Water Pollution X

WITPRESS

WIT Press publishes leading books in Science and Technology.
Visit our website for the current list of titles.
www.witpress.com

WITeLibrary

Home of the Transactions of the Wessex Institute.
Papers presented at Water Pollution X are archived in the WIT eLibrary in volume 135 of WIT Transactions on Ecology and the Environment (ISSN 1743-3541).
The WIT eLibrary provides the international scientific community with immediate and permanent access to individual papers presented at WIT conferences.
Visit the WIT eLibrary at www.witpress.com.

TENTH INTERNATIONAL CONFERENCE ON
WATER POLLUTION: MODELLING, MONITORING AND MANAGEMENT

Water Pollution X

Honorary Chairman

E. Andronescu
University Politehnica of Bucharest, Romania

Conference Chairmen

C.A. Brebbia
Wessex Institute of Technology, UK

A.M. Marinov
University Politechnica of Bucharest, Romania

International Scientific Advisory Committee

Organised by
Wessex Institute of Technology, UK
University Politehnica of Bucharest, Romania

Sponsored by
WIT Transactions on Ecology and the Environment

WIT Transactions

Transactions Editor

Carlos Brebbia
Wessex Institute of Technology
Ashurst Lodge, Ashurst
Southampton SO40 7AA, UK
Email: carlos@wessex.ac.uk

Editorial Board

Water Pollution X

Editors

A.M. Marinov
University Politechnica of Bucharest, Romania

&

C.A. Brebbia
Wessex Institute of Technology, UK

WIT*PRESS* **Southampton, Boston**

A.M. Marinov
University Politechnica of Bucharest, Romania

C.A. Brebbia
Wessex Institute of Technology, UK

Published by

WIT Press
Ashurst Lodge, Ashurst, Southampton, SO40 7AA, UK
Tel: 44 (0) 238 029 3223; Fax: 44 (0) 238 029 2853
E-Mail: witpress@witpress.com
http://www.witpress.com

For USA, Canada and Mexico

Computational Mechanics Inc
25 Bridge Street, Billerica, MA 01821, USA
Tel: 978 667 5841; Fax: 978 667 7582
E-Mail: infousa@witpress.com
http://www.witpress.com

British Library Cataloguing-in-Publication Data

A Catalogue record for this book is available
from the British Library

ISBN: 978-1-84564-448-2
ISSN: 1746-448X (print)
ISSN: 1743-3541 (online)

The texts of the papers in this volume were set individually by the authors or under their supervision. Only minor corrections to the text may have been carried out by the publisher.

Printed in Great Britain by Martins the Printers.

Preface

This volume contains most of the papers presented at the Tenth International Conference on Modelling, Monitoring and Management of Water Pollution held in Bucharest and organized by the University Politehnica of Bucharest and the Wessex Institute of Technology of the UK with the collaboration of the WIT Transactions on Ecology and the Environment.

The Conference, which started in 1991 has always been very successful and provided a continuous exchange of information between scientists, managers and academics from different countries, concerning water pollution problems.

A worldwide increase in demand for water by communities, industries, and agriculture, has placed great importance on water science and management practice. The future development of many countries will depend decisively upon how the increasing demand for water will be satisfied, and how the economic and environmental cost for that demand may be met.

The objective of the Water Pollution Conference is to create an international forum for inter-disciplinary discussion and dissemination of papers related to water quality and environmental problems caused by the increase of pollutant loads or by a bad management of water resources.

The topics of the papers published in the book are: Coastal areas, Lakes and rivers, Groundwater and aquifer issues, Environmental monitoring and sensing, Remediation, Experimental and laboratory work, Mathematical modelling, Oil spills, Wastewater treatment and management, Low cost technologies, Agricultural contamination, Pollution prevention, Emerging technologies for water treatment, Low cost technologies, Pharmaceutical and pesticides, Bioaccumulation, Micropollutant prevention in drinking water.

The Editors are indebted to Professor Ecaterina Andronescu, Rector of the University Politehnica of Bucharest, for her support of the meeting and having accepted to act as Honorary Chairperson.They are also grateful to all authors for the excellent standard of papers and to the members of the International Scientific Advisory Committee and other colleagues who helped to select contributions and in this manner to ensure the quality of the book.

The Editors
Bucharest, 2010

Contents

Section 6: Biosystems

Section 7: Wastewater treatment and management

Section 1
Water quality

What priority pollutants occur in stormwater and wastewater?

S. Zgheib[1], R. Moilleron[2] & G. Chebbo[1,3]
[1]*Université Paris-Est, Cereve, UMR-MA102 Ecole des Ponts ParisTech, AgroParisTech, France*
[2]*Université Paris-Est, Cereve, UMR-MA102 Université Paris-Est Créteil Val de Marne, AgroParisTech, France*
[3]*Faculté de Génie, Université Libanaise, Lebanon*

Abstract

After the implementation of the Water Framework Directive (WFD), there is an increasing need to monitor pollutant concentrations entering surfaces waters. This concerns all kinds of effluents, as well as urban waters (i.e., wastewater and stormwater). This study provides the first database on the occurrence of 88 individual substances in urban water and highlights direct discharges of these substances to the aquatic environment. Indeed, these hazardous pollutants, listed or not in the WFD, could be released without any treatment to watercourses during a storm.
Keywords: dissolved phase, particulate phase, priority pollutants, occurrence, wastewater, stormwater, urban catchments, water framework directive.

1 Introduction

Assessing concentrations of contaminants in domestic and industrial effluents has become an integral part of risk assessment and environmental regulation. Within this diverse research area there have been many works in the past on metals (Garnaud [1] and Gromaire-Mertz *et al.* [2]) and persistent organic chemicals, such as PAH (Gasperi *et al.* [3], Rocher *et al.* [4] and Grynkiewicz *et al.* [5]) and PCB (Chevreuil *et al.* [6]). More recently (Rule *et al.* [7, 8]), interest has been shown in priority pollutants of the Water Framework Directive (2000/60/CE), whose main objective is to achieve a "good status" for all the

WIT Transactions on Ecology and the Environment, Vol 135, © 2010 WIT Press
www.witpress.com, ISSN 1743-3541 (on-line)
doi:10.2495/WP100011

water bodies. The WFD requires Member States to gain a better understanding of priority substances entering surfaces waters. In fact, water bodies are a major recipient of an extensive array of contaminants produced by human activities. The latter can be discharged into watercourses after incomplete removal by wastewater treatment plants during dry and wet weather periods, by combined sewer overflows and by storm sewers after runoff on impervious urban surfaces.

In addition, Zgheib *et al.* [9] have noticed that the WFD list does not take into account some pollutants specific to urban areas. So, they extended the list of priority pollutants and defined 88 of them to ensure a good coverage of urban activities. Moreover, they have advised, as a general rule for monitoring water quality, to consider the analysis of contaminants on both the dissolved (D) and the particulate phases. They developed a monitoring or screening methodology taking into account all the precautions from sampling to analysis. All the analyses were performed by a certified French laboratory (Zgheib *et al.* [9]). This paper provides qualitative results on the occurrence of the 88 priority pollutants: (i) for three separate storm sewer systems with different land use patterns; (ii) for two domestic sewer systems very contrasted in terms of population density and land use features during dry weather periods.

2 Methodology

2.1 Experimental sites, sampling and analytical procedures

Five sites were investigated in this study to collect wastewater and stormwater as previously described by Zgheib *et al.* [9] and Zgheib [10]. Three sites, drained by separate storm sewer systems, were chosen to collect the stormwater samples. These sites differ by their land use pattern; there was a residential area in the suburb of Paris, namely Sucy-en-Brie (2.15 km^2, with an impervious surface coefficient (ISC) of 0.25), and two urban areas, one in Paris centre, ZAC Paris Rive Gauche (0.64 km^2, ICS 0.80) and another in the suburb of Paris, Noisy-le-Grand (2.30 km^2, ICS 0.65). The two remaining sites differ both in their sewer systems and their surface areas, and were chosen from which to collect wastewater during dry weather periods. Clichy was drained by a combined sewer system (120 km^2 with an impervious surface coefficient (ISC) of 0.50) and Sucy-en-Brie by a separate sewer system (2.15 km^2, with an impervious surface coefficient (ISC) of 0.25).

Ordinary water quality parameters, organic and mineral pollutants were analysed on each sample. For this reason, two different refrigerated automatic water samplers (+ 4°C) were used to preserve sample integrity. Water samples were collected at the outlet of each watershed. The automatic samplers were programmed to provide a mean rain event sample for stormwater and a mean day sample for wastewater. They allowed collecting 12 x 1 litre samples. They were equipped with two types of containers to avoid any contamination. Hence, for organic pollutants, samples were collected in Pyrex amber glass bottles with Teflon sampling tubes, whereas for mineral pollutants and ordinary water quality parameters, samples were collected in polyethylene (PE) bottles with PVC tube.

Once the whole water samples were fetched, they were then filtered with a succession of 90-mm pre-combusted glass filters from 2.7 µm (GF/D, Whatman) to 0.7 µm (GF/F, Whatman) in a glass filtration unit. Analyses were carried out on both the dissolved and the particulate phases in accordance with French (AFNOR) or international (ISO) standard methods by a certified laboratory.

2.2 Occurrence calculation approach

The results were expressed in mg/L for the dissolved phase (D) and the whole water sample or total phase (T), while they were expressed as mg/kg.dw for the particulate phase. Any substance was considered as present in a given sample, when its concentration was above the limit of detection (LOD) in, at least, one of the phases (D, P) or the whole water sample (T) for volatile organic compounds (VOCs). On the contrary, when the concentrations were below LOD for all phases, the substance was marked as absent in the given sample. This approach was applied to evaluate the occurrence of all pollutants in a given matrix, e.g., stormwater (SW) and wastewater (WW). The occurrence for all the contaminants was estimated for each matrix, SW or WW, and for each sampling location sample. The occurrence$_{ij}$ was defined as the percentage of presence for a substance (i) in a matrix or a site (j) as follows:

$$\text{occurence}_{ij}\ (\%) = \frac{s_i}{n_j} \times 100$$

where s_i is the number of times when the substance (i) was present in a sample for a matrix or a site (j) and n_j was the number of samples.

Similarly, the percentage of substances detected in the different phases (e.g., P, D or T) of a sample was:

$$\text{substances detected}\ (\%) = \frac{s_{detected}}{N} \times 100$$

where $s_{detected}$ is the number of substances detected for a site or matrix; N is the total number of investigated substances. This number (N) depends on the phase under investigation: N = 73 for both the dissolved and the particulate phases (D + P), N = 15 for the whole water sample (T) and, finally, N =88 for the total sample (i.e., D + P + T). Moreover, the substances were divided into three groups: the priority dangerous substances (sdp, n = 26), priority substances (sp, n = 17), with respect to WFD requirements, and the other substances more specific to urban pollution (others, n = 45).

3 Results and discussion

3.1 Occurrence of priority pollutants in stormwater

Thirteen samples of stormwater were collected from the three sites (Sucy-en-Brie, Noisy-le-Grand and ZAC Paris Rive Gauche). The screening revealed that stormwater samples contained 55 different individual substances or 62.5% of the substances investigated. An overview of both the type and the number of substances found in stormwater is summarized in table 1.

Among these substances, 64% belongs to the group of substances more specific to urban pollution. 36% of the substances derived from the WFD: 22% were priority substances and 14% were priority dangerous substances. Our results showed that only 5 VOCs were identified in stormwater and 50 organic substances or metals have been observed in at least one of the dissolved or particulate phases (D + P), while 33 substances have never been quantified. Overall, the total number of substances in stormwater whatever the site was comparable. In fact, the most frequently detected substances belong mainly to the following chemical families: metals, PAHs, PCBs, alkylphenols, organotins, phthalates and pesticides. Some chlorophenols and VOCs were less observed. No chloroalkane or PBDE and chlorobenzene were detected in stormwater (fig. 1).

Focussing on specific compounds to each site (fig. 1), one noted that the sites of ZAC Paris Rive Gauche and Noisy-le-Grand, representative of dense urban areas, were distinguished by the presence of certain VOCs (25% of samples), including methylene chloride (for 50% of samples), whereas these compounds

Table 1: Number of substances in SW and WW per sites, phases and types.

	N_{sample}	D	P	D+P	T	Sp	sdp	others	D+P+T
$N_{substances}$		73	73	73	15	26	17	45	88
ZAC Paris Rive Gauche	4	23	36	41	5	10	6	30	46
Noisy-le-Grand	4	22	38	44	1	10	8	27	45
Sucy-en-Brie	5	30	39	45	0	11	7	27	45
SW synthesis	**13**	**34**	**42**	**50**	**5**	**12**	**8**	**35**	**55**
Clichy – WW	4	29	34	47	7	12	7	35	54
Sucy-en-Brie-WW	4	21	30	36	5	10	7	24	41

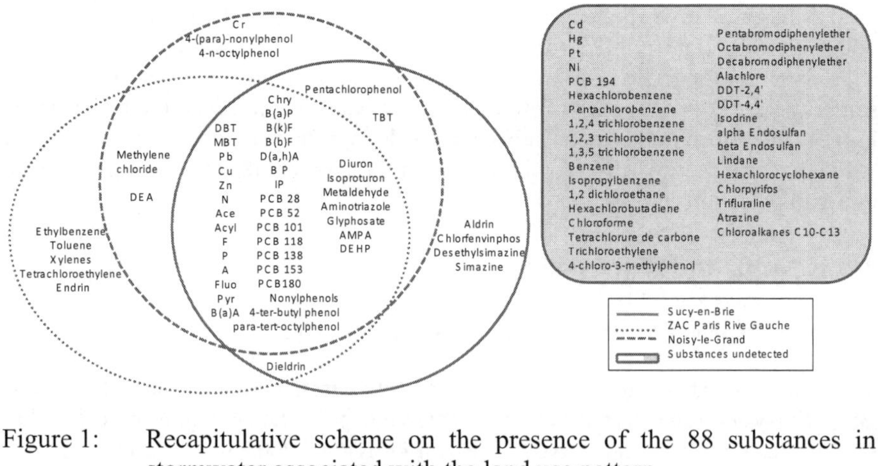

Figure 1: Recapitulative scheme on the presence of the 88 substances in stormwater associated with the land use pattern.

were absent in the samples from the residential area of Sucy-en-Brie. This latter was characterized by the presence of a large number of pesticides (aldrin, simazine, desethylsimazine, and chlorfenvinphos). Finally, Cr, 4-(para)-nonylphenol and 4-n-octylphenol were observed only on samples from Noisy-le-Grand. However, 33 substances were never detected whatever the site. Their concentrations were always below LOD. This was the case of some metals (Cd, Hg, Pt, Ni), PCB 194, chlorobenzenes (tri-, penta- and hexachlorobenzene), most VOCs (benzene, chloroform, trichlorethylene, isopropylbenzene, 1,2-dichloroethane, hexachlorobutadiene, carbon tetrachloride, trichlorethylene), 4-chloro-3-methylphenol, PBDEs, some pesticides (alachlor, 2,4'-DDT, 4,4'-DDT, isodrin, endosulfan, hexachlorocyclohexan, chlorpyrifos, trifluralin, atrazine, endosulfan, lindane,) and C10-C13 chloroalkanes. The absence of VOCs might be related to a bias in the sampling procedure, not suitable for the analysis of these substances. Indeed, samples were collected only at the end of the storm, increasing the risk of losses. Moreover, COVs are known as having a high volatility, which participate in their transfer to the atmosphere, and shortening of their residence time in samples. To improve the analysis of COV, a punctual sampling seemed to be more appropriate. Furthermore, many reasons can explain the non-detection of certain pesticides. However, the main reason would be the stop of their uses. Most of these pesticides are now banned from use in France. To confirm this, more pesticide monitoring would have been necessary during their application periods (in spring and fall) at all sites. Finally, some LODs were too high to enable the quantification of some substances (i.e., Cd, Ni, PBDE). This remains one of the main problems working with certified laboratories. Their LOD are somehow quite high for some contaminants, when they are compared to those generally achieved by research centres. Nevertheless, since in many countries, regulation lays down to sewer network managers to work with those organisms, they must face such constraints. The occurrence for every site is illustrated in fig. 2. Stormwater reveals the following characteristics. PAHs were ubiquitous; the 16 PAHs were systematically quantified in all samples for all sites. DEHP, a priority substance, was also present in all samples. Although the use of PCBs has been progressively restricted in France in 1979, with the prohibition of their use (in inks, adhesives, as additives in some oils...), then in 1987 with the prohibition of the sale, acquisition or marketing of equipment containing PCBs (transformers), all PCB congeners without PCB194 were observed for all sites.

For Noisy-le-Grand and ZAC Paris Rive Gauche, the dense urban areas, three metals, Cu, Pb and Zn were quantified in 100% of the samples. For Sucy-en-Brie, the trend was similar for Cu and Zn, whereas Pb was present in a lesser proportion (77% of cases). Cr, meanwhile, has been detected in only 67% of Noisy-le-Grand samples. For organotins, only DBT and MBT were common to all sites, but in proportions differing from one site to another: 100% for both substances at Noisy-le-Grand and ZAC Paris Rive Gauche. For Sucy-en-Brie, DBT was present in 83% of samples, while MBT was in 33% of them. TBT, a priority dangerous substance, has been observed on Noisy-le-Grand and Sucy-en-Brie in 100% and 17% of the samples, respectively.

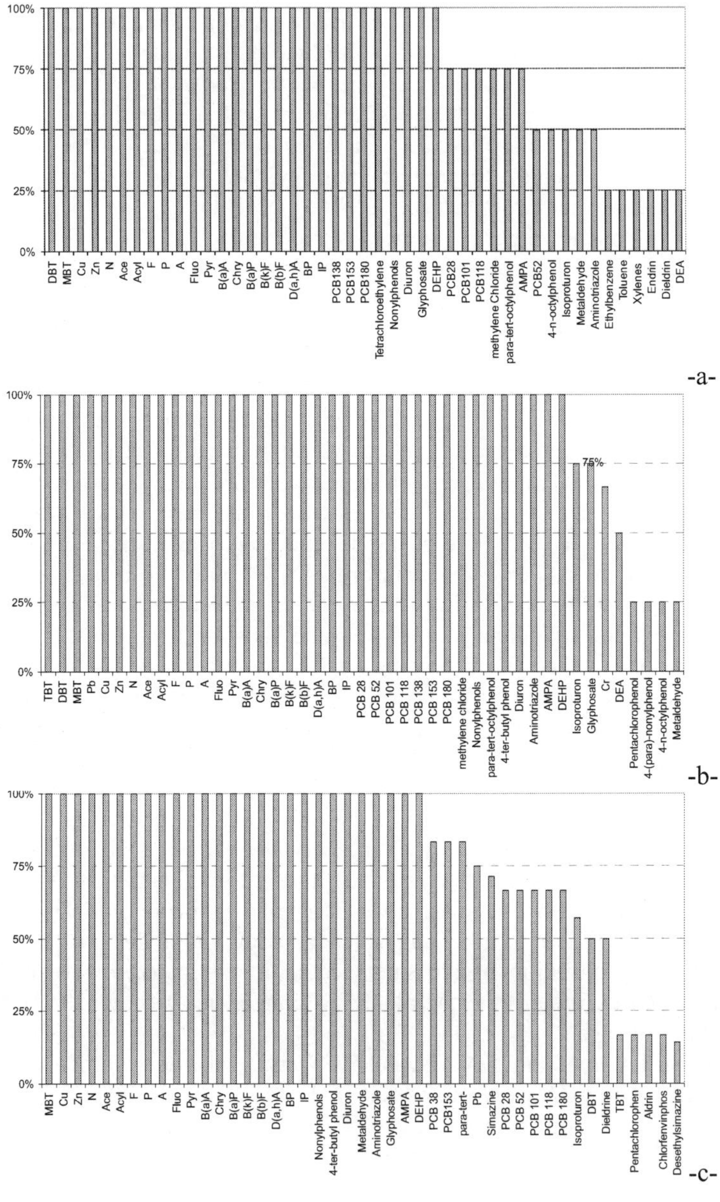

Figure 2: Occurrence of pollutants in stormwater in Paris ZAC Paris Rive Gauche (a); Noisy–le–Grand (b); and Sucy-en-Brie (c).

Two pesticides, namely isoproturon and diuron, were present respectively in 60 and 100% of stormwater samples. Moreover, all these samples were also contaminated with 4 other pesticides (metaldehyde, aminotriazole, glyphosate and its main degradation product AMPA). The degradation product of atrazine (a herbicide banned in France since 2003), the desethylatrazine (DEA), was common to both dense urban sites, while dieldrin (banned since 1994) was measured on the ZAC Paris Rive Gauche (25%) and Sucy-en-Brie (60%).

Three alkylphenols have been observed in more than 75% of samples, all sites combined: nonylphenol, para-tert-butylphenol and para-tert-octylphenol. Sucy-en-Brie was contaminated by all alkylphenols in more than 25% of cases. As previously mentioned, VOCs was the group of substances the least observed in stormwater, this seems to be in contradiction with what has been reported for wastewater releases in a previous study by Greaud-Hoveman *et al.* [11]. COVs were indeed present in 46% of samples, ranking just behind metals and PAHs Greaud-Hoveman *et al.* [11]. This underlines their strong domestic and industrial origins. Furthermore, stormwater samples from the dense urban site of the ZAC Paris Rive Gauche contained tetrachlorethylene (100% of cases). Tetrachloroethylene is mainly used as a solvent in dry cleaning and metal cleaning. It is also used for processing and finishing in the textile industry Rule *et al.* [8]. Finally, pentachlorophenol was present in 25% of samples from Noisy-le-Grand and Sucy-en-Brie. On the ZAC Paris Rive Gauche site, it was detected in one sample but not quantifiable (<LOQ). Pentachlorophenol is an ingredient of insecticides and fungicides Rule *et al.* [8]. This contaminant has not been observed in runoff in London, its absence was related to the prohibition of its use according to Rule *et al.* [8].

3.2 Occurrence of priority pollutants in wastewater

During dry weather periods, eight samples of wastewater were collected at the outlet of two different sewer systems (a combined and a separate one), they were analysed for the 88 substances. 54 substances were quantified in the combined sewer and 41 in the separated sewer, 61% and 47% of the 88 investigated substances, respectively (table 1). Although both domestic wastewater watersheds are contrasted in terms of land use patterns and type of sewer, many compounds were common in both sites. More precisely, 16 PAHs were observed in 100% of the samples from both sewer systems (fig. 3). This result is quite different from that previously reported for the analysis of PAHs in wastewater samples from the Parisian combined sewer network, during dry weather periods, by Gasperi *et al.* [12], where the occurrence for each PAH varied between 0 and 75%. However, analyses were carried out on the whole water sample in that study. This clearly shows that analyses on whole water samples with substantial amount of suspended matter are much less efficient for those substances than if performed on suspended matter itself, using suitable methods for solid phases.

For organotins, both sewer systems, exhibited the same trends, since DBT, TBT and MBT were observed in more than 50% of the samples. Two metals (Zn and Cu) were systematically observed in our samples in good accordance with previous studies from Garnaud [1], Rule *et al.* [7] and Gasperi *et al.* [12]. DEHP and alkylphenols, such as branched nonylphenol, para-tert-octylphenol and 4-ter-butylphenol, were also measured in all the samples. For VOCs, we noted the presence of chloroform in most of the samples along with toluene and xylene. The occurrence of pesticides varies, depending on the compound. However, diuron and AMPA were both present in 100% and 75% of cases, respectively. Glyphosate, aminotriazole, 2-4'-DDT, dieldrin, chlorfenvinphos, metaldehyde and DEA have been observed with pentachlorophenol, at least once in a sample. The sole PCB 28 was measured in one sample at Sucy-en-Brie. This contrasted with the samples from combined sewers, where all the PCBs were quantified at

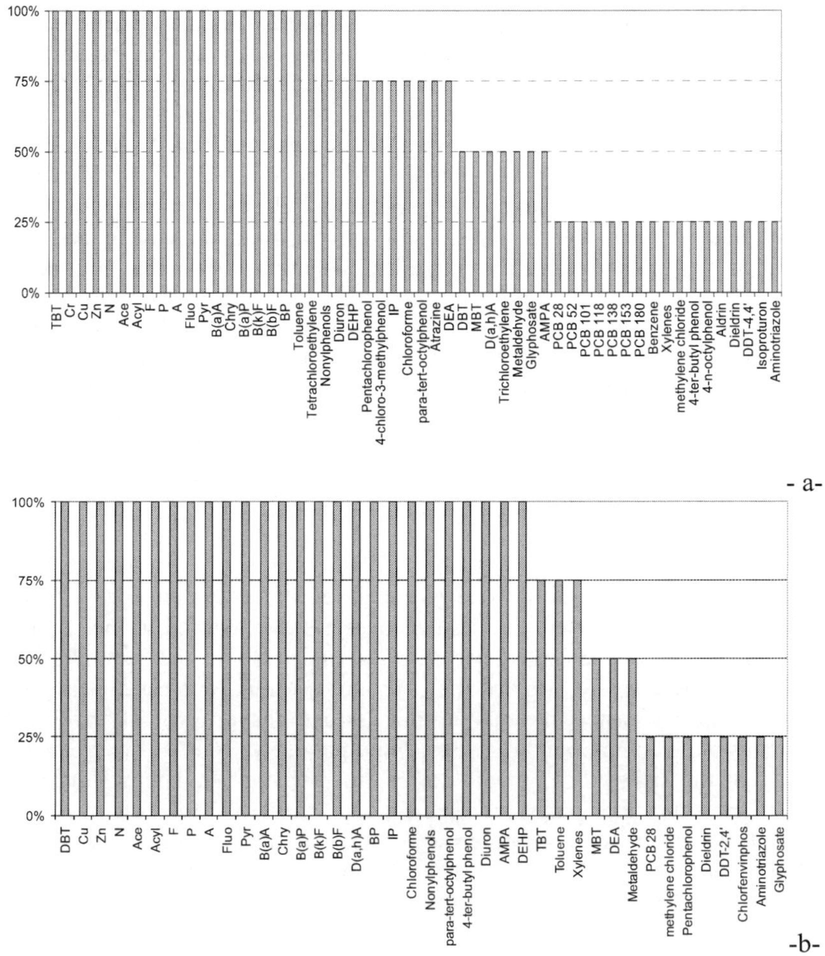

Figure 3: Occurrence of pollutants in wastewater from (a) a combined sewer system at Clichy and (b) from a separate sewer system at Sucy-en-Brie.

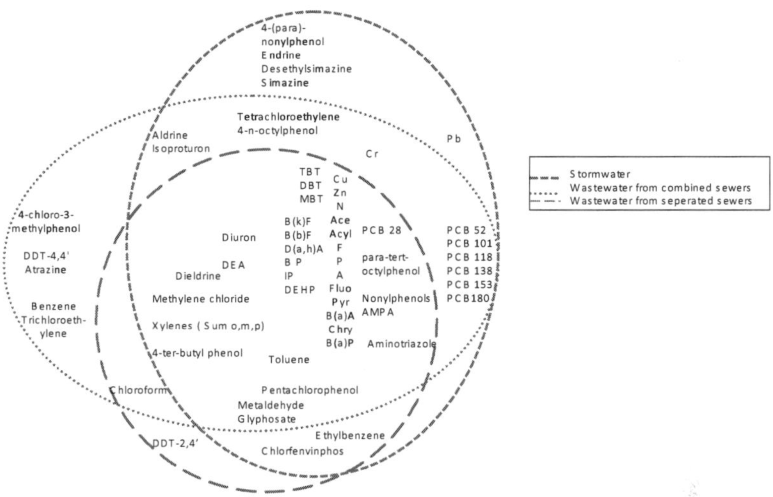

Figure 4: Recapitulative scheme on the presence of the pollutants in stormwater and wastewater.

least once (25-100%). Besides, the presence of ethylbenzene, 2-4'-DDT and chlorfenvinphos at Sucy-en-Brie can be related to their local use. Finally, some pollutants were never measured in wastewater from both the combined and the separate sewer systems: PCBs (congeners 52, 101, 118, 138, 153), some pesticides and Cr.

4 Conclusion

This paper describes the first results in Europe on the occurrence of priority pollutants in urban waters using an original methodology, called "screening". A list of 88 pollutants were analysed on 13 stormwater samples and 8 wastewater samples on both the dissolved and the particulate phases. Results showed that the number of substances in urban water was relatively comparable from one matrix to another (stormwater or wastewater). Although some substances were common to several matrices (28 substances), some other appeared to be more specific to a particular matrix (fig. 4). The wastewater was characterized by a greater number of VOCs and the stormwater by the presence of pesticides. Moreover, the number and the nature of pollutants were relatively comparable from one sample to another for different land use patterns, as observed for storm sewer systems, and for different type of sewer systems, as for wastewater.

Acknowledgements

The authors gratefully acknowledge the Seine-Normandy Water Agency (AESN), the Interdepartmental Association for Sewage Disposal in Paris

Conurbation (SIAAP), the Municipality of Paris, the Water and Sewage Disposal Departmental Administrations of Seine Saint Denis (DEA93), the Water and Sewage Disposal Departmental Administration of Val de Marne (DSEA94) and Regional Council of "Ile de France" (CRIF) for their financial support within the framework of the OPUR programme research.

References

[1] Garnaud, S. Transfert et évolution géochimique de la pollution métallique en bassin versant urbain. *Ph.D thesis, Ecole Nationale des Ponts et Chaussées*. 396p., 1999.

[2] Gromaire-Mertz, M.C., Garnaud, S., Gonzalez, A., and Chebbo, G., Characterisation of urban runoff pollution in Paris. *Water Science and Technology*. **39(2)**, p. 1-8, 1999.

[3] Gasperi J., Rocher V., Moilleron R. and Chebbo G., Review on the hydrocarbons fate within combined sewers: case of the "Le Marais" urban catchment (1994-2005). *Polycyclic Aromatic Compounds*. **27(2)**, p. 123-141, 2007.

[4] Rocher V., Azimi S., Gasperi J., Beuvin L., Muller M., Moilleron R. and Chebbo G., Hydrocarbons and metals in atmospheric deposition and roof runoff in central Paris. *Water Air and Soil Pollution*. **159(1-4)**, p. 67-86, 2004.

[5] Grynkiewicz M., Polkowska Z. and Namiesnik J., Determination of polycyclic aromatic hydrocarbons in bulk precipitation and runoff waters in an urban region (Poland). *Atmospheric Environment* **36(2)**, p. 361-369, 2002.

[6] Chevreuil M., Garmouma M., Teil M. J. and Chesterikoff A., Occurrence of organochlorines (PCBs, pesticides) and herbicides (triazines, phenylureas) in the atmosphere and in the fallout from urban and rural stations of the Paris area. *Science of the Total Environment*. **182**, p.1-3, 1996.

[7] Rule, K.L., Comber, S.D.W., Ross, D., Thornton, A., Markropoulos, C.K., and Rautiu, R., Survey of priority substances entering thirty English wastewater treatment works. *Water and Environment Journal*. **20**, p. 177-184, 2006.

[8] Rule, K.L., Comber, S.D.W., Ross, D., Thornton, A., Makropoulos, C.K., and Rautiu, R., Sources of priority substances entering an urban wastewater catchment--trace organic chemicals. *Chemosphere*. **63(4)**, p. 581-591, 2006.

[9] Zgheib S., Moilleron R. and Chebbo G. Screening of priority pollutants in urban stormwater: innovative methodology. *Water Pollution IX* **111**: p. 235-244, 2008.

[10] Zgheib S. Flux et sources des polluants prioritaires dans les eaux urbaines en lien avec l'usage du territoire. *Ph.D thesis, Ecole Nationale des Ponts et Chaussées*. 349p., 2009.

[11] Greaud-Hoveman L., Barré H., Houeix N., Lepot B., Lehnhoff C., Schneider M., Aguerre-Charol O. and Morin A., Les substances

dangereuses pour le milieu aquatique dans les rejets industriels et urbains. *Rapport d'étude d'INERIS*: 611 p., 2008.

[12] Gasperi J., Garnaud S., Rocher V. and Moilleron R. Priority pollutants in wastewater and combined sewer overflow. *Science of the Total Environment* 407(1): p. 263-272, 2008.

Hydrochemical evaluation of different spring waters for bottling

D. M. Bonotto, B. W. Tessari, L. Arthus, M. Santilli & L. Caprioglio
Departamento de Petrologia e Metalogenia, Instituto de Geociências e Ciências Exatas-UNESP, Brazil

Abstract

This investigation was carried out within the Paraná sedimentary basin in São Paulo and Mato Grosso do Sul States, Brazil, and involved the sampling of different spring water samples with the purpose of evaluating their quality for bottling. Several methods were utilized for acquiring the hydrochemical data, such as methyl orange end-point titration, potentiometry, ion selective electrodes, spectrophotometry, atomic absorption spectrophotometry and inductively coupled plasma spectrometry. The results obtained for the analyzed samples were compared with the guideline values established by the São Paulo State and Brazilian Health Ministry legislations for defining the potable water standards. The Brazilian Code for Mineral Waters that was established by Register 7841 published on 8 August 1945 was also taken into account for verifying if the spring waters are mineralized. The hydrochemical data allowed the identification of anthropogenic and geogenic inputs of some constituents in most of the samples, which affected the water quality and did not allow them to be utilized for commercial purposes (bottling). The waters of these springs can only become appropriate for human consumption after previous chemical treatment.

Keywords: drinking water quality, spring waters, hydrochemistry, mineral waters, bottling, contamination.

1 Introduction

In many countries, spring waters are extensively used for consumption purposes as an option to tap water, as many people believe they are healthy and/or can be

WIT Transactions on Ecology and the Environment, Vol 135, © 2010 WIT Press
www.witpress.com, ISSN 1743-3541 (on-line)
doi:10.2495/WP100021

utilized for health cures. Additionally, economic reasons have also favored their use as bottled waters, so the commercialization of mineral waters has widely increased, for instance, by about 15% per year in Brazil where circa 20 million consumers are involved [1]. The Brazilian Code for Mineral Waters was established by Register 7841 published on 8 August 1945 [2]. According to chemical aspects, the waters may be classified as follows:

-Radiferous: containing dissolved radioactive substances that sustain a permanent radioactivity;

-Bicarbonate-alkaline: containing dissolved alkaline compounds corresponding to a minimum of 0.2 g per liter of sodium bicarbonate;

-Earth-alkaline: containing dissolved earth-alkaline compounds corresponding to a minimum of 0.12 g per liter of sodium carbonate. They may be Ca-dominated if they contain at least 0.048 g per liter of Ca as calcium bicarbonate or Mg-dominated if they contain at least 0.030 g per liter of Mg as magnesium bicarbonate.

-Sulfated: containing dissolved sulfate corresponding to a minimum of 0.1 g per liter combined with sodium, potassium and magnesium;

-Sulfured: containing dissolved sulfur corresponding to a minimum of 0.001 g per liter;

-Nitrated: containing dissolved nitrate of mineral origin corresponding to a minimum of 0.1 g per liter;

-Chlorinated: containing dissolved chloride corresponding to a minimum of 0.5 g per liter of sodium chloride;

-Ferruginous: containing dissolved iron corresponding to a minimum of 0.005 g per liter;

-Radioactive: containing dissolved radon according to three different levels, i.e. (a) weakly radioactive (radon content between 5 and 10 Mache Unit per liter), (b) radioactive (radon content between 10 and 50 Mache Unit per liter) and (c) strongly radioactive (radon content higher than 50 Mache Unit per liter).

-Thoriferous: containing dissolved thoron corresponding to a minimum of 2 Mache Unit per liter;

-Carbogaseous: containing dissolved carbon dioxide gas corresponding to a minimum of 200 mL per liter at 20 °C temperature and 760 mmHg pressure.

The Brazilian mineral waters are also classified according to the temperature as follows [2]:

-Cold waters: temperature lower than 25 °C;

-Hypothermal waters: temperatures ranging from 25 to 33°C;

-Mesothermal waters: temperatures ranging from 33 to 36°C;

-Isothermal waters: temperatures ranging from 36 to 38°C;

-Hyperthermal waters: temperature higher than 38 °C.

Many mineral springs around the world do not contain strong concentrations of minerals, and are thus classified as lightly mineralized, or oligomineral waters [3]. This is also the case for many spring waters utilized for drinking purposes in Brazil and attention must be paid to diverse problems related to the interaction between the society and environment as it may occur the introduction of constituents able to cause modification on the desirable water quality for consumption purposes. Thus, the spring waters must also be properly managed like others hydrological resources, particularly if the intention is their use as bottled waters. This paper describes the results of some hydrochemical monitoring held in different aquifer systems in order to evaluate the water quality. Because it has been analyzed different spring waters with the purpose of bottle them for human consumption, the Brazilian legislation was utilized for providing guideline values for drinking water quality.

2 Analytical methods

Several standard analytical techniques were used for obtaining the parameters analyzed in the water samples, for example, dilution and incubation method, evaporation, methyl orange end-point titration, potentiometry, ion selective electrode, spectrophotometry, atomic absorption spectrophotometry and inductively coupled plasma spectrometry. Portable meters were used for *in situ* determinations of temperature, pH and dissolved oxygen (DO). The pH measurement was performed by a digital portable meter coupled to a combination glass electrode; buffer solutions equilibrated with the sample temperature were utilized to calibrate the equipment before the analyses. The DO was determined in a pointer meter recording the potential values generated by an O_2 sensible electrode consisting on a metallic wire covered by a thin layer of gold.

The true color for two water samples was determined by colorimetry (wavelength 455 nm) after filtering the suspended materials and using a program stored in the Hach DR/2000 spectrophotometer that was calibrated in color units based on the APHA-recommended standard of 1 color unit being equal to 1 mg L^{-1} platinum as chloroplatinate ion [4]. The same samples were analyzed for total coliforms through the fast ColiQuik Presence/Absence test [4] that utilizes a medium providing specific indicator nutrients, i.e. ONPG (o-nitrophenyl-β-d-galactopyranoside) and MUG (4-methylumbelliferyl-β-d-glucuronide).

A bench digital meter provided electrical conductivity readings obtained though a 1 cm^2 area platinum electrode calibrated with KCl standards. The dry residue (DR) (~total dissolved solids, TDS) content was evaluated on evaporating the filtrate to dryness in a weighed flask that was dried to constant weight at 180°C, with the increase in flask weight representing DR [5]. The alkalinity of the water samples was determined by titration using a titrator with sulfuric acid standard solution to an end point evidenced by the color change of a standard indicator solution [5]. The obtained values corresponded to alkalinity due to the presence of bicarbonate, carbonate and/or hydroxides.

The BOD (Biochemical Oxygen Demand) was measured by the dilution method, employing high purity distilled water provided by Barnstead Mega-Pure One Liter Water Still. For this purpose, the dissolved oxygen (DO) content remaining in six portions of each well-mixed sample transferred to separated 300-mL glass-stoppered bottles was evaluated after a five-day incubation period; the DO values measured potentiometrically were plotted against the mL of

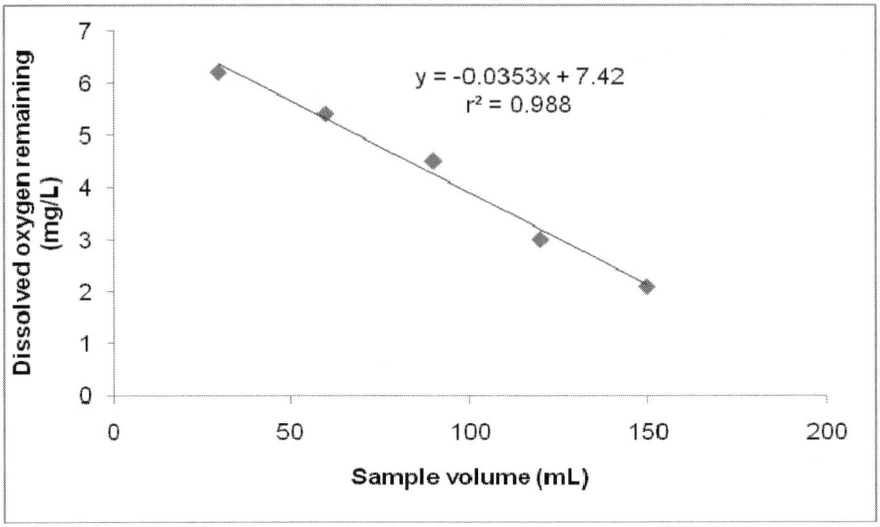

Figure 1: The dissolved oxygen values plotted against the sample volume for estimating the BOD (Biochemical Oxygen Demand).

sample taken, yielding a straight line that allowed obtain the BOD data [4]. Fig. 1 illustrates a typical straight line obtained for acquiring the results.

Calcium hardness (as $CaCO_3$) and magnesium hardness (as $MgCO_3$) of the water samples were determined by the colorimetric method (wavelength 522 nm) after chelating calcium with EGTA and calcium and magnesium with EDTA [4], parameters that allowed to evaluate Ca and Mg contents. The analyses of dissolved sodium were done by flame photometry, whereas potassium was determined by the tetraphenylborate method that is based on the combination of K with sodium tetraphenylborate to form an insoluble white solid read at 650 nm [4].

Chloride and fluoride were measured potentiometrically after adding a known amount of ionic strength adjustor to each sample, when necessary. Orion ion-selective electrodes coupled into a digital meter were used, where standards containing variable concentrations of chloride and fluoride were utilized for preparing calibration curves consisting on logarithmic straight lines involving the potential and concentration readings. Nitrate, nitrite, ammonium, phosphate and sulfate were determined by colorimetry [4] after adding reagents to the samples that are able to produce colored complexes read by a program stored in Hach DR/2000 spectrophotometer previously calibrated in variable concentrations at different wavelengths.

The metals Ba, Cr, Cr^{6+} and Fe were also measured colorimetrically. The amount of turbidity present in a fine white barium sulfate precipitate was read at 450 nm in order to supply the Ba concentration data. Cr^{6+} was determined on using 1,5-diphenylcarbohydrazide, which reacts to give a purple color read at 540 nm when Cr^{6+} is present. The 1,10 phenanthroline indicator formed an orange color in proportion to the iron concentration, which was read at 510 nm.

Other metals were measured by colorimetry, atomic absorption spectrophotometry and inductively coupled plasma spectrometry.

3 The selected sites at São Paulo and Mato Grosso do Sul States in Brazil

The methods described in previous item were applied to spring waters collected in areas situated at São Paulo and Mato Grosso do Sul States in Brazil. They are located at the following municipalities: Amambai in Mato Grosso do Sul State (site No. 1), Santa Rita do Passa Quatro in São Paulo State (Site No. 2) and São José do Rio Pardo in São Paulo State (Site No. 3) (Fig. 2). Two spring water samples were collected from different locations at São José do Rio Pardo municipality (Site No. 3). All sites are geologically situated at the Paraná sedimentary basin. The sedimentary sequence is almost undisturbed, with gentle dips towards the center of the basin, and covers since the Silurian-Devonian up to the Cretaceous periods. The major rock types [6] include sandstones, conglomerates, diamictites, siltstones, shales, mudstones, limestones, basalts and diabases that belong to Tubarão Group (Itararé Subgroup and Tatuí Formation), Passa Dois Group (Irati and Corumbataí Formations), São Bento Group (Pirambóia, Botucatu, and Serra Geral Formations), Bauru Group and different types of Cenozoic covers (Fig. 2).

Tietê River is the major drainage system at São Paulo State, which originates close to São Paulo city, cross the whole state, and discharges into Paraná River at the border with Mato do Grosso Sul State. Multiaquifer systems mainly comprising sandstones and basalts plus sediments from Passa Dois Group behaving as aquitards have been proposed to represent the hydrostratigraphy of the Paraná basin [7]. Groundwater occurs within the interflow zones and along cooling joints in basalts and diabases from Serra Geral Formation. The sandstones of Cretaceous age (Bauru Group) are moderately cemented and exhibit adequate properties to storage water. The Paleozoic sediments (Devonian-Permian age) also provide water that is relatively mineralized in the central parts of the basin and contains H_2S in some wells. The Guarani aquifer corresponds to sediments (Triassic-Jurassic age) consisting of silty and shaly sandstones of fluvial-lacustrine origin (Pirambóia Formation) and variegated quartzitic sandstones accumulated by eolian processes under desert conditions (Botucatu Formation). It has continental dimensions [7], extends over some 950,000 km^2 within the Paraná sedimentary basin, has an average thickness of 300-400 m, overlies formations ranging from the igneous basement to Paleozoic sediments of the Passa Dois and Tubarão Groups, and is covered by a thick (up to 1,500 m) basaltic package of Serra Geral Formation and Cretaceous sediments of the Bauru Group (Fig. 2).

Spring water samples for chemical analysis were collected at farms whose owners intended to bottle them for commercial purposes. Thus, they had to know the water quality for this purpose. The water discharge occurred through cracks, joints and faults from rocks outcropping in the Paraná sedimentary basin. The water samples (~5 kg) were stored in polyethylene bottles and, depending on the

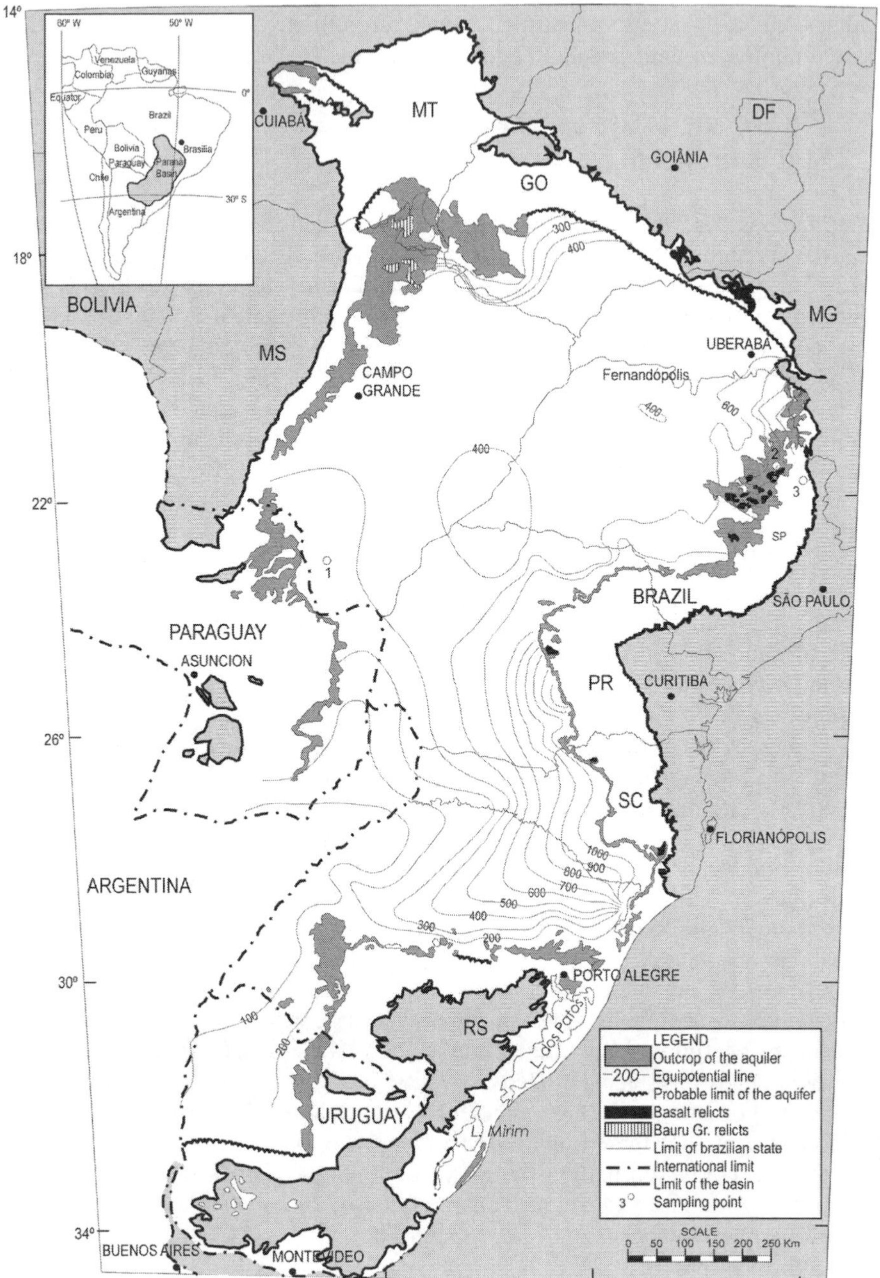

Figure 2: A simplified geological map of the Paraná sedimentary basin, the general groundwater flow in the Guarani aquifer according to [7] and the location of the sampling sites for spring waters.

Table 1: Results of the analyses of spring waters from the studied sites.

PARAMETER	UNIT	SITE No. 1	SITE No. 2	SITE No. 3(a)	SITE No. 3(b)
Sampling date	-	May 1995	Feb. 1995	July 1997	Aug. 1997
Discharge	L/day	-	17,280	-	-
Temperature	°C	-	26	-	-
Color	Pt-Co	2	2	10	0
Odor	-	Absent	Absent	Absent	Absent
Turbidity	NTU	0	0	2	0
pH	-	7.0	6.0	5.8	6.1
Conductivity	µS/cm	162	22.3	150	200
Free CO_2	mL	0.03	-	-	-
Dry Residue	mg/L	42	78	75	100
Settleable solids	mL/L	0.0	0.0	-	-
Dissolved Oxygen	mg/L	-	7.0	-	-
BOD[1]	mg/L	-	-	4.0	0.4
Total coliforms	n/100mL	-	16	Present	Present
Standard counting of bacteria	colonies/ mL	-	3,485	-	-

[1]BOD=Biochemical Oxygen Demand (5 days, 20°C).

requirements of the analysis, they were separated as unfiltered and unpreserved, filtered through 0.45 µm membranes and preserved with different acids, etc.

4 Hydrochemical analyses and water quality

The results obtained for the analyzed samples are reported in Tables 1 and 2 and compared with guideline values established by two different legislation (Table 3): Register 25 of the National Commission of Rules and Standards for Food of the Health Ministry published on 1976and Rule No. 12486 (NTA60) established in 20th October 1978 by São Paulo State for defining the potable water standards.

The data reported in Tables 1 and 2 for Site No. 1 were compared with the guideline values given in Table 3(a), as well with the Brazilian Code for Mineral Waters. The available data do not allow classify these spring waters as mineral waters, however, they are appropriate for human consumption (potable) and can be classified as “of natural spring”. The data for major ions in sample collected at Site No. 2 were plotted on a Piper [8] diagram. Fig. 3 shows that these waters are bicarbonated and mixed in terms of major cations. When the data obtained for these waters are compared with the guideline values given in Table 3(a) and also with the Brazilian Code for Mineral Waters, it is again not possible classify them as mineral waters. The reference values reported in Table 3(b) allow evaluate if these waters are potable and this is not the case in terms of the microbiological aspects as the number of total coliforms per 100 mL of sample is not absent. On the other hand, such water source could be chemically treated (for instance, by chlorination) in order to become potable, however, under such

Table 2: Results of the analyses of spring waters from the studied sites.

PARAMETER	UNIT	SITE No. 1	SITE No. 2	SITE No. 3(a)	SITE No. 3(b)
Sodium	mg/L	2.1	0.5	0.2	1.4
Potassium	mg/L	21.2	1.3	5.1	5.0
Calcium	mg/L	11.1	1.2	0.1	0.1
Magnesium	mg/L	1.9	0.4	0.7	0.7
Total Alkalinity	mg/L	8.5	12.0	20	52
Hydroxide alkalinity	mg/L	-	-	0	0
Carbonate alkalinity	mg/L	-	-	0	0
Bicarbonate alkalinity	mg/L	-	-	20	52
Chloride	mg/L	1.0	0.6	1.3	0.8
Nitrate	mg/L.	0.1	0.01	12.3	2.6
Nitrite	mg/L	-	-	0.02	0.01
Ammonium	mg/L	-	-	0.26	0.01
Sulfate	mg/L	0.1	0.03	0.1	0.1
Fluoride	mg/L	0.03	0.01	2.6	2.6
Aluminum	mg/L	0.2	0.05	-	-
Phosphate	mg/L	0.56	-	-	-
Barium	mg/L	0.02	0.06	-	1.0
Boron	mg/L	<0.02	<0.02	-	-
Cadmium	mg/L	<0.05	<0.05	-	-
Lead	mg/L	<0.10	<0.10	-	-
Copper	mg/L	<0.05	<0.05	0.08	0.13
Hexavalent chromium	mg/L	-	-	-	0.01
Total chromium	mg/L	<0.05	0.03	-	-
Iron	mg/L	<0.05	<0.05	0.48	0.16
Phosphorus	mg/L	0.2	0.008	-	-
Manganese	mg/L	<0.02	<0.02	0.68	0.96
Tin	mg/L	-	<0.01	-	-
Titanium	mg/L	-	0.08	-	-
Silicon	mg/L	5.7	6.11	-	-
Zinc	mg/L	<0.02	<0.02	2.1	2.1

circumstance, it no more could be considered as "of natural spring" for commercial purposes (bottling). Its discharge corresponding to 17,280 L/day is lower than that of several traditional bottlers in São Paulo State, but is higher than that of many others which had been already established since 1976 (Table 5).

The data reported in Tables 1 and 2 for Sites No. 3(a) and 3(b) were compared with the guideline values given in Table 3(a), as well with the Brazilian Code for Mineral Waters. The available data do not allow classify these spring waters as mineral waters. They are also not appropriate for human consumption (potable) as some compounds like nitrate, ammonium, manganese and fluoride exhibited values that exceeded the maximum established by the referred legislation. Additionally, the presence of total coliforms was also

Table 3: Parameters established by (a) Register 25 of the National Commission of Rules and Standards for Food of the Health Ministry published on 1976 and (b) Rule No. 12486 (NTA60) published in 20th October 1978 by São Paulo State.

PARAMETER	UNIT	VALUE (a)	VALUE (b)
Color	Pt-Co	< 5 mg/L	10 – 20 mg/L
Odor	-	Absent	Absent
Turbidity	NTU	< 3	2 - 5
Dry Residue	mg/L	< 1500	< 500
pH	-	from 4 to 9	from 5 to 9
Consumed Oxygen - acid	mg/L	< 4.5	< 2.5
N_2-ammonium	mg/L	< 0.03	-
Nitrate	mg/L	-	< 10
Chloride	mg/L	< 100	< 250
Fluoride	mg/L	< 1	< 1
Sulfate	mg/L	-	< 250
Arsenic	mg/L	< 0.05	< 0.05
Barium	mg/L	< 1	< 1
Cadmium	mg/L	< 0.01	< 0.01
Lead	mg/L	< 0.05	< 0.05
Cyanide	mg/L	< 0.2	< 0.2
Copper	mg/L	< 1	< 1
Hexavalent chromium	mg/L	< 0.05	< 0.05
Mercury	mg/L	< 0.001	-
Total iron	mg/L	-	< 0.3
Manganese	mg/L	< 0.05	< 0.05
Selenium	mg/L	< 0.01	< 0.01
Zinc	mg/L	< 5	< 5
Total coliforms	n/100mL	-	Absent
Escherichia coli	100 mL	Absent	-

identified and confirmed by the BOD data. Therefore, such water sources cannot be utilized for human consumption without previous chemical treatment, implying that they could not be considered as "of natural spring" for commercial purposes.

The presence of total coliforms, N-compounds, fluoride and manganese as pollutants in spring waters occurring in São Paulo State, Brazil, do not allow utilize them for commercial purposes (bottling). Despite some constituents may be typically geogenic, local anthropogenic influences due to non-appropriate management of the recharge area related to the springs may be responsible by the contamination of shallower aquifers. As a consequence of such diffuse pollution, the water quality becomes non-appropriate for human consumption without previous chemical treatment. This also implies on the water usage by bottlers and others users, which are requested to direct efforts on the protection and conservation of the soil covers and vegetation for avoiding the water pollution.

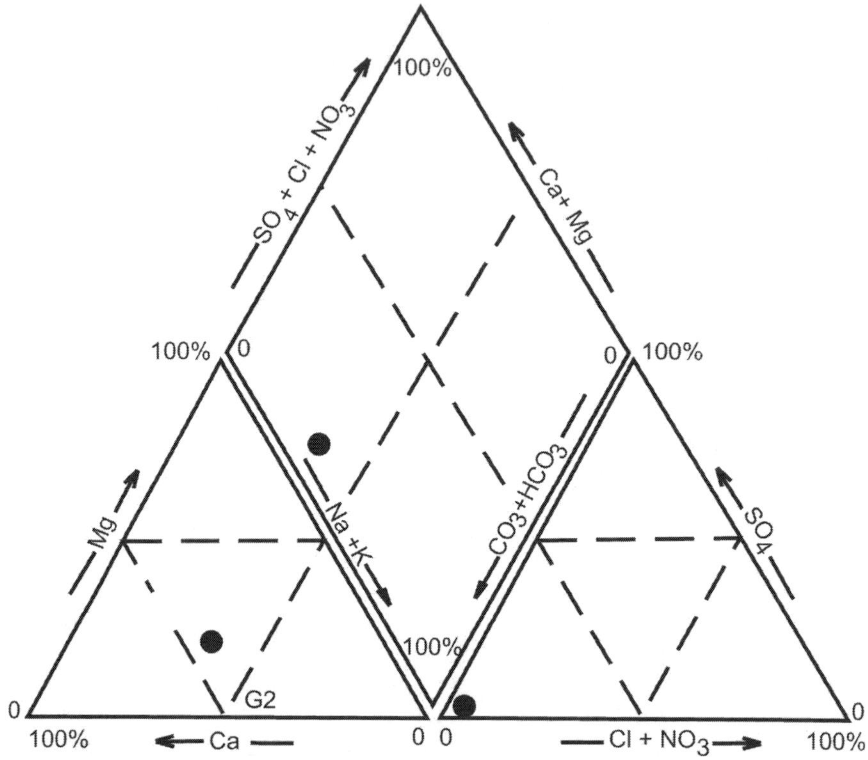

Figure 3: Major chemical data of the spring waters collected at Site No. 2 plotted on a Piper diagram.

Table 4: Discharge of springs located in São Paulo State that have had bottled waters since 1976 for supplying potable water classified as "of natural spring".

City	Location	Spring name	Discharge (L/day)
Analândia	Moinho Velho	Santo Antônio de Analândia	5,000
Oswaldo Cruz	Chácara Sugawara	Oswaldo Cruz	3,200
Poá	Sítio Primavera	Primavera	7,860
São José do Rio Preto	Jardim Conceição	No. 1	14,400
São Paulo	Vila Alípia	Petrópolis	8,000
São Paulo	Bairro Tremembé	Nova	3,500
São Simão	Jardim Conceição	No. 1	14,400

Acknowledgement

The CNPq (National Council for Scientific and Technologic Development) in Brazil is thanked for financial support of this investigation.

References

[1] Comércio de água mineral; SEBRAE, Online. www.sebrae-sc.com.br/ideais/default.asp?vcdtexto=31586&%5E%5E
[2] DFPM (Division for Supporting the Mineral Production), *The mining code, the mineral waters code and how applying research in a mineral deposit*. 8th edn., DFPM: Rio de Janeiro, 1966.
[3] Altman, N., *Healing springs*, Healing Arts Press: Rochester, 2000.
[4] HACH, *Water Analysis Handbook*, 2nd edn., Hach Company: Loveland, 1992.
[5] APHA (American Public Health Association), *Standard methods for the examination of water and wastewater*. 17th edn, Washington, 1989.
[6] Almeida, F.F.M. & Melo, M.S., A bacia do Paraná e o vulcanismo Mesozóico. *Mapa geológico do estado de São Paulo*, ed. IPT (Instituto de Pesquisas Tecnológicas do Estado de São Paulo), Promocet: São Paulo, v. 1, pp. 46-81, 1981.
[7] Araújo, L.M., Franca, A.B. & Potter, P.E., Hydrogeology of the Mercosul aquifer system in the Paraná and Chaco-Parana Basins, South America, and comparison with the Navajo-Nugget aquifer system, USA. *Hydrogeology Journal*, **7**, pp. 317-336, 1999.
[8] Piper, A.M., A graphic procedure in the geochemical interpretation of water-analyses. *Eos Trans. AGU*, **25**, pp. 914-928, 1944.

Groundwater hydrochemistry of an overexploited aquifer: the case of the Toluca Valley, Mexico

M. C. Jiménez-Moleón, R. M. Fuentes, M. T. Mota-González & M. Islas-Espinoza
Interamerican Center of Water Resources (CIRA), Autonomous University of the State of Mexico (UAEMéx.), Mexico

Abstract

The demographic, industrial and agricultural growth of recent decades in the State of Mexico has caused a considerable increase in groundwater extraction and the overexploitation is mainly due to drinking water, because it is the supply source for the two densely populated zones in the country: the metropolitan areas of Toluca City and Mexico City and its conurbation. In this work, 49 groundwater sources of drinking water were analyzed in two sampling periods (dry and rainy seasons). According to the Electric Conductivity data, 67% of the samples in the dry and 63% in the rainy season presented weak mineralization and the rest of the samples showed medium mineralization. In addition, the values of hardness showed that the water had good quality, with the exception of 3% of the wells in the rainy season, which had medium quality. The corrosion coefficient showed that two samples in the dry season and four in the rainy season had aggressive characteristics that might damage the extraction and distribution networks. The concentrations of major ions (measured by ionic chromatography) in both sampling periods showed this relative abundance: Na^+>Mg^{2+}>Ca^{2+}>K^+=HCO_3^->>Cl^->SO_4^{2-}. It was found that the HCO_3^- concentration increased with the groundwater flow in the area. Moreover, it was also possible to establish the principal species in solution and to corroborate the hydrochemical change in the area of study. Although there continues to be clear predominance of the Mg-Na-HCO_3 type, the Na-HCO_3 *facies* showed a progressive increase.

Keywords: drinking water, hydrochemistry, major ions, groundwater quality, Toluca Valley, Mexico.

WIT Transactions on Ecology and the Environment, Vol 135, © 2010 WIT Press
www.witpress.com, ISSN 1743-3541 (on-line)
doi:10.2495/WP100031

1 Introduction

The alteration of water quality is one of the most worrying problems in the world. At the time when water demands were modest, with limited industrial activity and unsophisticated agriculture, the discharges were not an overly serious problem. While in the current situation, in which the discharges have proliferated in number and pollutant load and various uses have withdrawn significant amounts of water, the problem has grown worse by increasing the number of pollutant sources. An example of the above is the valley of Toluca, fig. 1, an active industrial area with an accelerated urban and industrial growth, which is located in central Mexico, more particularly in the center of the State of Mexico, one of the valleys with greatest elevation in the Mexican highlands. In the 1970s, the valley had only 13 industries and currently the registered number exceeds 2500, which has led to overexploitation of their aquifers and pollution of their water bodies by sewage and agricultural runoff. Furthermore, in 1970, 230 wells were drilled to cover the deficiency of water in Mexico City; the subsequent exploitation of aquifers has caused negative effects, such as subsidence, cracking, reduced groundwater levels and hydrological imbalance of the basin.

The Toluca Valley is part of the upper watercourse of the Lerma River, with an average elevation of 2570 meters and a length of nearly 700 km^2, is located in

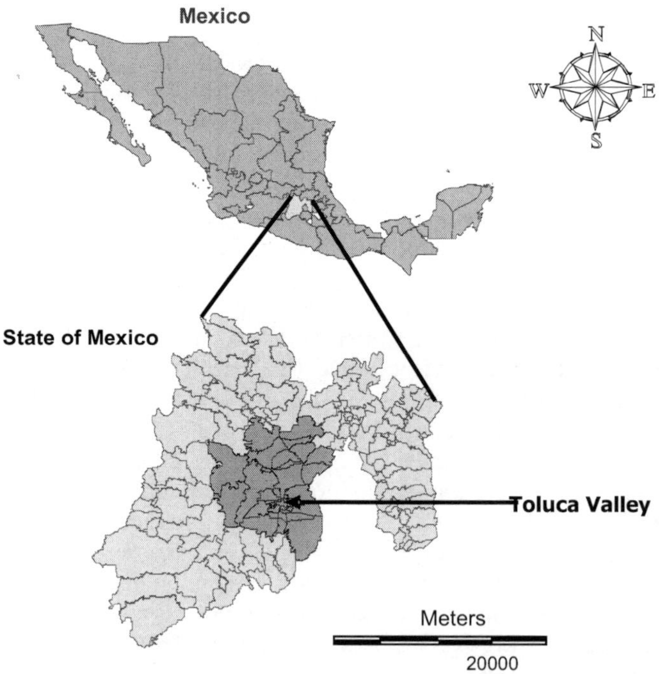

Figure 1: Toluca Valley location.

Figure 2: Direction of groundwater flow in the Toluca Valley (adapted from Franco-Plata [1]).

the physiographic province "Eje Neovolcánico Transmexicano", a region consisting of a strip of volcanic calcoalkaline type, whose age corresponds to the Upper Cenozoic. As general characteristics of this province, the presence of basalts with high permeability due to the degree of fracture should be noted. The groundwater flow has a direction similar to the flow of the superficial waters, which tends to move from the mountains surrounding the valley toward the center and north, to be drained into the Valley of Atlacomulco, fig. 2.

Antón [2] reported that the mineralization of the area coincided with what must have been groundwater in a volcanic environment: feldspars contributed with sodium, calcium and potassium, and pyroxene and biotite, with calcium and magnesium. The presence of sulfates and chlorides was low, although on average this was higher, indicating a recharge of surface water caused by rainwater infiltration. The water according to the Piper diagram was similar to that obtained in the campaigns conducted in previous decades regarding $NaHCO_3$ and/or $Mg(HCO_3)_2$, indicating that the groundwater hydrochemistry in this area had not undergone major changes until then. However, Esteller and Andreu [3] analyzed historical anthropogenic impacts and hydrogeochemical characteristics of the Valley of Toluca aquifer, and concluded that the intense exploitation of this aquifer had caused a change in the groundwater hydrochemistry. In addition to that, initially, the water extracted from the aquifer was linked to a local flow (recharge by rainwater and by lateral feeding from other aquifers), but, by 1998, the hydrochemistry became characteristic of a regional flow, as evidenced by the increase (from 26% to 35%) of the $Na\text{-}HCO_3$ type.

2 Materials and methods

Two samplings were performed, one between the months of May and June 2005 (dry) and one more in October and November of that year (rainy), to assess the possible seasonal influence on the water quality in the study area. Forty nine sampling points were considered in the municipalities of Xonacatlan, Lerma, San Mateo Atenco, Metepec and Toluca, the area most affected by anthropogenic activity in the Valley of Toluca, fig. 2. Sampling, preservation and transportation were made according to criteria established by the APHA-AWWA-WFPC [4]. In addition, for each of the samples the following were determined *in situ*: pH, temperature (T) and electrical conductivity (EC) with a multielectrode Conductronic PC18 portable. Major ions were analyzed by HPLC.

3 Results and discussion

The overall results obtained during the sampling program parameters *in situ* were (dry-rainy): T= 19.9±0.2° C - 18.4±0.1° C; pH= 7.12±0.02 - 7.60±0.01 and EC= 299±163 microS/cm - 311±157 microS/cm. So it was concluded that seasonal variation was not significant, only in the case of EC was the spatial variation significant, tending to increase towards the central part of the area of study. It means that it is parallel to the groundwater flow, indicating that mineralization increases with the contact time in the aquifer. An individual analysis of the wells determined that the wells that were subject to greater exploitation also have a higher degree of mineralization.

The degree of mineralization was determined according to [5], obtaining that in the dry period, 10% of the wells showed a weak mineralization, 57% medium accentuated mineralization and 33% medium mineralization, while in the rainy season the range of conductivity values narrowed, but increased the mineralization, since 12% of samples showed weak mineralization, 51% medium accentuated mineralization and the rest were medium.

The concentrations of the major anions and their relationship to each other can be seen in fig. 3, while table 1 shows the minimum, maximum and average values of both sampling periods. The average of all these anions hardly varies seasonally, following the order: $HCO_3^- >> Cl^- > SO_4^{2-}$. However, reviewing the limits of the range of variation, it can be seen that while the bicarbonate ion tended to increase in the rainy season, the sulfate ion was concentrated during the dry season. The ranges of variation of the most abundant anions, and their averages, are consistent with those reported in previous studies by CONAGUA, the National Water Commission [6].

Table 1: Minimum, maximum and average values of the major anions in the Toluca Valley wells.

Concentration	Minimum		Maximum		Average	
(mg/L)	Dry	Rainy	Dry	Rainy	Dry	Rain
HCO_3^-	49.67	70.87	241.6	257.2	138±34	140±44
Cl^-	7.66	6.93	32.27	23.71	14.17±0.38	12.80±0.24
SO_4^{2-}	0.84	1.04	12.28	12.41	4.52±0.19	4.00±0.10

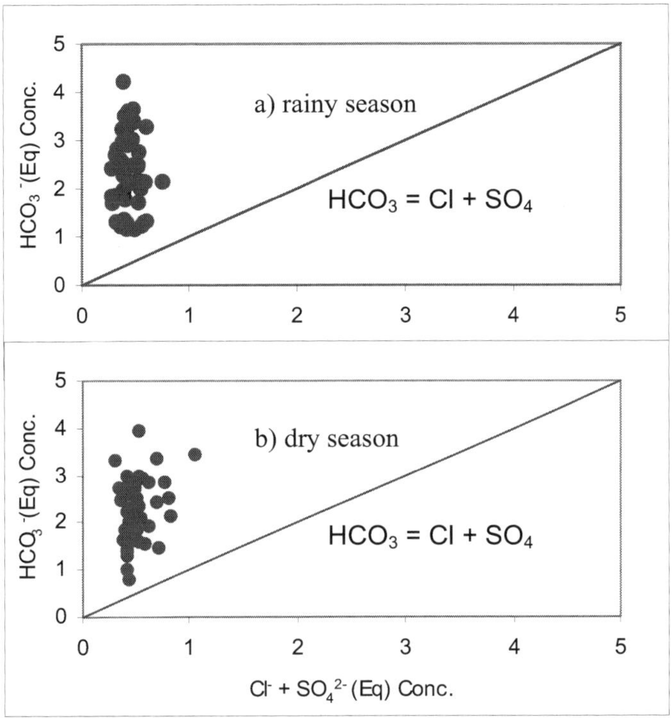

Figure 3: Relation HCO_3^- versus $Cl^- + SO_4^{2-}$.

In fig. 3, as described by the data in table 1, in both sampling periods a single group was identified in the groundwater of the Valley of Toluca, which clearly had the bicarbonate ion as predominant. Thus, the low presence of sulfate and chloride in the Toluca Valley wells is logical, because the soil in the area contains these ions in small quantities. In fig. 3 it also can be seen that while the sum of the concentrations of Cl^- and SO_4^{2-} showed a little spatial variation, the HCO_3^- concentration increased considerably from one well to another. Given the location of the wells, the low content of bicarbonate found in the eastern zone (Municipality of Ocoyoacac), i.e., in the areas closest to the recharge zones, and that the content of HCO_3^- increased parallel to the groundwater flow is remarkable. Therefore, the increase in the residence in the aquifer determined that water could dissolve more and thus, increase its content of HCO_3^-.

Furthermore, it is noteworthy that according to results of the groundwater balance presented by [7], the natural vertical recharge in the valley of Toluca has been increasing year by year: by 21% during 1992-1996, and by 81% during 1996-2000. While the recharge from the mountains (Nevado de Toluca and Sierra de las Cruces) increased by just 0.11% in 2000, i.e., by 2000, the vertical natural recharge in the center of the Valley was 12% higher than the recharge in the mountains. This effect might be associated with the increase in bicarbonate

Table 2: Concentration (mg/L) of major cations in Toluca Valley wells.

Well	Dry				Rainy			
	Na	K	Mg	Ca	Na	K	Mg	Ca
1	20.80	2.78	14.51	12.16	22.28	3.25	14.91	14.82
2	23.25	3.20	18.25	13.24	23.60	2.87	18.07	12.74
3	17.55	3.21	14.17	11.59	20.45	4.72	19.72	18.65
4	17.69	4.81	20.34	19.30	21.19	5.24	21.49	20.80
5	19.74	4.19	16.76	16.30	20.13	4.34	17.37	17.21
6	15.79	3.24	11.77	11.70	14.19	3.34	11.02	11.29
7	19.94	2.94	13.78	13.36	20.34	2.49	13.79	11.82
8	15.21	3.37	11.70	11.61	15.33	3.21	11.11	11.20
9	20.72	3.87	11.15	12.81	19.40	2.58	14.06	11.85
10	21.97	4.82	13.42	14.27	20.27	4.52	14.63	14.98
11	19.82	2.17	11.80	10.67	19.83	1.51	12.72	10.88
12	14.73	2.21	10.76	9.11	16.35	2.33	11.14	9.61
13	16.15	3.76	13.21	13.27	16.75	3.88	15.67	14.22
14	14.68	3.46	10.18	10.20	15.71	4.53	10.13	11.20
15	14.22	2.02	8.29	9.15	14.78	1.74	8.40	9.32
16	30.32	2.58	21.81	16.10	32.87	2.72	22.94	21.26
17	21.25	2.15	12.85	10.74	20.46	1.90	13.26	11.29
18	21.22	2.23	14.40	12.77	19.67	1.87	13.16	11.92
19	26.92	2.15	17.41	13.47	26.35	1.77	17.56	13.22
20	22.27	3.93	15.87	11.29	24.58	3.95	13.84	10.20
21	15.23	2.36	9.62	11.06	16.17	3.29	9.93	10.53
22	46.72	6.68	8.94	11.97	49.90	4.89	9.22	12.26
23	24.57	3.62	9.72	13.32	24.39	4.26	11.38	15.32
24	22.78	4.83	6.94	8.88	17.61	2.64	10.24	12.72
25	16.36	3.38	8.65	9.91	17.23	2.74	9.82	12.45
26	27.62	3.66	15.12	14.35	27.69	3.77	15.34	14.17
27	28.22	2.62	12.39	14.14	28.89	2.97	12.46	13.73
28	24.39	2.65	11.25	11.90	19.14	2.63	11.84	12.13
29	18.91	2.08	11.36	10.52	19.74	1.91	11.86	10.70
30	18.33	2.12	10.94	10.43	18.38	2.27	10.78	10.87
31	21.69	3.21	10.34	10.29	18.35	3.56	15.90	15.28
32	21.34	1.88	12.78	11.82	19.38	1.44	11.20	10.77
33	19.85	3.27	18.85	17.51	27.32	4.57	31.44	27.26
34	26.85	2.55	10.79	11.19	21.78	2.83	13.12	13.25
35	19.35	3.65	15.04	14.97	20.05	3.72	16.26	15.59
36	19.45	3.14	8.20	8.62	19.77	2.81	13.20	14.90
37	18.74	2.80	12.62	14.18	19.19	2.73	13.19	14.90
38	33.67	6.64	15.34	18.87	19.97	2.84	13.11	14.97
39	15.28	3.93	7.28	9.75	14.70	3.68	6.27	7.80
40	11.64	3.32	5.92	8.70	14.36	3.52	5.92	7.54
41	16.16	2.56	10.22	10.92	14.69	1.48	10.75	9.33
42	13.87	1.44	10.57	9.74	13.96	1.31	11.14	9.73
43	16.63	1.19	10.76	10.22	17.64	0.84	11.82	9.92
44	22.90	1.53	8.05	8.81	24.89	1.56	7.96	8.93
45	18.25	2.06	13.59	11.17	27.36	3.58	17.24	15.27
46	16.61	1.65	13.10	10.89	19.37	5.60	11.83	9.52
47	18.09	0.92	11.59	9.34	19.26	0.87	11.57	9.28
48	18.33	0.83	11.19	9.20	18.26	3.54	9.50	13.62
49	18.76	3.38	13.91	12.07	18.34	3.65	9.23	11.98

concentrations observed in 49% of the sources of supply during the rainy season. However, it must be noted that the 51% of the remaining sources of supply evidenced a dilution phenomenon as an effect of the rainfall, which made them sources of a potential risk of contamination.

Subsequently, the water tendency to corrosion was determined with the Larson index, k, [8, 9]: $k= (2[SO_4^{2-}]+[Cl^-])/[HCO_3^-]$. According to [8], a k-value ≤ 0.25 indicates resistance to corrosion, while higher values are generally more corrosive and can damage the water supply systems. The k obtained data indicated that 2/49 wells in the dry season and 4/49 in the rainy season showed aggressive characteristics.

Table 2 shows the concentrations obtained from the major cationic constituents quantified in the sources of supply in the Toluca Valley area during both sampling periods.

Once the concentrations of the major components in both sampling periods were obtained, the ionic balance was made to all the points under analysis, finding that the error was always less than 5%.

The average values of Ca^{2+} in the dry and rainy seasons were 11.99 mg/L and 12.92 mg/L, respectively. Spatially analyzing the data, it was observed that Ca^{2+} concentration increased towards the central part of the study area. The Mg^{2+} values were slightly higher than those of Ca^{2+}: with averages of 12.39 mg/L, in the dry season, and 13.23 mg/L in the rainy season, following a spatial distribution very similar to Ca^{2+}. Sodium, with average concentrations of 20.50 mg/L and 20.65 mg/L in the dry and rainy season, respectively, became the main cation; while the average concentrations of K^+ in the dry and rainy seasons were 3.0 mg/L and 3.02 mg/L, respectively, confirming (as expected) the conservative value of these parameters.

From the values obtained in this study, it could be concluded that the hydrogeochemical feature of the water in the area of study is the predominance of bicarbonate and sodium ions, which is consistent with other studies in the Toluca Valley [3, 6, 10, 11]. The relative abundances of the major ions in water were found in the following order: $Na^+>Mg^{2+}>Ca^{2+}>K^+=HCO_3^->>Cl^->SO_4^{2-}$.

The total hardness (TH) was determined by stoichiometric calculations from the concentrations of calcium and magnesium, so the quality of the water in the study area could be classified [5] as very good in the dry season (100% of well samples with values of TH < 122 mg $CaCO_3$/L), while in the rainy season, 98% of the samples had between 43 and 48 mg $CaCO_3$/L (good quality) and 2% had a value of 199 mg $CaCO_3$/L (medium quality).

The graphical treatment of major ion chemistry [12] was analyzed. Thus, the affinity of $Na^+ + K^+$ versus Cl^- was studied, plotting their equivalent concentrations, because the Cl^- ion tends normally to be associated more with alkali metals than with Ca^{2+} and Mg^{2+}. In fig. 4, it is deduced that Na^+ and K^+ might be related to Cl^- ions and there could be an excess available to join other anions. Fig. 4 corresponds to the rainy season, but the same trend was observed in the dry season. In addition, the natural affinity of the relation between $Ca^{2+} + Mg^{2+}$ and HCO_3^- was graphically evaluated, fig. 5, where it is shown that in most cases, even after forming the aqueous species $Ca\text{-}HCO_3$ and $Mg\text{-}HCO_3$, there was an excess of HCO_3^- that could be related to other relatively abundant ions (Na^+ and K^+). Fig. 5 corresponds to the dry season; however, the same trend occurred in the rainy season. In conclusion, the relationships found by considering the affinity among ions (figs. 4 and 5), the aqueous species Na-HCO_3, Ca-HCO_3, Mg-HCO_3, Na-Cl and K-Cl, were likely to occur in the Toluca Valley groundwater system.

Finally, the geochemistry of water was analyzed through the Piper trilinear diagram (fig. 6); the presented figure corresponds to the rainy season, although the trend is the same in both sampling periods. The samples were mainly of the NaMg-HCO_3 type, but also defined were MgNa-HCO_3 and MgCa-HCO_3, indicative of a volcanic environment. Given the location of the wells, in the north and northeast of the area (Lerma, Ocoyoacac and Xonacatlan), the water mostly belonged to the Na-HCO_3 family, while in the south and southwest, the Mg-HCO_3 and Ca-HCO_3 facies were observed. With respect to cations, the water was classified as mixed, while for anions, the water was totally bicarbonated.

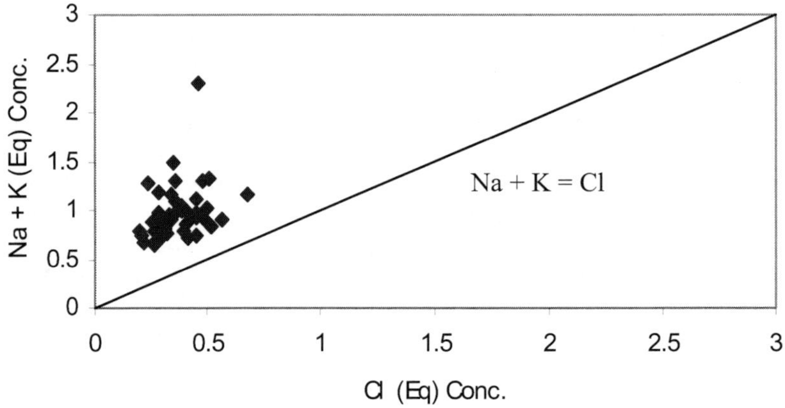

Figure 4: Relation alkali versus Cl^- (rainy season).

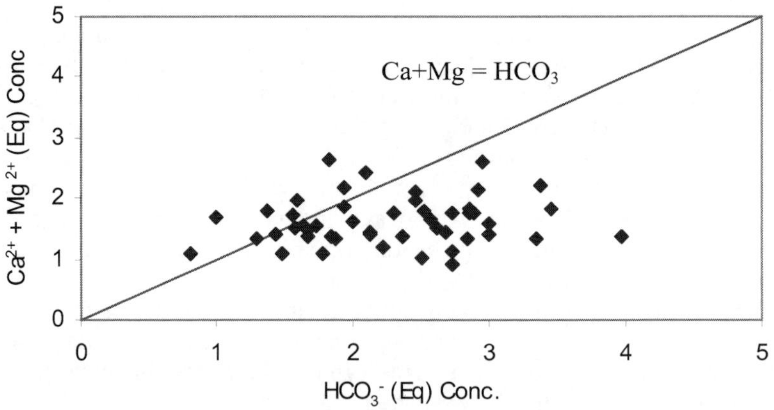

Figure 5: $Ca^{2+} + Mg^{2+}$ versus HCO_3^- (dry season).

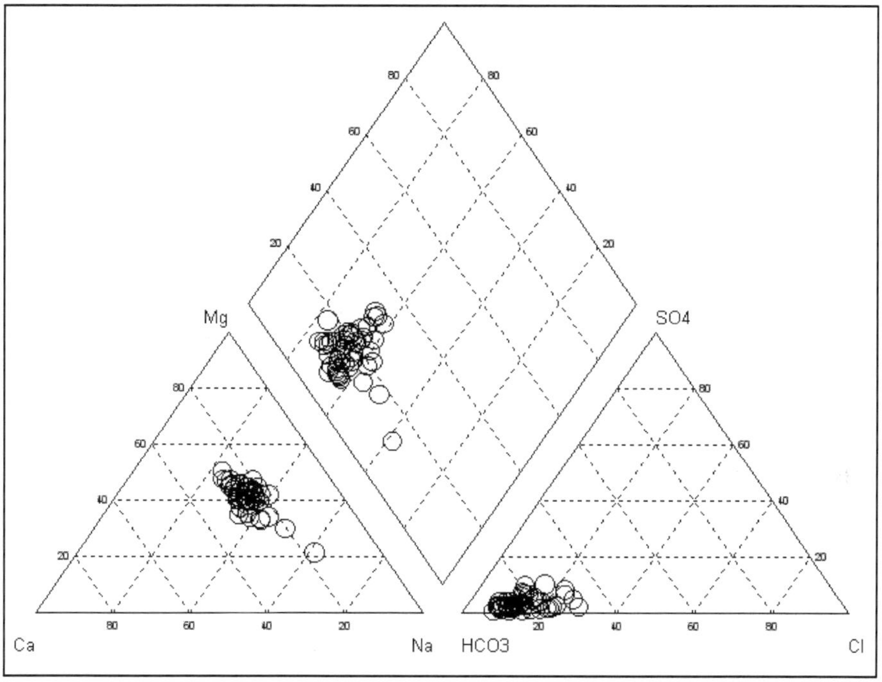

Figure 6: Piper diagram for the Toluca Valley wells (rainy season).

Historically, the researches on the area of study concluded that the hydrochemical facies of the Toluca Valley would not have presented changes until 1986 (MgNa-HCO_3, NaMg-HCO_3) [7]; however, by 1998 a change in the hydrochemistry was reported, which began to show an increase in the Na-HCO_3 facie, reaching 38% [3]. This trend was confirmed and accentuated in the present study, where this facie reached 47% of the samples. The progressive increase of the sodium cation in the water is probably due to the dominating soil type in the area (rhyolites and trachytes) and to the overexploitation that has been exerted on the resource.

4 Conclusions

After treatment and analysis of results, the following conclusions were reached.

The physicochemical parameters, T, pH and EC showed no great seasonal variation, although in the case of EC, the range of spatial variation is quite large. According to EC, the 33% and 37% of the samples in the dry and rainy seasons, respectively, showed marked mineralization.

The chemical composition of water did not show a significant seasonal change. The relative abundance of ions was: $Na^+ > Mg^{2+} > Ca^{2+} > K^+ = HCO_3^- >> Cl^- > SO_4^{2-}$. In addition, it was found that the main anion concentration (HCO_3^-)

increased in parallel to groundwater flow, i.e., when the residence time of water in the aquifer was greater.

With regard to the total hardness, the water had good quality, except for 3% of the wells in the rainy season that had an average quality.

The coefficient of corrosion (k) showed that two samples in the dry season and four in the rainy season showed aggressive characteristics that could damage the extraction and distribution network.

Chemical affinity studies identified that the more likely species in solution were $NaHCO_3$, $Ca(HCO_3)_2$, $Mg(HCO_3)_2$, NaCl and KCl.

The hydrochemical classification of the water through the Piper diagram showed a clear predominance of the MgNa-HCO_3 *facies*. In addition, the hydrochemical change in the drinking water sources due to the gradual increase in the Na-HCO_3 type was confirmed.

Acknowledgement

This publication was made possible thanks to the financial support of the Autonomous University of the State of Mexico, through the Secretary of Research and Advanced Studies, Projects 2005/2005 and 2506/2007U.

References

[1] Franco-Plata, R. Adecuación del método de evaluación de vulnerabilidad de acuíferos (DRASTIC), mediante el uso del trazador de partículas en el acuífero del curso alto del río Lerma, Estado de México, M. Sc. Thesis CIRA-UAEMéx., 2002.

[2] Antón, D., Las aguas subterráneas (Chapter 6). *Sequía en un mundo de agua*, ed. Piriguazú Ediciones/CIRA-UAEMéx.: San José/Toluca, pp.67-100, 2000.

[3] Esteller, M.V. & Andreu, J.M., Anthropic effects on hydrochemical characteristics of the Valle de Toluca aquifer (Central Mexico). *Hydrogeology Journal,* **13(2)**, pp.378-390, 2005

[4] APHA-AWWA-WFPC, *Standard Methods for Examination of Water and Wastewater*, 17th edition, American Public Health Association & Water Pollution Control Federation: Washington, 1992.

[5] Rodier, J., *Análisis de aguas. Aguas naturales, aguas residuales, agua de mar*, Omega S.A.: Barcelona, 1998.

[6] CONAGUA, *Programa Hidráulico Nacional*, Comisión Nacional del Agua: México D.F., 1998.

[7] CONAGUA, Determinación de la disponibilidad de agua en el acuífero Valle de Toluca, Comisión Nacional del Agua: México D.F., 2002.

[8] Özcan, H., Hüseuin, E., Alper, B., Yasemin, K., Orhan, Y. & Yiğini, Y., Assessment of the water quality of Troia for the multipurpose usages. *Environmental Monitoring and Assessment*, **130(1-3)**, 389-402, 2007.

[9] Zhang, Y. & Edwards, M., Anticipating effects of water quality changes on iron corrosion and red water, *Journal of Water Supply: Research and Technology- AQUA*, **56(1)**, 55-68, 2007
[10] Esteller, M.V. & Díaz-Delgado, C., Environmental effects of aquifer overexploitation: A case study in highlands of Mexico. *Environmental Management*, **29(2),** 266-278, 2002.
[11] Espinoza, R. Relación entre factores antrópicos y naturales y la concentración de nitrato en agua subterránea del valle de Toluca. M. Sc. Thesis, Faculty of Chemistry-UAEMéx., 2005.
[12] Umar, R., Muqtada, M., Khan, A. & Absar, A., Groundwater hydrochemistry of a sugarcane cultivation Belt in parts of Muzaffarnagar district, Uttar Pradesh, India, *Environmental Geology,* **49(7)**, 999-1008, 2006.

Volatilization of benzene from gasoline: the effect of ethanol blends

F. Fedrizzi[1], E. Lovatel[2], N. Vieceli[2], E. Cardoso[2], J. Cagliari[2], A. Finotti[2] & I. Nascimento[2,3]
[1]*Environmental Engineering Course, University of Caxias do Sul, RS-Brazil*
[2]*Environmental Sanitation Institute, University of Caxias do Sul, RS-Brazil*
[3]*Center of Exact Sciences and Technology, Environmental Sanitation Institute, University of Caxias do Sul, RS-Brazil*

Abstract

In this work experiments were performed with vapor collection in columns to assess the evaporation profile of benzene from gasoline-ethanol blend fuels. The vapors from two columns simulating gasoline-contaminated soils (with and without ethanol) were monitored for 77 d. The instrumental analysis was performed by Gas Chromatography (GC) with a Flame Ionization Detector (FID). Compound identification was based on the GC retention times of standard BTX (benzene, toluene and xylenes). The concentration of benzene in the vapor phase decreased by 89.09%, considering the entire experimental period, while the toluene and xylene concentrations were increased by 239.34 and 251.78%, respectively. These results suggest that the benzene evaporation behavior was affected by the interactions among ethanol and other aromatic compounds. These results are particularly important, since ethanol is an alternative to gasoline blends. Furthermore, benzene (a well known carcinogen) was retained in the liquid phase and, in this way, can reach the underground water sources. These findings can point out necessary changes in traditional risk models for soil spills that are based on compound concentrations in the environment.

Keywords: gasoline, ethanol, benzene, BTX, evaporation.

WIT Transactions on Ecology and the Environment, Vol 135, © 2010 WIT Press
www.witpress.com, ISSN 1743-3541 (on-line)
doi:10.2495/WP100041

1 Introduction

After the first oil crisis in the 1970s, with growing environmental pollution and the need for clean energy production, several nations around the world developed programs to support the use of alternative fuels, including ethanol and gasohol (which is a gasoline-ethanol blend) as automobile fuels. The main advantages of the ethanol blend fuels are the reduction of greenhouse gas emissions and enhancement of the octane number of gasoline. The ethanol content in Brazilian gasoline ranges from 24 to 25%.

Brazil has two national programs to insert ethanol in its energy matrix. In the first one combustible anhydrous ethyl alcohol (AEAC) is mixed with gasoline in volume percentages varying from 20 to 25%. The mixture volume is established in a law enacted by the government and depends on market and economic variables. In the other program, combustible hydrated ethyl alcohol (AEHC) is sold as fuel to be used in vehicles powered exclusively by ethanol or by flex fuel. Anhydrous ethyl alcohol has been used in the country since 1939 as a gasoline additive with percentages varying from 5 to 25%. As a consequence, Brazilian ethanol production has grown from 31,499 m^3 of AEAC and 62,215 m^3 of AEHC in 1939/1940 to 8,400,000 m^3 of AEAC and 13,900,000 m^3 of AEHC in 2007/2008. The technical process utilized in AEAC makes this product very competitive on the domestic and international markets.

Despite the mixture benefits, the presence of ethanol in gasoline can affect the BTX (benzene, toluene and xylenes) volatilization behavior and these compounds can evaporate from spills of underground storage tanks or by accidental releases raising a vapor plume just above the contaminated soil. In the soil spill situation, the rate of evaporation is a function of temperature, wind-speed, atmospheric conditions, solar radiation, dimension of the oil spill and the volatility and diffusion characteristics of the fuel [1].

Compounds with four to six carbon atoms volatilize rapidly. In the gasoline case BTX are the chemicals of greatest concern because of their toxicity and benzene carcinogenic activity [2]. In gasohol the presence of ethanol can modify the dissolution capacity of BTX and increase the groundwater pollution. Some of these changes reported in the literature are variations of free phase measurements in monitoring wells, co-solvency effects that render the BTX dissolution capacity higher in presence of ethanol, preferable biodegradation of ethanol in gasohol plumes, which causes anoxic conditions for BTX degradation [3, 4] and effects of gasohol contamination [5, 6]. Other studies focus on the thermodynamic properties of ethanol and gasoline-blended fuel, behavior of entrapped petroleum products in the subsurface, and soil vapor extractions systems [7-9] or measurements of the vapor pressure enhancement of alcohols-gasoline blends [10]. Thus, despite the well-known harmful effects of BTX in the atmospheric environment, the quantitative analysis of the influence of the ethanol on the evaporation rates of each individual chemical is still lacking in the literature.

The main objective of the present study was the quantitative analysis of the influence of ethanol on the evaporation behavior of BTX, compared to neat gasoline. With this purpose, two columns were prepared and filled with selected sand. One column was contaminated, with neat gasoline and other with a gasoline-ethanol mixture.

2 Materials and methods

2.1 Samples and solvents

The gasoline samples, without ethanol, were kindly provided by Motors Laboratory of the Alberto Pasqualini refinery – Brazilian Petroleum (REFAP – PETROBRAS). Anhydrous ethanol (25 ml) was mixed with 75 ml of neat gasoline. All reagents and solvents used in this work were of analytical grade. The gasoline samples presents 30.63 (% mass) of aromatics, 11.72 (% mass) of alkanes, 17.90 (% mass) of substituted alkanes, 39.75 (% mass) of other components, and vapor pressure of 51.5 kPa. The BTX standards were purchased from Merck (São Paulo, SP, Brazil) and used with no further treatment.

2.2 Experimental columns

Two plexiglass columns were contaminated with neat gasoline and gasoline-ethanol. The columns present the following characteristics: height: 0.97 m; internal diameter: 0.35 m; total volume: 93.32 L. The columns were filled with 115 kg of washed river sediment with the following characteristics: particle diameters ranging from 0.25 to 0.59 mm, constituted mostly by quartz, uniformity coefficient and coefficient of curvature of 1.08 and 0.89, respectively. A bubbling pressure of a 14.8 centimeter water column was used in this work. The columns were filled with water, from bottom to top to avoid air bubbles and allowed to rest for one week. After this, the columns were left to drain until the water level was 0.5 m high from the base. Initial conditions of the columns were obtained by introducing 1.15 L of neat gasoline in the upper layer of column 1 and the same volume of a gasoline-ethanol mixture (25% v/v of ethanol) in the upper layer of column 2. Oil Red dye (Aldrich 200 mg/L) was added to the gasoline and the gasoline-ethanol mixture in order to monitor the gasoline liquid plume variation. This kind of dye is largely used in laboratory experiments with petroleum hydrocarbons as reported by MacDowell et al. [11]. After the contamination, the columns were closed and the junction between the column body and the cap was sealed with silicon to avoid vapor loss. Gas bags were placed at the top of both columns to collect volatile substances. The Columns were sealed to avoid the influence of the external atmosphere such as wind and pressure variation. The environmental temperature was monitored daily and showed an average value of 20.3 °C.

Plexiglas has previously been used for studies with organic compounds such as the study of biofiltration of BTX [12] and the study of naphthalene bioavailability in a Plexiglass glove box [13]. Baek et al. [14] studied the benzene sorption in sand and conducted a test to verify the influence of Plexiglass® in the results. In a separate experiment concentrations of benzene solution were measured with time for glass and Plexiglass containers under a closed system. They found that the concentrations of benzene solution in the glass container were lower than those in the Plexiglass container, the greatest difference being 9.8%. This indicates that the use of Plexiglass for the column test would not greatly affect the experimental results on the transport and sorption of benzene in their study.

2.3 Instrumental analysis

The instrumental analysis was performed with a gas chromatograph PerkinElmer (model Autosystem–XL) with flame ionization detection. The samples were injected in the splitless mode. A 30 m Elite column (film thickness 0.25 μm, internal diameter 0.25 mm) was temperature-programmed from 40 °C (held for 10 min) to 220 °C (held for 0 min) at 5 °C/min. The carrier gas (Helium) flux was 1mL/min. The quality assurance/quality control of analysis was developed on the basis of the ANR Analytical Lab Quality Assurance Manual (August 25, 2006, University of California, USA).

3 Results and discussion

Figure 1 shows the weekly BTX concentration in the neat gasoline and gasoline-ethanol columns, during the sampling period. Although the samples in the present work cannot be classified as soil, these results partially disagree with enhancement of the BTEX (benzene, toluene, ethyl benzene, and xylenes) retention in soil by ethanol reported by Österreicher-Cunha et al. [15] that studied the degradation and distribution processes of ethanol-amended gasoline in tropical soil under unsaturated conditions. Considering that the conditions in the neat gasoline and gasoline-ethanol columns are the same and that some biodegradation occur, the benzene concentration must be higher in the gasoline-ethanol column, since that to Österreicher-Cunha ethanol could delay BTEX degradation because of its constitutive degradation by soil microbiota. However, this was not observed in our results, suggesting that, compared to the biodegradation, co-solvency effects must play a more important role it the reduction of the benzene concentration in the headspace.

In the present work benzene, toluene, and xylenes had their maximum concentrations in the gasoline-ethanol column, at weeks 6, 4, and 11, respectively. Figure 2 shows the relative volatilization of BTX, considering the final and initial times of the entire experimental period where the decrease of the benzene concentration and the increase of the toluene and xylenes volatilization, in the vapor phase can be better observed. Toluene and xylenes shows the highest relative volatilization in the gasoline-ethanol column (239.34 and 257.78%, respectively). These results can be better explained by the cosolvent theory approach.

According Chen et al. [16] the addition of polar organic solvents (such as ethanol) that are completely-miscible or highly-soluble in water to a mixture of hydrocarbons (such as gasoline) and water initiates a cosolvent effect. Water-miscible cosolvents may reduce the net polarity of the oxygenate-water

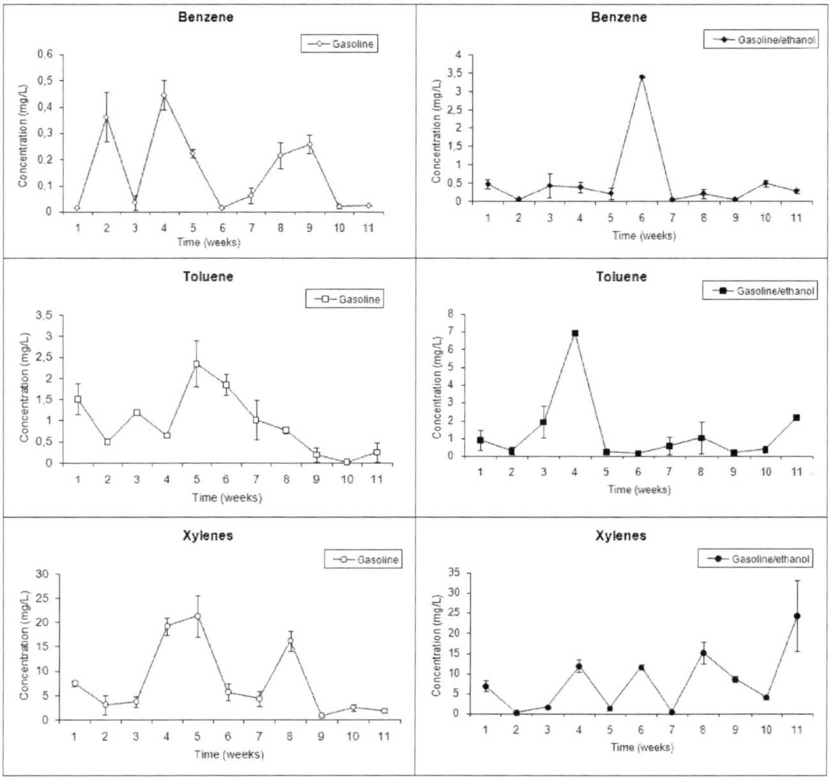

Figure 1: Weekly variation of the benzene concentration in the neat gasoline and gasoline-ethanol columns. Error bars represent ± standards deviations.

solvent. When a miscible constituent is added to water, it increases the quantity of nonionic organic compounds that can dissolve in the mixed solvent. In this way, while the ethanol increases the water solubility of toluene and xylenes due to the co-solvency effect, the distance among the molecules of those compounds is increased as well their volatilities. Although attractive intermolecular forces are not expected to occur among ethanol and benzene molecules, the decrease of the benzene concentration in the vapor phase of the ethanol-gasoline column should be due to the enhancement of the toluene and xylenes volatility.

Many works have been published regarding the cosolvency promoted by gasoline-ethanol mixtures. According Yu et al. [17] the proportions of BTX in ethanol-amended gasoline are changed due to the cosolvent behavior of ethanol in the presence of an aqueous phase, as well as a reduction in the interfacial tension between the aqueous/nonaqueous phases. The addition of ethanol to the gasoline hydrocarbons can generates a mixture with a boiling point lower than the original compounds. To our knowledge, the present work is the first one reporting measurements of the volatilization of BTX in laboratory sand

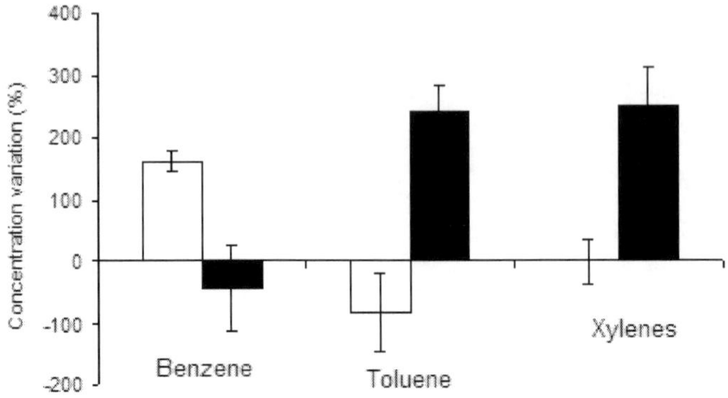

Figure 2: Relative volatilization of benzene toluene and xylenes on the experimental columns, considering the final and initial experimental times. Error bars represent ± standards deviations. Neat gasoline (□); gasoline/ethanol (■).

Table 1: Some physical properties of ethanol and BTX.

Compound	Dipole moment (D)	Density ($g\ cm^{-3}$)	Water solubility ($mg\ L^{-1}$)
Ethanol	1.69	1.49	∞
Benzene	0	0.87	175
Toluene	0.37	0.86	535
Xylenes*	0.91	0.61	167

*Average value of the *o*, *m* and *p* isomers.

columns. As such, published results cannot be compared. Pasteris et al. [18] reported similar gas phase concentrations of toluene and *m*-xylene, evaluating the vapor phase transport and biodegradation of gasoline compounds. Those authors studied a large-scale field lysimeter representing a 2.3 m thick sandy unsaturated zone over a gravel aquifer. However, no comparisons were made with gasoline-ethanol fuel.

The co-solvency effects are appreciable among polar (ethanol) and nonionic organic compounds of low polarity (gasoline hydrocarbons). In the present work, the high relative volatilization of toluene and xylenes can be due to the permanent dipole forces that occur among ethanol-toluene and ethanol-xylenes. Table 1 show that toluene and xylenes presents the highest dipole moment.

To Kanai et al. [19] the intermolecular forces are strong enough to allow the chromatographic separation of *meta* and *para* isomers of xylenes.

4 Conclusions

Ethanol may decrease the volatilization of benzene in gasoline-ethanol column. On the other hand, concentrations of toluene and xylenes in the vapor phase of the gasoline-ethanol column increases drastically. Besides the presence of ethanol, other aspects concerning the BTX volatilization, like adsorption on the sand grains and biodegradation must be addressed.

On the basis of the results of the present study, we can infer that if ethanol promotes the displacement of the contaminant gasoline plume down the contaminated sand, and if benzene is retained in the liquid phase due to the cosolvency effects, a higher concentration of this contaminant can reach the underground water and pose a risk to this important water source.

Acknowledgements

The authors would like to thank CNPq, Pronex Fapergs/CNPq, and University of Caxias do Sul for financial support and REFAP PETROBRAS for the material support.

References

[1] Mackay, D., & Matsugu, R., Evaporation rates of liquid hydrocarbon spills on land and water. *Canadian Journal of Chemical Engineering*, **51**, pp. 434-439, 1973.

[2] Hsieh, C., Chang, K., & Kao, Y., Estimating the ozone potential formation on volatile aromatic compounds in vehicles tunnels. *Chemosphere*, 39, pp. 1433-1438, 1999.

[3] Courseuil, H.X., & Alvarez, P.J., Natural bioremediation perspectives for BTEX-contaminated groundwater in Brazil: effect of ethanol. *Water Science and Technology*, **34**, pp. 311-318, 1996.

[4] Courseuil, H.X., Aires, J.R., & Alvarez, P.J., Implications of the presence of ethanol on intrinsic bioremediation of BTX plumes in Brazil. *Hazardous Waste Hazardous Materials*, **13**, pp. 213-221, 1996.

[5] Powers, S.E., Rice, S., Dooher, B., & Alvarez, P.J., Will ethanol-blended gasoline affect groundwater quality? *Environmental Science and Technology*, **35**, pp. 24A-30A, 2000.

[6] Powers, S.E., Hunt, C., Heerman, S., Courseuil, H., Rice, D., & Alvarez, P., The transport and fate of ethanol and BTEX in groundwater contaminated by gasohol. *Critical Reviews on Environmental Science and Technology*, **31**, pp. 79-123, 2001.

[7] Nan, Z., & Tan, Z.C., Thermodynamic properties of ethanol and gasoline blended fuel. *Energy and Fuels*, **19**, pp. 2432-2437, 2005.

[8] Dror, I., Gerstl, Z., Prost, R., & Yaron, B., Abiotic behavior of entrapped petroleum products in the subsurface during leaching. *Chemosphere*, **49**, pp. 1375-1388, 2002.

[9] Yang, J.W., Cho, H.J., Choi, G.Y., & Lee, S.H., Cost-effective monitoring for a soil vapor extraction (SVE) system. *Environmental Monitoring and Assessment*, **70**, pp. 201-210, 2001.

[10] Pumphrey, J.A., Brand, J.L., & Scheler. W.A., Vapour pressure measurements and predictions for alcohol – gasoline blends. *Fuel*, **79**, pp. 1405-1411, 2000.
[11] McDowell, C.J., Buscheck, T., Powers, S.E., Behavior of gasoline pools following a denatured ethanol spill. *Groundwater*, **41**, pp. 746–757, 2003.
[12] Abumaizer, R.J, Kocher, W., & Smith, E.H., Biofiltration of BTEX contaminated air streams using compost-activated carbon filter media. *Journal of Hazardous Materials*, **60**, pp. 111- 126, 1998.
[13] Burgos, W.D., Berry, D.F., Bhandari, A., & Novak, J.T., Impact of soil–chemical interactions on the bioavailability of naphthalene and 1-naphthol. *Water Research*, **33**, pp. 3789–3795, 1999.
[14] Baek, S.S., Kim, S.B., & Kim, D.J., Irreversible sorption of benzene in sandy aquifer materials. *Hydrological Processes*, **17**, pp. 1239–1251, 2003.
[15] Österreicher-Cunha P, Vargas EA Jr, Guimarães JRD, Lago GP, Antunes FS, Silva MIP. 2009. Effect of ethanol on the biodegradation of gasoline in an unsaturated tropical soil. *Int Biodeterior Biodegrad* 63:208-216.
[16] Chen, C.S., Rao, P.S.C., & Delfino, J.J., Oxygenated fuel induced cosolvent effects on the dissolution of polynuclear aromatic hydrocarbons from contaminated soil. *Chemosphere*, **60**, pp. 1572–1582, 2005.
[17] Yu, S., Freitas, J.G., Unger, A.J.A., Barker, J.F., & Chatzis J., Simulating the evolution of an ethanol and gasoline source zone within the capillary fringe. *Journal of Contaminant Hydrology*, **105**, pp. 1–17, 2009.
[18] Pasteris, G., Werner, D., Kaufmann, K., & Höhener, P., Vapor phase transport and biodegradation of volatile fuel compounds in the unsaturated zone: a large scale lysimeter experiment. *Environmental Science and Technology*, 36, pp. 30-39, 2002.
[19] Kanai, H., Inouye, V., Yazawa, L., Goo, R., & Wakatsuki, H., Importance of Debye and Keesom interactions in separating m-Xylene and *p*-Xylene in GC–MS analysis utilizing PEG stationary phase. *Journal of Chromatographic Science*, **43**, pp. 57-62, 2005.

Section 2
Pollution prevention and control

On experiments on oil-spill booms in three Atlantic coastal zones: La Rochelle, Lorient and Brest

F. Muttin
Engineering School EIGSI La Rochelle, France

Abstract

The intrinsic performance of the protection of coastal zones of the Atlantic against oil-spills by using floating booms needs to be improved in the current conditions. To define a contingency plan in advance, a one-dimensional numerical code has been used for many years. A three-dimensional membrane finite-element method is proposed to give more complete results on boom geometry and boom stress. To improve the booms' performances and validate numerical methods three experiments are described. The first experiment is conducted in the open-sea harbour of La Rochelle, with current conditions higher than 3 knots. In the Lorient oil wharf an experiment attempt shows the constraints due to the uncertainties of a coastal zone with harbour infrastructure under extension and intensive maritime traffic. In the Elorn estuary, Brittany, France, a third experiment is described with current conditions bounded by 0.9 knots. The environmental performance of a floating boom is also evaluated using an economic approach. The budgets of these three experiments are evaluated and analyzed. These budgets estimate the technical and human efforts needed to use a boom in such coastal zones. The knowledge of both environmental and economical performances must be added to the theoretical performances given by numerical computations.

Keywords: spilled oil, floating boom, environmental evaluation, economic evaluation, numerical methods, scale-one experiments.

1 Introduction

The protection of the maritime environment against oil pollution is governed by several international conventions (MARPOL, Boon Agreement, ERIKA

WIT Transactions on Ecology and the Environment, Vol 135, © 2010 WIT Press
www.witpress.com, ISSN 1743-3541 (on-line)
doi:10.2495/WP100051

packages…) and by many international establishments (REMPEC, EMSA, OSRL…), as well as the initiative action of the World Bank [1] in terms of material and response.

Around the world many naval exercises, or drifts, are achieved every year to improve oil spill responses. These experiments target both operational and technical goals. During 2009, we can take as examples the drift involving China, Russia, Japan and Korea (NOWPAP, December 2009), and the EPC 09 exercise in the Mediterranean Sea between Bulgaria, Georgia, Greece, Latvia and the United States. During spring 2009, an exercise named Euronyme 2009 gathered French and Spanish vessels, and EMSA, in the Mediterranean Sea.

Maritime environmental protection against oil discharges is the objective of contingency pre-planning, such as the POLMAR plans in France, or oil industry response plans. These plans frequently use floating barriers, named booms. These structures are flexible devices, connecting mooring points, which are anchoraged along the sea bed, near a shore, or in a harbour. A boom can also be towed by a ship, in an estuary, or in high sea conditions. This eco-technology is used to contain, or deviate, floating oil pollution, before recovery by a skimmer.

The project BAR3D [2] aims to validate a three-dimensional (3D) numerical model used to improve the installation plan of floating oil barriers (Muttin *et al.* [3] and Muttin [4]). The numerical computations verification uses necessary experiments. For that purpose several experiments at scale-one have been performed in three locations on the French Atlantic coast. The measurements on the boom geometry and on the boom stress can be compared with the numerical results.

The numerical code is developed to predict the geometry and the stress inside the boom device and its mooring system. The 3D computation is initialized by a one-dimensional (1D) computation done with the software FORBAR. This initialisation step permits one to reduce the computation time by starting the solution from an approximation on a coarse grid (Adams [5] and Nobile [6]).

The main objective of this paper is the description of the three scale-one experiments of oil barriers, in coastal, estuary and harbour waters. Experiments permit the improvement of boom installation as well as an evaluation of the technical and human efforts involved. These means are presented in various coastal environments of the Atlantic.

The paper is organized as follows. The La Rochelle, Lorient and Brest experiments are described in sections 2, 3 and 4, from the environmental, numerical and economical points of view. Section 5 presents the discussion of the economical and environmental dimensions.

2 La Rochelle harbour sea opening

This experiment is conducted in the deep sea harbour of La Rochelle. The experimental zone is located inside the Anse Saint-Marc, in the vicinity of the viaduct of the La Rochelle oil terminal. Between 5 and 13 May 2008, a floating boom was installed in the sea opening between Ré Island, the oil terminal and the harbour infrastructure, which is under extension on the sea. The boom device is 100 m long. It is anchoraged within five mooring points. The boom device is composed of rigid floats. The South end-point of the boom plan is anchoraged on a buoyancy coffer connecting a mooring chain on a dead mass. The North end-point is attached to a buoy connecting a mooring chain on another dead mass. Three intermediary points are anchoraged on both sides of the boom using Brittany anchors weighting 35 kg. A total of six anchors are moored. Each mooring line is perpendicular to the side of the boom, e.g. the South-North end-points of the boom plan. The installation period corresponds to a high tide coefficient.

Figure 1 shows the boom plan in the South of the oil terminal viaduct. Two artificial viaduct pile foundations are near the boom plan North end-point.

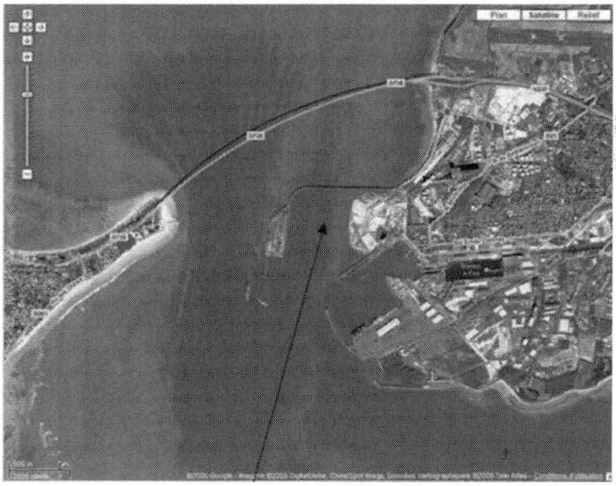

Figure 1: Location of the boom plan 100 m long of the La Rochelle harbour experiment.

The cardinal sides of the experimental zone are respectively, a sea opening (South), an artificial dock (West), a sea opening with viaduct piles (North) and a cliff (East) with the new harbour infrastructure. The sea bottom morphology is an estuarine mudflat, surrounding a rocky flat.

The experiment is conducted by the partners of the BAR3D national project, with the main contributions of the National Harbour of La Rochelle and the "Phares et Balises" Subdivision. The boom is delivered by the company TOTAL SA.

The following figure shows the boom plan. At that moment the sea current reverses after its zero value of a tide-down period end. The low tide is back to this moment. Figure 2 is a view from North to South, taken from the oil terminal viaduct.

The main result observed in figure 2 is the four adjacent curves take by the boom device. A geometrically singular point corresponds to the tension of a mooring line. Note that these three singular points are not aligned with the end-points of the plan. It is a consequence of the installation of the six intermediary

Figure 2: The boom plan under a small sea current at the beginning of a tide-up period.

Table 1: Budget repartitions between the partners for the La Rochelle scale-one experiment.

	research (€)		administration & industry (€)		total (€)	
head	63519 86%	59%	43650 62%	41%	107169	74%
material means & equipments	10454 14%	28%	26943 38%	72%	37397	26%
total	73973	51%	70593	49%	144566	

mooring lines during an opposite current orientation corresponding to a tide-down period. During that installation period the entire boom device is stressed by the current and is attached solely by its end-points, e.g. boundaries. The smooth shape of the boom at that moment can be recognized in figure 2 by looking at the end-points and the three singular points.

Another mechanical limitation of this boom plan is the absence of additive buoyancies at the three singular points. Here, the buoyancy comes solely from the boom rigid floats. The tension of a mooring line increases when the sea current increases. Consequently, the boom buoyancy can become insufficient to balance the vertical component of the mooring line tension. It is observed during this experiment the sinking of the boom locally, when the sea current is high. This effect can be reduced by using inflated tubes instead of rigid floats.

Here follows the economic evaluation of this experiment. The repartitions of the budget between the participants are set by a contingency table (Greenacre [7]). The horizontal direction of the table gives the natures of the budget, while the vertical repartition concerns the partnership kinds.

In table 1 we note the high proportion of the equipments provided by the administration and the industry, showing the numerous material means involved. It concerns the boom device, its mooring system, the three pneumatic boats, the mooring boat service, the towing vedette and the instrumentations. The principal budget comes from the head resources involved by the research partners. This includes the preparation of the test and the measurements during the experiment. The preparation contains the numerical computations of the boom plan by using a 1D model. The measures done are for example the sea current velocity by an ADCP sensor and the boom device tension by a dynamometer on the coffer at the South end-point.

3 Lorient petroleum wharf

This experiment is a test attempt. The experiment has been delayed because of the presence of much public transportation shipping in the experiment zone, the extension of the concerned oil terminal quay and several uncertainties on the marine current and the water bathymetry.

The test has been expected for the second semester of 2008. The boom device floats are composed of inflated tubes. The boom device is a single section of 100 m, or 70 m long, for either former or new quay infrastructure. On one end the boom section moves vertically on a stressed cable. It permits the boom plan respecting the sea level variation during tide. It is located near the tanker quay. On its other end the boom is anchoraged on a buoyancy coffer moored with a chain on a dead mass. It is located near the shore line.

The cardinal sides of the experimental zone are respectively, the Lorient harbour channel (South), a fishing harbour quay and a public transportation quay (West), a rocky coastal structure (North) and a two-rivers estuary (East). The sea bottom is an estuarine tidal mudflat. It exists along the quay an artificial bed allowing tanker vertical displacement with tide. The figure 3 is a view from East to West of the zone, showing the boom plan on the former oil wharf.

Figure 3: The boom plan of the experiment attempt on the former oil wharf of Lorient harbour (view from East to West).

Table 2: Budget repartitions between the partners for the Lorient experiment.

	research (€)		administration & industry (€)		total (€)	
head	9347 96%	44%	12040 86%	56%	21387	90%
material means & equipments	400 4%	17%	1900 14%	83%	2300	10%
total	9747	41%	13940	59%	23687	

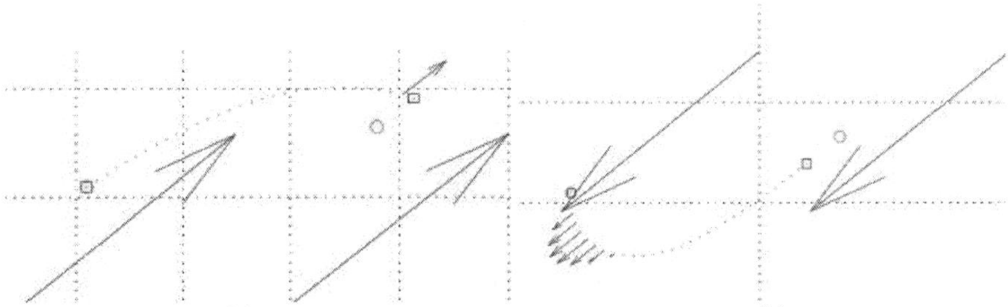

Figure 4: Lorient oil wharf boom plan computations: (a) midterm tide-down period; and (b) midterm tide-up period.

The preparation of this experiment is conducted by the partners of the BAR3D project, the Oil Storage Facilities of TOTAL SA in the Lorient Harbour and the local authorities, CCI Lorient and the Regional Council of Brittany. Let us give the economic evaluation of the experiment. The repartitions of the budget between the partners are given in the contingency table 2.

The budget repartition underlines the principal contribution of the administration and the industrial partner. Despite the fact that the experiment is an attempt, the contribution of research work is significant. During this attempt, we do not involve any water pollution response material and responders.

Figures 4(a) and (b) show the results of the computations of the boom plan at both midterm tide-up and midterm tide-down periods. The computation solves the tension cable equilibrium equation being a first order non-linear differential equation. A finite-difference scheme is used with a fixed-point method. The iterations proceed on the tension on one end of the boom until the other end converges to the desired anchorage location. The sea current is taken parallel to the tanker quay.

The main observations in the figures 4.a and 4.b are on their right sides. We can see the large displacement of the boom plan end, attached on a coffer, given by a small square on the figures. The coffer is moored by a chain on a dead-mass, represented by a small circle. This large amplitude is a consequence of a practical constraint on the mooring chain length. This one must be strictly higher than the sea level change of the tide periods considered. This displacement can be reduced solely by a reduction of the mooring chain length. Note that a reduced mooring chain length renders it more vertical during high tide periods. In practice, the mooring chain length is taken as three times the maximal water height. The constraint comes from the vertical component of the traction of the mooring chain which can favour the drag of the dead mass.

4 Brest Elorn estuary

The experimental zone is located near the right bank of the Elorn estuary, at the level of the small city of Relecq-Kerhuon, located at the East of the Brest bay, Iroise sea, at the North-East of the Iroise Bridge. Between 2 and 9 November 2009, a 210 m long boom is installed between four anchorages points, 60 m equidistant. The boom device floats are inflatable tubes. The installation period corresponds to the beginning of the winter season.

The cardinal sides of the experimental zone are respectively: the Elorn estuary (South), a mooring zone for yachting (West), a rocky coastal structure (North) and the Elorn River and its natural channel (East). The sea bottom of the experimental zone is an estuarine tidal mudflat. The natural channel is near the boom plan East end.

The test period corresponds to a high tide coefficient. The Elorn river debt is heavy during that period, as a consequence of rains. The Elorn flows to SW. The sea current is NE, when tide goes up. The maximal sea level change is [1.29, 7.04] m corresponding to around a 6 m tide height. Figure 5 shows the Elorn estuary and the boom plan.

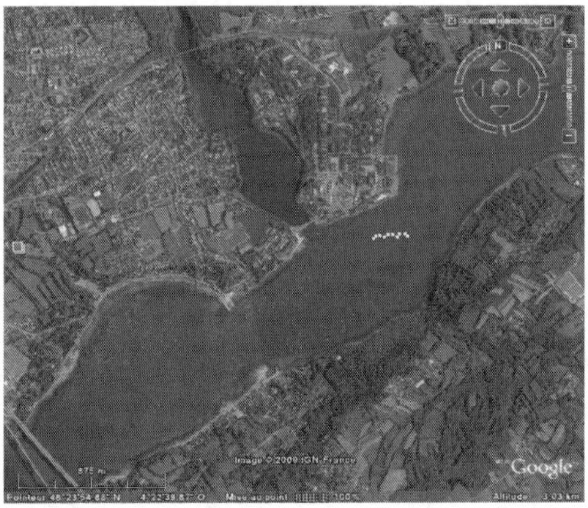

Figure 5: The boom plan (the boom is 210 m long) for the Elorn estuary experiment.

Figure 6: The third section of the boom plan for the Elorn estuary during a tide going-up period (view from west to east).

The experiment is conducted by the partners of the BAR3D project, with the valuable contributions of the “Stock POLMAR” of Brest, the “Phares et Balises Subdivision du Finistère”. The boom is delivered by the “Stock POLMAR”.

Figure 6 shows the third section of the Elorn boom plan, during a tide going-up period. This third section is the East side of the boom plan. Three measurement light bars are installed at the quartile lengths of this boom section. Each bar is attached vertically on the bottom part of the sub-sea skirt of the boom device. It indicates locally the vertical angle of the boom skirt bottom. It is admitted that skirt angle amplitude higher than 10° is inefficient to contain floating oil (Violeau *et al.* [8]).

A result observed during the experiment is the large amplitude of the boom skirt angle, which is higher than 20°. We note that the angle measured is on the tangent at the skirt bottom and the vertical. As a consequence it cannot be concluded that the boom is inefficient. For boom efficiency, the angle of the skirt to be taken into account is the cord between the skirt top-bottom and the vertical. Our measure can be improved by using the bars attached on the full height of the skirt.

The next figure shows the stress map in the third boom device. It is computed by using a 3D membrane finite-element method (Muttin [9]). The mesh uses 21 sections of nodes given by the initialization stage. Each nodal position of the boom mesh at the sea surface is initialized by using a 1D numerical model. The current velocity is 0.4 m/s. The angle between the boom device end-points and the current is 38.70°. On the two coffers visible in figure 6, the computed tensions of the boom device are 193 kg for the first coffer, and 201 kg for the far East coffer.

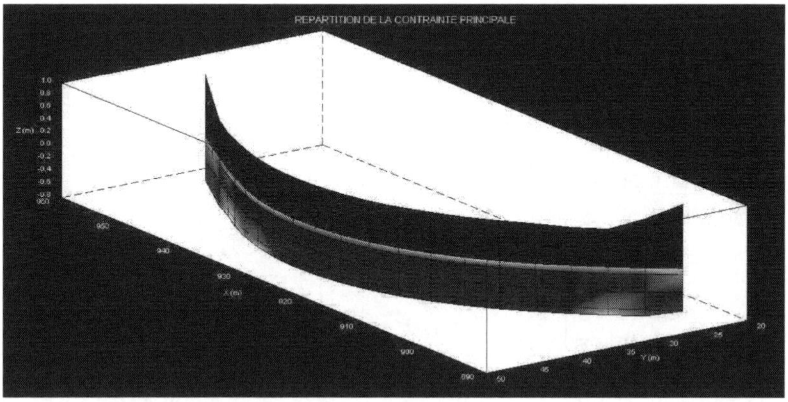

Figure 7: The stress map in the third section of the boom plan for the Elorn estuary (view direction similar in figure 6).

Table 3: Budget repartitions between the partners for the Elorn estuary experiment.

	research (€)		administration & industry (€)		total (€)	
head	45302	53%	39800	47%	85102	61%
	71%		53%			
material means & equipments	18763	34%	35711	66%	54474	39%
	29%		47%			
total	64065		75511		139576	
		46%		54%		

The map in figure 7 indicates a concentration of the stress in the leach located at the upper part of the skirt under the boom float.

Let us now look at the economic evaluation of this experiment. The budget repartitions between the partners are showed using a contingency table.

The budget repartition of the Elorn estuary experiment is similar with the La Rochelle experiment. The head for the research activity is the principal component. The material means and equipments are mainly involved by the administration and the industry. That suggests an invariant budget structure for such experiment. The total amount level shows the effort to install floating boom for coastal protection against oil pollution.

5 Discussion

5.1 Economic analysis

An economic analysis is given on the following table. It shows some economic indicators, as the human effort needed, given in man*months and the ratio

Table 4: Some economic indicators versus the three oil boom experiments in Atlantic coastal zones.

	La Rochelle	Lorient	Brest Elorn
human effort (man*month)	12	3	9
public/private budgets (%)	29% / 71%	18% / 82%	47% / 53%
cumulative ratio of research funds	4,3	5,0	4,9

between the public budgets and the private ones. The cumulative effect of the state funds is given. It concerns the BAR3D project part dedicated to experiments.

We underline a quasi constant cumulative effect of the research funds. It appears as quasi invariant in the three experiments. This factor is closed to 5. It proves the high interest of the involved partners for such experiments. It suggests that the environmental technology properties verification and testing must contain that kind of economic indicators.

Concerning the two performed experiments (La Rochelle, Brest Elorn) the ratio between the public and private budgets appears lower or higher than the targeted 1/3-2/3 repartition suggested by national research policies. It is expected that such research activities on environmental performance verification and improvement are compatible with a sustainable development goal from an economic point of view (Seager *et al.* [10]).

5.2 Environmental analysis

The principal environmental uncertainty during the experiment of La Rochelle comes from the amplitude of the sea current. The current can be higher than the existing measures. It is expected to be a consequence of the recent embanking of the sea water zone for the harbour infrastructure extension. The strait between the Ré Island and the coast is reduced and the current will increased according to a Venturi effect. It suggests before any boom experiment to record the sea current in the zone concerned.

Another environmental uncertainty results from the Elorn experiment. In an estuary two flows are superposed. The river flows down, while the tides make the sea water flow up or down. In the Elorn estuary experimental zone, the current of the water surface does not take opposite directions when the tide goes up or down. It may be a consequence of the water stratification and the estuary geomorphology, with a bending geometry of the channel in the vicinity of the zone.

The experiment attempt at the oil wharf of Lorient can be interpreted as the difficulty to perform a test at scale-one in a harbour where the geomorphology contains uncertainties, and where the port infrastructures are under transformation. It can be note the difficulty to install during several days a boom plan and a measurement unit in a harbour zone where the maritime circulation can be intensive (Gottinger [11]).

6 Conclusion

The performances evaluation and improvement of oil booms are a risky task from two points of view.

6.1 Technical performances

A main conclusion from the La Rochelle and Elorn estuary experiments is the lack of knowledge in the sea current conditions. Our main recommendation is an in-situ recording of the environmental conditions before the experiment will be made, even if existing measures are available. A lack of quality in the variables used in the computations will result in imprecise numerical results, even being 1D and 3D, for decision making. The pre-planning of response plan can be improved in that direction (Harrald [12]).

6.2 Economical performances

The conclusion is a valuable invariance of the budget structure for the performed experiments of La Rochelle harbour and Elorn estuary. It suggests that when an experiment is envisaged this budget structure can be adopted. The economic performance evaluation appears as a monitoring process of a boom plan. Nevertheless, the attempt of the Lorient experiment shows that a significant budget can be spent without the certainty that the experiment can be performed. It suggests that even if the budget structure of an experiment is quite invariant the performing of an experiment is uncertain.

Acknowledgements

This research is supported by the French Research Agency ANR, and ADEME, under grant n° BAR3D/0694C0067, project n° ANR-06-ECOT-003-01. The author acknowledges the French framework, PRECODD, on Environmental Technologies and Sustainable Development. The network “Recherche et Innovation Technologique RITMER”, “Pollutions Marines Accidentelles et leurs Conséquences Ecologiques” and “Pôle Mer Bretagne” are gratefully acknowledged.

References

[1] World Bank, Implementation Completion Report, Report n° 34859, December 29, 2005.
[2] BAR3D project, http://bar3d.eigsi.fr

[3] Muttin, F., Nouchi, S., Numerical Study of Oil Spill Containment Boom by the Finite-Element Method, *Proc. of Eighth International Conference on Modelling, Monitoring and Management of Water Pollution*, Water Pollution VIII Conference, Bologna, Wessex Institute of Technology, pp. 245–254, 2006.

[4] Muttin, F., Oil spill boom modelling by the finite-element method, *Proc. of Ninth International Conference on Water Pollution: Modelling, Monitoring and Management*, Water Pollution IX, WIT Transactions on Ecology and the Environment, Wessex Institute of Technology and Alicante University, Vol 111, pp. 383–392, 2008.

[5] Adams, M. F., Parallel multigrid solvers for 3D unstructured finite element problems in large deformation elasticity and plasticity, *International Journal for Numerical Methods in Engineering*, **48(8)**, pp. 1241–1262, 2000.

[6] Nobile, F., Coupling strategies for the numerical simulation of blood flow in deformable arteries by 3D and 1D models, *J. Math. Comput. Modelling*, **49(11–12)**, pp. 2152–2160, 2009.

[7] Greenacre, M. , *Correspondence Analysis in Practice*, Interdisciplinary Statistics, Chapman & Hall, Taylor & Francis Group, 2nd edition, 2007.

[8] Violeau, D., Buvat, C., Abed-Meraim, K., De Nanteuil, E., Numerical modelling of boom and oil spill with SPH, *Coastal Engineering*, **54**, pp. 895–913, 2007.

[9] Muttin, F., Structural analysis of oil-spill containment booms in coastal and estuary waters, *Applied Ocean Research*, 30(2), pp. 107–112, 2008.

[10] Seager, T.P., Tuler, S.P., Kay, R., Linkov, I., Satterstrom, F.K., Environmental Performance Metrics For Oil Spill Response, Technical Report, Coastal Response Research Center, University of New-Hampshire and NOAA, January 31, 2006.

[11] Gottinger, H.W., Monitoring pollution accidents, *Eur. J. Oper. Res.*, **104**, pp. 18–30, 1998.

[12] Harrald, J.R., Contingency planning: building the infrastructure for crisis decision making, *International Journal of Mass Emergencies and Disasters*, **8(2)**, pp. 137–150, 1990.

Pollution of drinking water transported by corroded metallic pipelines

A.-Kh. M. Hussain, A. A. Sanoussi & H. A. M. Hussain
Civil Engineering Department, Engineering College,
El-Fateh University, Libya

Abstract

Corrosion of metallic water pipelines is a function of the pipe's surrounding environment (soil and water). This means that it is important to know the physical and chemical characteristics of the pipe's surrounding environment. The corrosion of externally unprotected metallic water pipelines, particularly ductile iron pipes, in localities with aggressive soil conditions is becoming a significant problem: repair and replacement costs are high, and failures cause supply interruptions to customers and inconvenience to the public. Anti-corrosive protection for metallic water pipelines, their fittings and accessories is very important, because they may be attacked by corrosion with time. The tendency of a metallic substrate to corrode is a function of the surface characteristics of the metal and of the metal/protective film interface, the physical, electrical and electrochemical properties of the film, and the nature of the environment in which the pipelines system is placed.

In this work the authors have looked at the corrosion problems of metallic water pipelines and their control. The corrosive properties of water and soil environments are reviewed. Parameters affecting corrosion are discussed. The pollution of drinking water transported by metallic pipes caused by corrosion is discussed. The purpose of this work is to provide guidelines for materials selection in water and soil environments, and how the water pipelines can be protected against metallic corrosion.

Keywords: corrosion of metallic water pipelines, drinking water pollution, corrosive water, metallic water pipes.

www.witpress.com, ISSN 1743-3541 (on-line)
doi:10.2495/WP100061

1 Introduction

Corrosion is one of the most important problems facing the water services industry and in the metallic piping systems. It is the most common threat to the performance and life of pipeline systems. Corrosion takes an even more significant toll in human life and safety [1–3]. It is the destruction of metal or metal alloy by chemical change (chemical reactions between a metal or metal alloy and its environment), the chemical reaction between a metal and acid, electrochemical change, or physical dissolution. The result of corrosion is the eating away of the metal surfaces of hydraulic components. Crevice corrosion occurs for steel bolts, and for drilled and tapped stainless steel pipes in aerated salt water. Copper pipeline systems carrying water can be attacked by pitting corrosion. It is more rapid and generalized when the PH-range is less than 7. Heavy metals in drinking water, such as lead (Pb^{++}), Copper (Cu^{++}), Zinc (Zn^{++}), and Iron (Fe^{++}), come from the corrosion of internally unprotected metallic water pipelines [4]. Municipal water must not only be treated for portability, but also for minimum corrosivity towards the distribution lines and the home plumbing system [4–6]. This can cause excessive leakage (losses of water quantity) past the sealing surfaces of the affected parts.

Accumulated corrosion and oxidation products on pipelines reduce their product, their strength, life and performance. Oxidation is the conversion of the surface portions of metal to oxide on heating in air or oxygen. It is caused by the chemical reaction of oxygen from the air with particles of water. Most products of oxidation are soluble in water and are acidic in nature, which can cause corrosion of parts throughout the system. Metal atoms in nature are present in minerals. The same amounts of energy needed to extract metal atoms from their minerals are emitted during the chemical reactions that produce corrosion. Corrosion engineering is the design and application of methods to prevent corrosion.

Losses of water from leaking pipelines have a high cost. These leaks have a corresponding hazardous effect on the health . Thousands of leaking pipelines must be repaired or replaced to prevent recurring losses of water and contamination of flowing groundwater. Corrosion of pipeline system built below groundsurface is a function of the nature of the surrounding soil and groundwater [7]. The soil in direct contact with the pipe will have the greatest influence on its behaviour over time which highlights the importance of knowing the physical and chemical characteristics of the pipe's immediate environment.

2 Metallic pipe materials

A pipe can be defined as a rigid or semi-rigid tube fabricated by many different methods and in many different materials (metallic and non-metallic). Types of metallic water pipes are Aluminium tubes, Copper tubing, Grey cast iron pipes, Ductile cast iron pipes, Steel pipes, Stainless steel pipes, Titanium tubing and Tungum tubes. The Aluminium, Copper, Stainless steel, Titanium and Tungum

Figure 1: Corroded pipelines removed for replacement.

pipes are resistant to general corrosion. However, the Grey cast iron, Ductile cast iron and steel pipes have no resistance to corrosive atmosphere [8] as shown in Fig. 1. Ductile cast iron pipes are the type most widely used because of their rigidity, strength and durability in many kinds of soils. It is available in a wide range of tensile strengths. It is commonly used for hydraulic lines.

3 Stages of corrosion

The first stages of corrosion are indicated by rust spotting or the appearance of a few small blisters. Rust spotting is the very earliest stage of corrosion and in many cases is left unattended. Standards have been established for evaluating the degree of rust spotting and may be found in ASTM D 610 – 68 or (Steel Structures Painting Council Vis – 2). If the rust spots go unattended, a mechanism for further corrosion is provided. The definition of corrosion rate is the total metal loss divided by the period of time over which the metal loss occurred.

4 Types of corrosion

Corrosion is often identified with atmospheric rusting of iron base alloys. Corrosion attacks metallic water pipelines and can cause leaks and explosions in the pipe. Corrosion may be in the forms of pitting corrosion, definite holes, pockets, cavities, crevice corrosion, environmentally induced cracking, hydrogen damage, inter-granular corrosion, de-alloying, erosion – corrosion and graphitic corrosion. Different types of corrosion are seen in Fig. 1. Many metals and alloys have their own individual corrosion problem. Each corrosion problem has its own cure. Some corrosion can be controlled by electrical means (cathodic protection) or the application of a non-metallic barrier (coating). Such barriers should be barrier (100% pore free) when used in highly aggressive water, and since this is rarely the case, they do not form a universal cure for

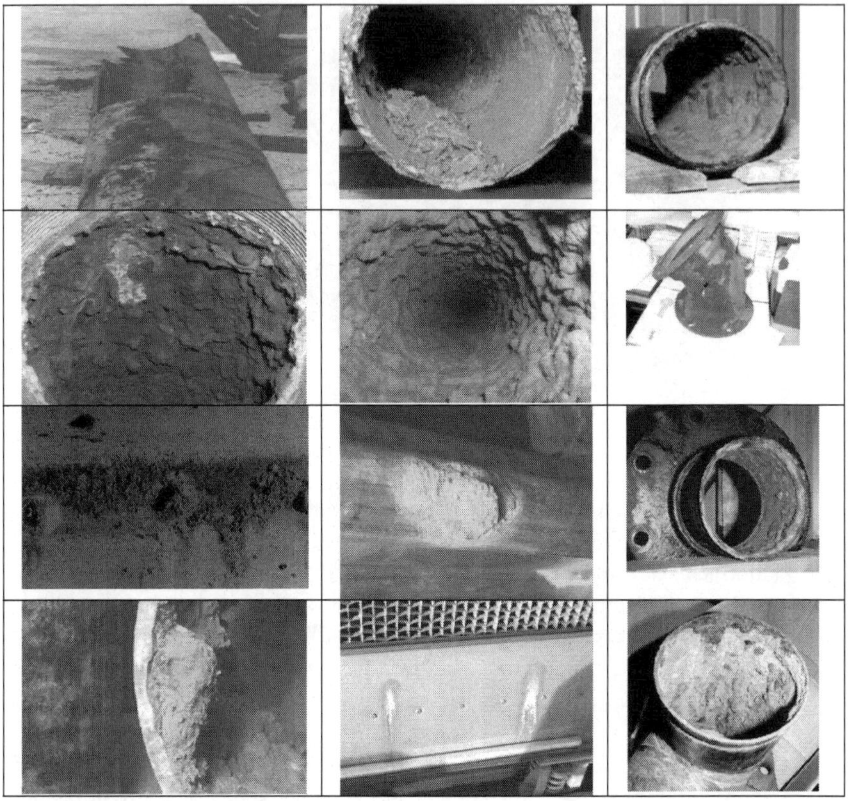

Figure 2: Different types of corrosion of metallic water pipelines.

metallic corrosion. Sometimes the metal generates its own barrier from corrosion products (metal oxide), the metal is then in the passive condition. Specific types of corrosion are discussed in the following paragraphs.

4.1 Galvanic corrosion

The main cause of deterioration in buried pipelines is galvanic corrosion. It is caused by the electrolytic action of a galvanic cell. It occurs for welded pipes, for carbon steel pipe flanges near the weld, and near the junction between dissimilar alloys. It decreases at points farther from the junction due to higher resistance through a longer electrolyte path. A local galvanic action is set up when there is a difference of potential between the areas on a metallic surface that is an electrolyte, as shown in Fig. 2. The resulting electrochemical reaction develops a potential difference between these metals. This causes one metal to be negative (anodic) and the other metal to be positive (cathodic). In a dry cell battery, the zinc case is the anode and the carbon rod the cathode. Now, when an external electrical circuit is completed, current flows from the zinc case into the electrolyte, taking with it particles of zinc.

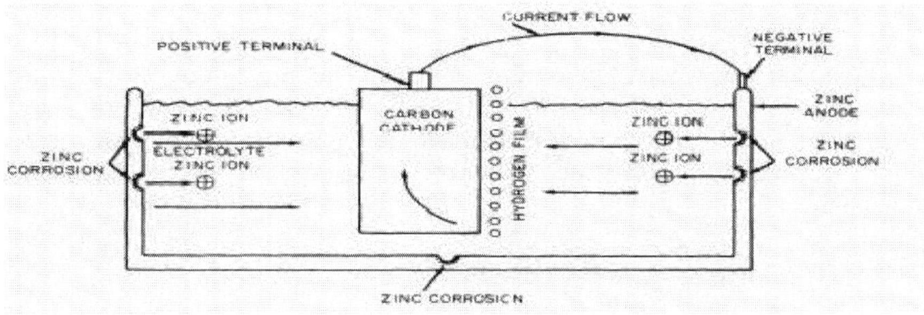

Figure 3: Galvanic cell showing internal galvanic action.

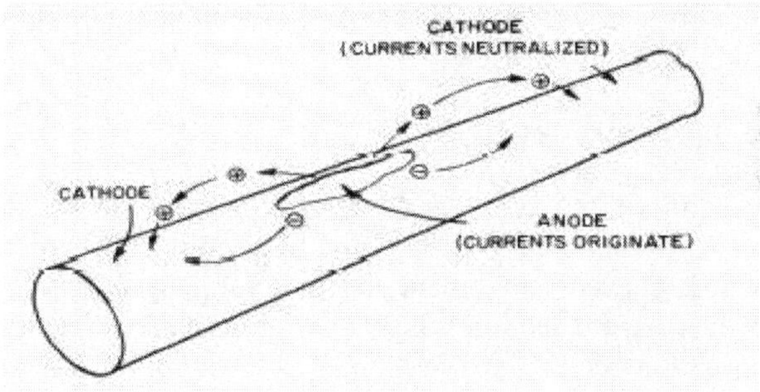

Figure 4: Pipe with corroding (anode) and noncorroding (cathode) areas.

4.2 Corrosion due to mill scale

The mill scale embedded in the walls of iron pipe during its manufacture is one cause of pipe corrosion. It actually becomes the cathodic area, the iron pipe the anodic area, and the moist soil the electrolyte as seen in Fig. 3. Current leaves the iron pipe wall and passes through the electrolytic soil to the mill scale. This electrochemical action causes severe pitting of the pipe metal at the anodic areas. Continued action of this type will eventually weaken the pipe to the extent of failure.

4.3 Corrosion due to cinders

Another type of corrosion occurs when iron pipe is laid in a cinderfill in direct contact with the cinders. The cinders and the iron pipe make up the dissimilar metals. The pipe forms the anodic area, the cinders form the cathodic area, and the highly ionized soil serves as the electrolyte. The current leaves the pipe through the soil to the cinders and returns to the pipe. Severe corrosion occurs at the points where the current leaves the pipe.

4.4 Corrosion due to dissimilarity of pipe surface

This type of galvanic corrosion occurs when there are bright or polished surfaces on some areas of the pipe walls in contact with suitable electrolytic soil. These bright surfaces become anodic to the remaining pipe surfaces. In highly ionized soil, the polished surfaces corrode at an accelerated rate, thus weakening the pipe at that point.

4.5 Corrosion due to different soil conditions

This is a general corrosion problem, especially prevalent in highly alkaline areas. Corrosion currents leave the pipe wall and pass into compact soils and enter the pipe wall from light sandy soils. The intensity of the corrosion currents and the resulting rate of corrosion at the anodic areas of the pipe are directly proportional to the conductivity of the soil.

4.6 Corrosion due to stray currents

Direct current circuits that pass in and out of an electrolyte usually cause stray currents, many of which are a direct cause of corrosion. Corrosion does not occur at the point where the current enters the structure, because it is catholically protected. However, at the section where the current leaves the structure, severe stray current corrosion occurs.

4.7 Corrosion due to bacteria

Microbiological corrosion is another distinct type of corrosion caused by electrolytic or galvanic cell action [9, 10]. It is the deterioration of metals by corrosion processes that occurs as either a direct or an indirect result of the metabolic activity of bacterial organisms, particularly in water or soil environments. Microbiological corrosive action in the soil is due to physical and chemical changes in the soil caused by the presence of these organisms close to the metal surface in stationary slimy deposits produce corrosive substances, such as, carbon dioxide, hydrogen sulphide, ammonia, and organic and inorganic acids. These corroding substances are found only in the locality of the colony and may be undetected in the surrounding water or soil. Bacteria that cause corrosion in this way need to produce only small amounts of corrosive products for localized attack. However, colonies of bacteria that do not produce corrosive products may act as a protective film around the metal, causing unequal distribution of electrical potential, which gives rise to local anodes and cathodes. In this way, the production of local cells will cause increased corrosive action. Biological corrosion is extremely difficult to control, since the organisms are very resistant to normal methods of sterilization. Probably the most logical method to reduce microbiological corrosion is by the use of some barrier coating between the environment and the metal. Some bacteria are responsible for the production of active galvanic cells. These bacteria are mostly found in highly waterlogged, sulphate-bearing, blue clay soils. The bacteria concentration, as well as the corrosion rate, varies considerably with the different seasons of the year. Cast-iron and steel pipes are corroded mostly by sulphide production.

4.8 Dezincification

This is a selective type of corrosion that occurs in copper and zinc alloys. When alloys of this kind (brasses) are exposed to this type of corrosion, the zinc dissolves out of the alloy and leaves only the copper.

4.9 Graphitization

Another type of compositional corrosion is graphitization or graphitic softening. It is a form of disintegration that attacks buried grey cast iron water pipes and their fittings after many years of exposure in soil and groundwater. Cast iron is an alloy made of iron and carbon, the carbon being in the form of graphite. When cast iron with such a composition is subjected to graphitization, the iron dissolves out and leaves only the graphite. This action leaves cast-iron pipes and other similar equipment weakened mechanically. However, after graphitization corrosion occurs, the graphite pipe may last for many years if it is not subjected to any mechanical forces or sudden pressures. The action of this type of corrosion is similar to dezincification.

4.10 Hydrogen embrittlement

This is a term applied to metal that becomes brittle because of the action of some form of corrosion that causes the formation of hydrogen on its surface. When hydrogen forms on the surface of steel, the action of the hydrogen may form blisters or actually embrittle the metal. Hydrogen liberated near the surface of steel in an electrolyte will diffuse into the metal quite rapidly. The hydrogen picked up by the steel is in an atomic state and causes the steel to become brittle. When the production of atomic hydrogen on the surface of the metal stops, the hydrogen leaves the metal in a few days and the metal again regains its original ductility.

4.11 Stress fatigue of metals (corrosion affects metals that are under stress)

The action caused by stresses on a pipeline or structure is due to the shifting of the various rocks and soils of the earth. Usually a complete pipeline is not under stress; certain sections are under stress while adjacent sections are not. Because of these pressures and strains, localized electrochemical action takes place. The section of the pipe or structure under stress becomes anodic, whereas the unstressed sections become cathodic. In this way, the pipe under stress begins to corrode and weaken because of the action of corrosion.

5 Pipe performance and selection

A range of performance parameters have to be considered when selecting a pipeline material, so that the final decision based on economic and availability factors can be made. These parameters are compatibility of pipe with media and environment (its resistance to corrosive atmosphere), pipe bore size, strength of pipe for internal and external pressure, temperature of media and environment, flexibility and bend radius, weight and compactness, expansion and compatibility of pipe with jointing method. Where steel pipes have to be used from a strength requirement, a greater use of stainless steels is being made in the smaller pipe bore sizes. Where larger steel pipes have to be used they can be lined with a coating of PTFE (Polytetra- fluorocthylene), which can be extended over the flange faces, so that the media doses not come in contact with the steel pipe [12–14]. Typical manufacturer's range of pipe sizes showing the thickness of the PTFE lining in the range of 2.8 mm to 5 mm. For environmental protection of the external surfaces of the pipes, the external surfaces of the pipe can be coated with a material which is resistant to the working environment, before being installed. Specially developed plastic coatings have greatly increased the range of basic materials which can be used for water pipelines in a wide range. The main benefits obtained from such coatings are protection against corrosion and chemical attack, increased abrasive resistance and non- toxicity. Plastic coatings also lessen the risk of impact damage, reduce erosion and in certain instances improve the flow characteristics of some piping systems. Coatings for general use include Polyvinyl chloride (PVC), Polyethylene (PE), and Polypropylene and nylon.

6 Conclusion

It is known that more than half of all corrosion problems could be avoided by direct application of established knowledge. Each corrosion problem has its own cure. To prevent excessive corrosion, good inspection procedures and preventative maintenance practices are required. Proper design consideration are also necessary, as well as, selection of the proper coating system. Regular inspections of coating should be conducted. Since corrosion of substrates under coatings takes place in stages, early detection will permit correction of the problem thereby preventing ultimate failure. Sanitary engineers are responsible for minimizing the direct and indirect costs and risks to the public from corrosion. Causes of corrosion have to be known. Methods of prevent and minimize its effects have to be practically followed.

References

[1] Walker, R., Corrosion Is Pollution. Uni-Bell PVC Pipe News, 1995, 18 (2).
[2] Rajani, B., McDonald S. and Felio G., Water Mains Break Data on Different Pipe Materials for 1982 and 1993. NRC Canada Report, August 1995.
[3] USEPA, Water on Tap. A Consumer's Guide To The Nation's Drinking Water, US Environmental Protection Agency, Office of Water (4601), July 1997.

[4] Lee R.G., Becker, W.C. and Collins, D.W., Lead at the Tap. Sources and Control J. Am. Water Works Assoc. 81(7), 1989, pp. 52-62.
[5] Plishka M.J. and Shenkiryk M., Chemical Cleaning Process for Water Systems. Water Engineering & Management, March 1996, 143(3).
[6] Reiff F., Biofilm in Water Distribution Systems. Drinking Water & Health, A Publication of the Public Health Advisory Board to the Chlorine Chemistry Council®, Fall 1996, 3(4).
[7] Jakobs J.A. and Hewes F.W., Underground Corrosion of Water Pipes in Calgary. Canada, Materials Performance, May 1987, 26(5).
[8] Millette J.R., Hammonds A.F., Pansing, M.F., Hanson E.C. and Clark A., Aggressive water: Assessing the Extent of the Problem. *J.Am.* Water warks Assoc. 72(5) (1980): 262-266.
[9] Simgley J.E., The Search for a Corrosion Index. *J.Am.* water Works Assoc. 73(11) (1981): 579-582.
[10] Kriewall G., Tracking Bacteria Not Easy. Times Herald, August 22, 1997.
[11] Ruda T., Microbial Regrowth and Distribution System Management. Opflow- American Water Works Association, August 1997, 23(8).
[12] Merrill D.T. and Sanks R.L., Corrosion Control by Deposition of $CaCO_3$ Films. J. Am. Water Works Assoc. 69(11) (1977): 592- 599; 69 (12) (1977): 634- 640; 70(1) (1978): 12-18.
[13] Langelier W.F., The Analytical Control of Anticorrosion Water Treatment. J. Am. Water Works Assos. 28 (1936): 1500.
[14] Schock M.R., Understanding Corrosion Control Strategies for Lead, J. Am. Water Works Assoc. 81(7) (1989): 88-100.

Effects of water soluble extract from the dead leaves of deciduous trees on the growth control of toxic cyanobacteria like *Microcystis aeruginosa*

K. Shimada[1], N. Kitamura[2], Y. Yoshida[2], H. Matsushima[2] & Y. Asada[3]
[1]*Civil Engineering Department, Graduate School of Science and Technology, Nihon University, Japan*
[2]*Civil Engineering Department, College of Science and Technology, Nihon University, Japan*
[3]*General Education Department, College of Science and Technology, Nihon University, Japan*

Abstract

Because of the toxic microcystin from *Microcystis aeruginosa* (*M.a.*) in eutrophicated lakes and reservoirs, effective counter measures against its excessive growth have been widely called for. Recent research on the growth control of *Microcystis* was conducted especially concerning the effects of straw and emergent plants. The authors have discussed that the water soluble extract (WSE) from the dead leaves of deciduous trees like cherry (*Purunus Yedoensis*) and maple (*Acer buergerianum Miq.*) bring about a clear effect on the growth control of *M.a.* The WSE solutions from the dead leaves of cherry and maple trees were added to an *M.a.* culture solution to set the initial condensed tannin loading intensity (CTLI) per unit cell at about 26 pg-CT/cell, and then, the daily change of the *M.a.* cell concentration was observed. Accordingly, the WSE addition resulted in controlling the cell concentration and its influence was clearly affected by the initial CTLI level. The condensed tannin (CT) in the WSE solution was preferentially separated to produce a CT fraction (CTF) solution by use of casein/acetone, and thus the CTF solution also caused a good growth control. The WSE fractioning treatment resulted in separating the CT substance from the WSE solution. As for the growth control tests using the

WIT Transactions on Ecology and the Environment, Vol 135, © 2010 WIT Press
www.witpress.com, ISSN 1743-3541 (on-line)
doi:10.2495/WP100071

WSE and CTF solutions, the initial CTLI level was set at between 13 to 18 pg-CT/cell, and accordingly the CT effect was confirmed in terms of the growth control similarly in both the WSE and the CTF addition cases. Therefore, the results imply that the CT extracted from the dead leaves is the major growth control factor of *M.a.,* and thus the dead leaves of deciduous trees function as an effective growth control agent.

Keywords: water soluble extract, condensed tannin, dead leaves, growth control, Microcystis aeruginosa.

1 Research background

Eutrophication in inland lakes and reservoirs induces the excessive growth of phytoplankton under relatively high water temperature conditions from late spring through summer seasons. Along with the advance of this phenomenon that is called algal blooming or "Aoko" in Japanese, toxic cyanobacteria such as *Microcystis* and *Anabaena* especially grow and produce toxic substances like microcystin and anatoxin. Consequently, in the lakes where the algal blooming occurred, there were deaths of animals and fish and even incidents of human death due to lake water use in Brazil in 1996. These examples reflect the urgent need of counter measures for controlling the growth of the toxic cyanobacteria in regard to water environment preservation and water resource protection. Practical counter measures against algal blooming, include direct sediment/algal removal, copper sulfate addition, supersonic treatment [1], and electric processing [2]. These have been implemented although each process was not always technically effective, economically practical, or environmentally sustainable. Otherwise, there are some other reports on studies about the growth control of toxic cyanobacteria through the use of the natural plants like barley straw [3, 4]. In respect to the growth control of *Microcystis*, Ridge et al. [5] reported about the effects of leaf litter, and Nakai et al. [6] examined the influence of emergent plants like *Myriophyllum spicatum.* In these reports, it was supposed that the polyphenol (PP) of the natural organic compounds acts as an inhibitory factor to control the growth of the toxic cyanobacteria. Polyphenol has been known to hold medicinal actions [7, 8]. Antibacterial or bactericidal action has also been reported about the effect of catechin contained in tea, which is due to condensed tannin such as catechin that has been polymerized in several molecule units [9–11]. Toda et al. [12] examined the antibacterial action of catechin to E. coli forms, and Augustin et al. [13] reported on the bactericidal action of condensed tannin.

2 Research objectives

The authors have conducted a series of experiments about the effects of water soluble extract (WSE) from the dead leaves of deciduous trees like cherry (*Purunus Yedoensis*) and maple (*Acer buergerianum Miq.*) on the growth control of *Microcystis aeruginosa* (*M.a.*). Consequently, the growth control was clearly observed on occasions of the WSE solution addition in comparison with the control in which the WSE solution was not added. The results show that condensed tannin (CT) was detected in the WSE solution and is believed to be one of the important factors affecting the growth control [14]. The major objective this research aimed at finding was whether or not the CT functions as a growth control/inhibitory factor of *M.a.* cells when using the WSE solution containing the CT from the dead leaves of cherry trees and the CTF solution coming from the WSE solution that has been fractioned by the use of casein/acetone. That is, the CT in the WSE solution could be preferentially separated to hold the CT substance in the solution, which is defined as the CTF solution. The remaining solution after the fractioning treatment of the WSE solution to separate the CT substance would contain no condensed tannin or less condensed tannin than the WSE solution itself or the CTF solution. This is defined as the WSE-remaining solution (WSE-R). These three solutions of WSE, CTF and WSE-R were used for experimentally evaluating the effects of the CT from the dead leaves on the growth control of toxic cyanobacteria like *Microcystis aeruginosa.*

3 Experimental procedures for *M.a.* growth control

3.1 Preparation of the WSE, CTF and WSE-R solutions

In order to prepare the WSE solution containing the CT from dead leaves, dead leaves of deciduous trees like cherry and maple were collected at several parks in metropolitan Tokyo during late autumn and early winter. The dead in-room-dried leaves were first pulverized by a food mixer at 7,000 rpm for 30 seconds, and the pulverized sample of 1.0 g was immersed in 100 mL of D.I. water under laboratory room conditions for the seven days following the pulverization. Then, the supernatant portion was filtered to prepare the WSE solution. The WSE fractioning treatment to obtain the CTF solution was based on the procedure as in the flow chart in Figure 1 [15].

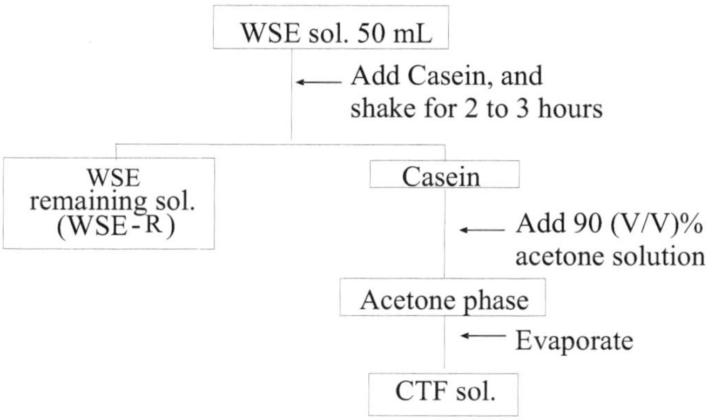

Figure 1: WSE fractioning treatment for CTF solution.

WIT Transactions on Ecology and the Environment, Vol 135, © 2010 WIT Press
www.witpress.com, ISSN 1743-3541 (on-line)

The WSE solution of 50 mL was first prepared as above, into which 2g of casein (Wako Pure Chemical Industries, Ltd., Japan) was added and agitated vigorously for 2 to 3 hours. Then, the casein was removed after the solution had sat for several minutes. Into the removed bulk of the casein was added a 50 mL 90 (v/v) % acetone solution, and it was air-stopped with a Parafilm (Pechiney Plastic Packaging Incorporation, USA). After agitating for one night, the acetone portion was removed by a rotary evaporator. In this remaining solution the CT substance itself was preferentially separated from the original WSE solution, and this is defined as the CTF solution. Furthermore, after the WSE fractioning treatment, the remaining solution would presumably contain less or no condensed tannin than the original WSE solution or the CTF solution. This is defined as the WSE remaining solution (WSE-R). Growth control tests using the WSE-R solution resulted in observing no growth control effect because of the less or no CT substance existing.

3.2 Condensed tannin analysis in WSE, CTF and WSE-R solutions

In general, the CT reacts with a 1% ferric chloride solution to produce a dark green color for cherry WSE solution and a dark blue color for maple WSE solution. The concentration analysis on the WSE and CTF solutions was based on the vanillin-sulfuric method for condensed tannin [16] and the Folin-Denis method for polyphenol [17]. Both the CT and PP concentrations have been expressed in catechin equivalent. For the CT and PP quantitative analysis, the pure chemical of (+)-Catechin Hydrate was purchased from Spectrum Chemical MFG Corp., USA. The PP naturally contained CT.

3.3 Pre-incubation of *M.a.*

The *Microcystis aeruginosa* culture (NIES 102) for the present experiments was originally transferred from the National Institute for Environmental Studies (NIES) in Japan. Before use for the growth control tests, the *M.a.* culture was pre-incubated at 28±1℃ under a light intensity of 2,500 to 3,000 lux for 16 hours light and 8 hours dark condition for the seven days following the initiation of pre-incubation. For the *M.a.* culture pre-incubation, a M-11 culture medium solution which had been pre-autoclaved for sterilization at 121℃ for 15 minutes was used for the growth control tests as in Table 1.

Table 1: Nutrients in the M-11 culture medium solution.

$NaNO_3$	100.0mg	Na_2CO_3	20.0mg
K_2HPO_4	10.0mg	$FeSO_4 \cdot 7H_2O$	1.0mg
$MgSO_4 \cdot 7H_2O$	75.0mg	2NA(EDTA · 2Na)	1.0mg
$CaCl_2 \cdot 2H_2O$	40.0mg	Deionized Water	1,000 mL

pH=8.0

3.4 Growth control tests using WSE and CTF solutions

Procedures for the *M.a.* growth control tests are briefly summarized. A 150 mL M-11 culture medium solution that was previously sterilized as above was separated into a test bottle of 500 mL Erlenmeyer flask, into which a NIES 102 pre-incubated culture was added to set up the initial cell concentration at about 300,000 cells/mL. Then, the WSE or CTF solution that was pre-filtered through a 0.22 μm membrane filter (Millipore Corporation, Japan) was added into the test bottle. The legend of the control in each figure is when neither the WSE nor the CTF solution was added into the test bottle. The growth control effect was examined on the basis of observing the daily change of cell concentration in the mixture solution of the test bottle. After the cyanobacterial colony in the mixture solution inside the test bottle was scattered by a supersonic treatment, a small amount of the mixture solution was quickly taken into a Hemacytometer Thoma Type Counting Chamber to directly measure the cell number by a microscope. The cell number measurement was conducted five to ten times for each of the five samples from the test bottle mixture solution.

4 Results and discussions

4.1 CT and PP analysis on WSE solution

Two kinds of dead leaves from cherry and maple trees were used to prepare the WSE solutions, and then the CT and PP concentrations on each WSE solution were analyzed and listed in Table 2. The CT and PP concentrations are expressed in mg-CT/mL and mg-PP/mL as catechin equivalent.

Table 2: CT and PP analysis on WSE solution.

	CT (mg-CT/mL)	PP (mg-PP/mL)	CT/PP Ratio
cherry	0.13	0.54	0.24
maple	0.32	0.95	0.34

According to Table 2, the CT extractions from the cherry and maple trees were 0.13 mg-CT/mL and 0.32 mg-CT/mL respectively, and the PP extractions were 0.54 mg-PP/mL and 0.95 mg-PP/mL. As a result, the CT/PP ratios were 0.24 and 0.34 respectively for each kind of dead leaf, which means that the CT and PP extraction amounts from the maple trees were higher than those of the cherry. Although it is not clearly understood, it probably is due to the kind of deciduous trees, the sunlight intensity during the photosynthesis light reaction, or the differences in sample-collecting time or location of the dead leaves.

4.2 CT extraction from dead leaves

Incidentally, the extraction volume for the WSE solution preparation was actually 100 mL for 1 g of the dry dead leaves, and the total extraction per unit weight of the dead leaves was estimated to be 13 mg-CT/g for the cherry and 32 mg-CT/g for the maple. The CT substance elutes easily into water from the dead leaves, and the amount of the CT extracted from the cherry tree after seven days immersion under room temperature was 13 to 20 mg-CT/g for D.I. water and 11.9 to 14.8 mg-CT/g for lake waters. As for the solvent extraction from the dead cherry leaves, the possible amount of the CT extracted by a 50% ethanol solution (ethanol:D.I. water = 1:1 v/v) was 18.6 to 21.9 mg-CT/g. When using a 50% ethanol solution, the CT extraction process was actually repeated three times, and a coloring reaction was not observed for the supernatant solution when a drop of ferric chloride solution was added. Therefore, the CT contained in the dead leaves of the cherry tree seemed to be completely removed by the 50% ethanol solution. Accordingly, the CT extraction rate by the D.I. water would be about 0.60~0.91 (average 0.72) when compared with that of the 50% ethanol solution, which means that the CT substance was easily extracted with a relatively high recovery rate only by immersing the dead leaves into water.

4.3 Effects of CT on *M.a.* growth control

Using the two kinds of WSE solutions from the dead cherry and maple leaves, the *M.a.* growth control tests were conducted and the results were plotted in Figure 2.

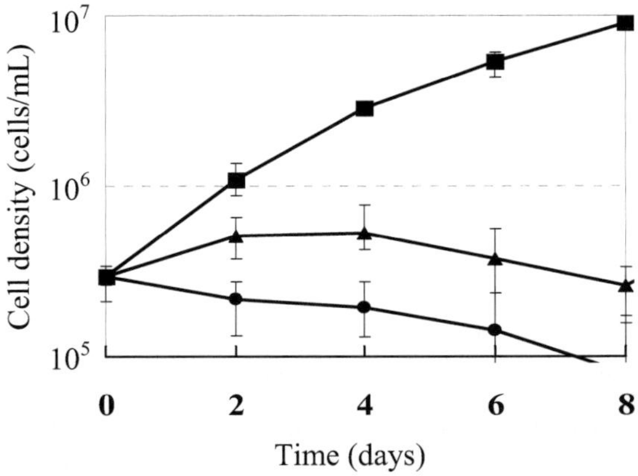

Figure 2: Effects of WSE solution on *M.a.* growth control under CTLI of 26 pg-CT/cell.

The CT was supposed to be the major growth control factor. Based on the pre-determined CT concentration in each WSE solution, the addition of the cherry WSE solution was decided to be 10 mL, and that of the maple WSE solution was 4 mL because of the CT extraction difference between the two kinds of dead leaves. Accordingly, the initial CT concentration was 0.008 mg-CT/mL in both test bottles, while the PP concentration was 0.032 mg-PP/mL for the cherry WSE solution and 0.024 mg-PP/mL for the maple WSE solution. Therefore, the CTLI per unit *M.a.* cell in each test bottle was initially adjusted at 26 pg-CT/cell. In this case, the initial *M.a.* cell concentration was arranged to be about 300,000 cells/mL. As in Figure 2 showing the *M.a.* cell concentration change in the control test bottle, the *M.a.* cell seemed to be growing during the first 8 days of the test period. Distinct growth control/inhibition was observed in the WSE added-test bottles. This implies that some water soluble substances like the condensed tannin were very effectively acting on the growth control of the *M.a.* cells. Although the CTLI was set up initially at nearly the same level for the test bottles, the effect on the growth control was a little bit higher when using the maple WSE solution than when using the cherry WSE solution. The growth control seems to have been affected by the composition and/or the amounts of the PP of the organic substances in the WSE solution.

4.4 Effect of fractioning treatment on CT separation

Table 3 shows the CT and PP amounts and the CT/PP ratios in the WSE and CTF solutions obtained from the dead cherry leaves. The CT and PP concentrations of Sample 1 were 0.09 mg-CT/mL and 0.40 mg-PP/mL for the WSE solution, and 0.05 mg-CT/mL and 0.06 mg-PP/mL for the CTF solution. The CT/PP ratio of Sample 1 was 0.23 for the WSE and 0.83 for the CTF respectively. This implies that the WSE fractioning treatment using casein/acetone results in an increase in the CT/PP ratio of the CTF since the fractioning treatment aims to preferentially separate the CT substance from the WSE solution to obtain the CTF solution with a higher CT/PP ratio. According to Table 3, the CT/PP ratios of the other samples ranged from 0.19 to 0.25 for the WSE solutions and from 0.29 to 0.83 for the CTF solutions. The WSE fractioning treatment by use of casein/acetone would seem to effectively but not completely separate the CT substance itself.

Table 3: Amounts of CT and PP in WSE and CTF solutions.

Sample #	Solutions	CT (mg-CT/mL)	PP (mg-PP/mL)	$(CT/PP)_{WSE}/(CT/PP)_{CTF}$
Sample 1	WSE/CTF	0.09/0.05	0.40/0.06	0.23/0.83
Sample 2	WSE/CTF	0.12/0.07	0.59/0.15	0.20/0.47
Sample 3	WSE/CTF	0.14/0.10	0.56/0.21	0.25/0.48
Sample 4	WSE/CTF	0.10/0.11	0.45/0.16	0.22/0.69
Sample 5	WSE/CTF	0.10/0.07	0.54/0.22	0.19/0.32
Sample 6	WSE/CTF	0.07/0.02	0.35/0.07	0.20/0.29

4.5 Effects of WSE and CTF solutions on *M.a.* growth control

Using both the WSE and CTF solutions, the *M.a.* growth control tests were conducted similarly to those done using the WSE solution as in Figure 2. The data obtained are shown in Figure 3 when using Sample 1 and in Figure 4 when using Sample 2. Both Sample solutions are almost the same in CT/PP ratio for the WSE solutions, but the CT/PP ratios differ a little for the CTF solutions because of the effect of the fractioning treatment. As is shown in Figures 3 and 4, a typical log growth phase of the *M.a.* cells was observed in the control during the first four days from the starting, while the growth control/inhibition tendency was clearly observed especially for the first six days on occasions of adding the WSE or CTF solution with the initial CTLI level of 13 to 16 pg-CT/cell. This would be mainly due to the effect of the CT of the organic substances contained in the dead leaves. Considering along with the results in Figure 2, it is assumed that the very effective growth control/inhibition would be sustained for at least 6 to 8 days when the initial CTLI per unit *M.a.* cell were adjusted to higher than 14 pg-CT/cell.

4.6 Effects of WSE-R solution on *M.a.* growth control

The results in Figures 3 and 4 show that the CTF solution also keeps a good growth control effect on the *M.a.* cells. Assuming that the growth control effect is due to other factor than the condensed tannin, no growth control effect would be observed when using the WSE-R solution since it presumably contains no or less condensed tannin substance. Therefore, if a growth control effect takes place when using WSE-R solution, there would be other factors affecting the growth control. As for the WSE, CTF and WSE-R solutions for the growth control tests,

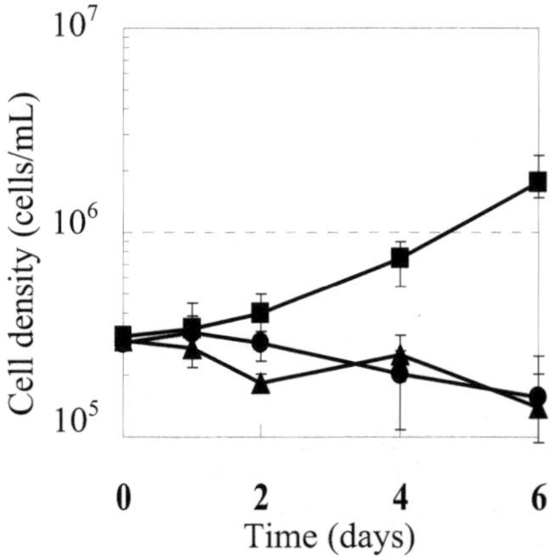

Legends:
■:Control ▲:WSE(14pg-CT/cell)
●:CTF(13 pg-CT/cell)

Figure 3: Effects of WSE and CTF solutions on *M.a.* growth control at various CTLIs in cases using Sample 1.

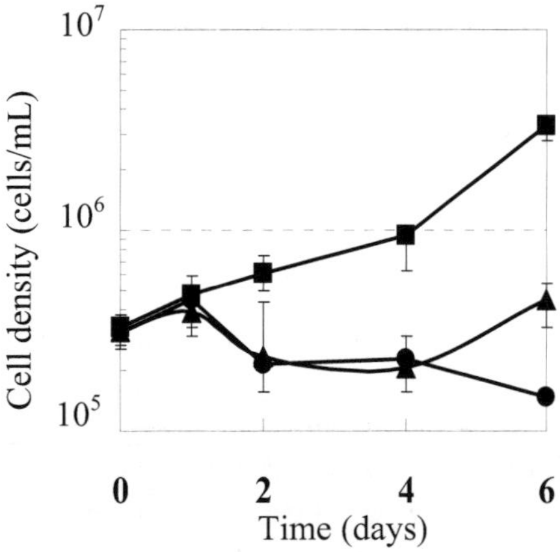

Legends:
■:Control ▲:WSE(16pg-CT/cell)
●:CTF(15 pg-CT/cell)

Figure 4: Effects of WSE and CTF solutions on *M.a.* growth control at various CTLIs in cases using Sample 2.

the CT concentrations and the initial CTLIs are given in Table 4. By the WSE fractioning treatment, the CT concentration in the WSE-R solution decreased down to about 14% of the original WSE solution. In this case, the rate of the CT recovery from the WSE solution to the CTF plus WSE-R solutions was about 86%.

Table 4: CT concentrations and CTLIs in WSE, CTF and WSE-R solutions.

Solution	CT (mg-CT/mL)	CTLI (pg-CT/cell)
WSE	0.14	18
CTF	0.10	13
WSE-R	0.02	2.2

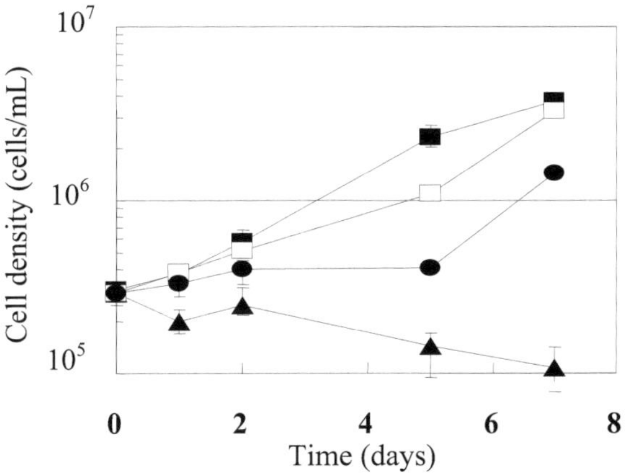

Legends: ■ Control ▲WSE(18pg-CT/cell)
● CTF(13pg-CT/cell) □WSE-R(2.2pg-CT/cell)

Figure 5: Effects of WSE, CTF and WSE-R solutions on *M.a.* growth control.

In Figure 5, the results are shown about the growth control/inhibition tests using the WSE, CTF and WSE-R solutions. The amount of each solution added was 4 mL, and thus the initial CTLI in each solution was 18 pg-CT/cell for the WSE solution, 13 pg-CT/cell for the CTF solution and 2.2 pg-CT/cell for the WSE-R solution. As is shown, the growth control/inhibition effect due to the solution addition seems to be very clearly affected by the initial CTLI level, while there was no clear growth control effect observed when using the WSE-R solution. Therefore, it is assumed that the condensed tannin would function as a growth control factor to the *M.a.* cells.

5 Conclusions

Widely in temperate zones or cold district regions, the dead leaves of deciduous trees like cherry and maple contain organic substances like polyphenol and especially condensed tannin, and the condensed tannin can easily elute into water. Growth control tests have been conducted to evaluate the effect of the condensed tannin on toxic *M.a.* cells by using WSE solutions obtained from dead cherry tree leaves. The results above imply that the condensed tannin extracted from the dead leaves of the deciduous trees functions to be very effective for the growth control/inhibition of *M.a.* cells. A counter measure using the dead leaves would be very rational and effective for the growth control of the toxic cyanobacteria. Some important conclusions are summarized as follows.

1) The CT amount extracted from the dead leaves of cherry trees after seven days immersion at room temperature was 12.7 to 20 mg-CT/g for D.I. water, and the possible amount of the CT extracted by a 50% ethanol solution (ethanol:D.I. water = 1:1 v/v) was 18.6 to 21.9 mg-CT/g. Accordingly, the CT extraction rate by D.I. water was 0.60~0.91 (average 0.72) when compared with that by a 50% ethanol solution.
2) The *M.a.* cells seem to be growing during the first 8 days of the test period, while the distinct growth control/inhibition was observed in the WSE added-test bottles. This apparently implies that some water soluble substances like condensed tannin would very effectively act for the growth control of the *M.a.* cells
3) The growth control/inhibition tendency was clearly observed on occasions of the WSE or CTF addition at the CTLI at 13 to 16 pg-CT/cell especially for the first 6 days. This would be mainly due to the effect of the CT of the organic substances contained in the dead leaves to function as the growth control/inhibition factor.
4) There was no growth control effect observed when using the WSE-R solution with its initial CTLI at 2.2 pg-CT/cell.

The condensed tannin extracted from the dead leaves could be a major growth control factor on *M.a.* although the mechanism has not been proven definitely in regards to the growth control of the *M.a* cell. Our research has been focusing on disclosing the possible mechanism such that the condensed tannin affects on the photosynthesis light reaction activity of the *M.a.* cells to decrease their energy production function [18, 19]. As described above, the counter measure against the growth control of *Microcystis* using the dead leaves of natural plants is supposed to be very effective and rational.

Abbreviations:
CT: Condensed Tannin
CTF: Condensed Tannin Fraction
D.I. water: deionized water
M.a.: *Microcystis aeruginosa*
PP: Polyphenol

WSE: Water Soluble Extract
WSE-R: Water Soluble Extract remaining

Acknowledgements

The present research work has been performed by many students of the Nihon University, College of Science and Technology, Civil Engineering Department Aqua-Enviro System Laboratory. The authors would like to express their sincere gratitude to all students who have contributed to the project over the last three years.

References

[1] Guangming Z., Panyue Z., Hong L., & Bo W., Ultrasonic damages on cyanobacterial photosynthesis. *Ultrasonic Sonochemistry*, 13, pp.501-505. 2006

[2] Yunfeng X., Ji Y., Mingming O., Yalin W., & Jinping J., Study of *Microcystis aeruginosa* inhibition by electrochemical method, *Biochemical Engineering Journal*, 36 (3), pp.215-220, 2007

[3] Ferrier M.D., Butler B.R. Sr., Terlizzi D.E., & Lacouture R.V., The effects of barley straw (*Hordeum vulgare*) on the growth of freshwater algae. *Bioresource Technology*, 96, pp.1788-1795, 2005

[4] Gibson M. T., Welch I. M., Barrett P. R. F, & Ridge I., Barley straw as an inhibitor of algal growth II: laboratory studies. *Journal of Applied Phycology*, 2, pp.241-248, 1990

[5] Ridge I., Walters J., & Street M., Algal growth control by terrestrial leaf litter: a realistic tool?. *Hydrobiologia*, 395/396, pp.173-180, 1999

[6] Nakai S., Inoue Y., Hosomi M, & Murakami A., *Myriophyllum spicatum*-released allelopathic polyphenols inhibiting growth of blue-green algae *Microcystis aeruginosa*. *Water Research*, 34 (11), pp.3026-3032, 2000

[7] Nakayama M., Suzuki K., Toda M., Okubo S., Hara Y., & Shimamura T., Inhibition of the infectivity of influenza virus by tea polyphenols. *Antiviral Research*, 21, pp.289-299, 1993

[8] Sakanaka S., Sato T., Kim M., & Yamamoto T., Inhibitory effects of green tea polyphenols on glucan synthesis and cellular adherence of cariogenic streptococci. *Agric. Biol. Chem*, 54 (2), pp.2925-2929,1990

[9] Junli Z., Takahashi K., Kono Y., Suzuki Y., Takeuchi S., Shimizu T., Yamaguchi I., Chijimatsu M., Sakurai A., Sato Y., & Kitamura H., Bioactive condensed tannins from bark; chemical properties, enzyme inhibition and anti-plant viral activities, *J. Pesticide Sci.*, 15, pp.585-591,1990

[10] Hattori M., Kusumoto I., Namba T., Ishigami T., & Hara Y., Effect of tea polyphenols on glucan synthesis by glucosyltrans ferase from Streptococcus mutans. *Chem. Pharm, Bull.* 38 (3), pp.717-720, 1990

[11] Ikigai H., Nakae T., Hara Y., & Shimamura T., Bactericidal catechins damage the lipid bilayer. *Biochimica Biophysica Acta*, 1147, pp.132-136, 1993
[12] Toda M., Okubo S., Ikigai H., Suzuki T., Suzuki Y., & Shimamura T., The protective activity of tea against infection by vibrio cholerae 01. *Journal of Applied Bacteriology*, 70, pp.109-112, 1991
[13] Augustin S., Antimicrobial properties of tannins. *Phytochemistry*, 30 (12), pp.3875-3883,1991
[14] Kitamura N., Nagakawa T., Yoshida Y., Matsushima H., & Asada Y., Studies on the growth control of toxic Cyanobacteria by use of the water-extracts from the dead leaves of deciduous trees. *Environmental Engineering Research, JSCE*, 43, pp.543-549, (in Japanese) 2006
[15] Tanno K. & Nonomura H., Antibacterial substances in green tea extract. *Japanese Society for Food Sci. and Tech.* 21 (9), pp.445-449. (in Japanese) 1974
[16] Richard B. B. & William T. J, Analysis of condensed tannin using acidified vanillin. *J. of the Science of Food and Agriculture*, 29, pp.788-794, 1978
[17] Folin O. & Denis W., On phosphotungstic-phosphomolybdic compounds as color reagents. *J. Biol. Chem.* 12, pp.239-243,1912
[18] Shimada K., Kitamura N., Nomoto S., Yoshida Y., Matsushima H., & Asada Y., Effect of the condensed tannin from the dead leaves of deciduous trees on the photosynthesis reaction inhibition to *Microcystis aeruginosa. Proc. of EMECS 8th Conf.* pp. 132-133, 2008
[19] Shimada K., Nomoto S., Kitamura N., Yoshida Y., Matsushima H., & Asada Y., Mechanism of growth control of toxic cyanobacteria by use of the dead leaves of roadside trees. *The 3rd ASPIRE-IWA Conference and Exhibition.* oral-STD8-0-2, 2009

Waste water treatment plants as redistributors of resistance genes in bacteria

M. Gómez[1], M. T. Díaz[1], M. Araujo[2], R. Sueiro[2] & J. Garrido[2]
[1]*Labaqua S.A., Santiago de Compostela, Spain*
[2]*Lab. Microbiology (Instituto de Investigación y Análisis Alimentarios), Universidad de Santiago de Compostela, Spain*

Abstract

The aim of this study was to assess the number of both extended-spectrum β-lactamases (ESBL) producing *Enterobacteriaceae* and vancomycin-resistant enterococci (VRE) present in the inflow (hospital and urban sewage) and effluent waste water treatment plant (WWPT) and receiving waters.

The results show that the average counts of ESBL producing *Enterobacteriaceae* in hospital sewage were much higher (more than 2.5 logs) than those observed in urban waste water. On the other hand, the sewage treatment plant studied did not perform well in removing pathogenic microorganisms. Although *Enterobacteriaceae* concentrations were reduced in the final effluent by almost two log units as compared with the crude sewage concentrations, a high number of *Enterobacteriaceae* were detected in the effluent, representing a risk for microbiological pollution of water resources. A strong increase in ESBL-producing bacteria in the receiving water was detected after effluent discharge WWPT.

The results obtained have also demonstrated that the hospital sewage releases a high concentration of VRE (> 3.0×10^5 CFU/100 ml), which represents 25% of the total population of enterococci presents in these waters. The hospital sewage contains a proportion of VRE 10 times higher in comparison with that detected in the urban sewage. The VRE proportion in the urban sewage was very low (2.6%) and constant along the different samplings. In the other hand, sewage discharged from municipal sewage treatment plant effluent also increased the VRE proportion in the river Sar water (1.5% of EVR in upstream to 3.8% downstream from the plant). In conclusion, sewage (particularly hospital sewage) contains a high proportion of ESBL producing *Enterobacteriaceae* and vancomycin resistant enterococci, implying a potential risk of spreading these

WIT Transactions on Ecology and the Environment, Vol 135, © 2010 WIT Press
www.witpress.com, ISSN 1743-3541 (on-line)
doi:10.2495/WP100081

resistant bacteria and resistance genes to the environment and possibly to the human and animal population.

Keywords: β-lactamases, drug resistance, ESBLs, hospital and urban sewage, Enterobacteriaceae, vancomycin-resistant enterococci.

1 Introduction

During the last fifty years the utilization of antimicrobial drugs has steadily increased. Antimicrobials are extensively used in human medicine to treat or prevent bacterial infectious diseases. Antimicrobial resistant bacteria and antimicrobial residues are excreted from patients into the hospital wastewater daily [1]. Consequently, the hospital wastewater may be a hot spot for the propagation and dissemination of antimicrobial resistant and pathogenic bacterial isolates. Some reports showed that high resistance rates to several antimicrobial agents are also observed in commensal and environmental bacteria [2, 3]. Thus, the emergence and dissemination of antibiotic resistant bacteria in hospitals and environments is an emerging challenge to humankind.

Strains of *Enterobacteriaceae* that produce extended-spectrum β-lactamases have emerged as significant pathogens. First reported in the mid-1980s, they were mainly found in *Klebsiella pneumoniae* and *Escherichia coli*, although they can now be found in many other species [4]. The emergence of multiresistance in the *Enterobacteriaceae* family is a question that needs attention, because these are important causative agents of hospital infections, typically associated with pneumonias, blood stream infections, urinary tract infections, bacteremia and other intra-abdominal infections [5-8]. Reports of infection or colonization with ESBL-producing *Enterobacteriaceae* strains have focused mainly on hospitalized patients or nursing home residents [4]. However, ESBL-producing *Enterobacteriaceae* have also been described in animals and in patients in the community, with and without chronic conditions [9, 10]. Additionally, studies have demonstrated the ESBL-producing bacteria are isolated with increasing frequency in many parts of the world [7, 11-17].

Although many resistant bacteria have been isolated from environmental samples, there are only limited studies that have addressed the detection of ESBL-producing bacteria from these types of samples [18, 19]. Prado *et al* [19] observed the presence of *Klebsiella pneumoniae* producing ESBL in the effluent and sludge of hospital sewage plants. Therefore, the hospital sewage or the low efficacy of hospital sewage-treatment plants could result in the dissemination of ESBL-producing bacteria. In addition, it is generally accepted that waste water treatment plants play an important role in redistributing and disseminating antimicrobial resistance properties to bacteria. In fact, the almost permanent presence of detectable antimicrobial levels in the waters of waste water treatment plants could strongly influence the environment, creating a selective pressure that would be ultimately responsible for the dominance of populations of resistant microorganisms among those present in the habitat.

Although it was investigated the presence of ESBL-producing bacteria in hospital sewage and environmental samples, we lack data on the quantitative importance of these bacteria in these environments, consequently the aim of this study was to quantify ESBL-producing *Enterobacteriaceae*, present in the inflow (hospital and urban sewage) and effluent waste water treatment plant (WWPT) and receiving waters.

By the other hand, enterococci have emerged as one of the clinical challenges when identified as the cause of serious or life threatening infections. Isolates of *Enterococcus faecalis* and *Enterococcus faecium* are the third- to fourth-most prevalent nosocomial pathogen worldwide [20]. Antimicrobial resistance in strains causing nosocomial infections is a growing problem and VRE in particular are considered a serious threat in hospitals around the world [21]. Vancomycin is often used as a last resort in treatment of antibiotic resistant gram-positive bacterial infections caused by organisms such as multi-resistant enterococci and methicillin resistant staphylococci. Infections due to VRE are associated with higher treatment costs, prolonged morbidity, and greater mortality rates [22].

In the USA, the prevalence of VRE is mainly documented as nosocomial infection in humans [23, 24]. The situation regarding VRE in Europe is diverse with prevalences ranging from <1 to >40% and many aspects of VRE acquisition and spread are still unknown [20]. In any case, acquired vancomycin resistance appears to be a serious and growing therapeutic challenge among enterococci all over Europe. According Werner *et al* [20], some EU countries have experienced an increasing VRE trend over time (e.g., Ireland, Germany, Greece); in other countries VRE prevalence is still low (e.g., in Nordic countries, the Netherlands); a few EU Member States showed decreasing VRE rates (e.g., Austria, Portugal, Italy).

The spread of these organisms from the hospital environment or other sources to environmental waters through discharged sewage or other means could increase the prevalence of these strains in the human population and become a potential risk to human health [25-27]. Several studies have detected VRE strains in hospital-impacted sewage [26, 28], sewage [29] and in receiving waters [28].

The second aim of the present study was to assess the number of vancomycin-resistant enterococci present in the hospital sewage and receiving waters.

2 Materials and methods

2.1 Water samples collection

Samples were collected from sewage effluent University Hospital of Santiago de Compostela (Spain), both wastewater and municipal sewage treatment plant of Santiago de Compostela, River Sar (upstream and downstream of sewage treatment plant discharge, sampling sites 1, 2 and 3, respectively) and estuary of Arousa (sampling site 4, shellfish cultivation area). Site 1, is at 0,2 Km upstream of WWPT, site 2 is at 7 Km downstream of WWPT and site 3 is at 15 Km downstream of WWPT.

Sampling was carried out in April 2004, September 2004, February 2005 and June 2005, over 2 weeks in each month. Wastewater and water samples were collected into sterile bottles. The wastewater samples were collected during a 24h period, harvesting 50 ml every 60 min, refrigerated and analysed within 24h.

2.2 Isolation and enumeration of *Enterobacteriaceae*

The *Enterobacteriaceae* count from waste water samples were obtained by inoculating one ml aliquots of the appropriate dilution in chromogenic Coli-ID agar (supplied by bioMerieux, France). All plates were incubated aerobically for 24 $\pm$ 2h at 36.0 $\pm$ 1.0ºC. All pink/violet and blue/grey colonies on Coli-ID agar were counted as presumptive *Enterobacteriaceae.* Typical colonies were picked off for confirmation as *Enterobacteriaceae* based on the following tests: Gram stain, oxidase and glucose fermentation [30].

The *Enterobacteriaceae* count from River Sar and estuary of Arousa were analysed by standard membrane filtration technique, using the same culture media and incubation conditions.

The same methods were employed for ESBL-producing *Enterobacteriaceae* counts, supplementing Coli-ID agar with 2 μg/ml of ceftazidime, cefotaxime, ceftriaxone and aztreonam. Confirmation of β-lactamase production by strains isolated from agar supplemented Coli-ID was carried out using the test specified by Steward *et al* [31].

2.3 Isolation and enumeration of enterococci

At the laboratory, wastewater samples (0.1 ml aliquots of the appropriate dilution) were directly streaked onto Slanetz & Bartley Medium for enterococci count. Water samples from River Sar and estuary of Arousa were analysed by membrane filter technique [32]. VRE were isolated on M-enterococcus containing 32 μg /ml vancomycin and were incubated for 48 $\pm$ 3h at 36.0 $\pm$ 1.0°C.

3 Results and discussion

The average counts of *Enterobacteriaceae* and ESBL-producing *Enterobacteriaceae* obtained from all wastewater samples (hospital and urban sewage, influent and effluent WWPT) collected throughout the study are shown in Figure 1. The results show that among the urban and hospital sewage there is little difference in the total level of enterobacteria. On the contrary, it is noteworthy that the average counts of ESBL producing *Enterobacteriaceae* in hospital sewage were much higher (more than 2.5 logs) than those observed in urban wastewater. This clearly highlights the importance of hospital sewage as a major source of antibiotic-resistant bacteria.

One of the recurring concerns because of the presence of ESBL-producing bacteria in these environments is associated with the transfer of conjugative plasmids, which also carry genes of resistance to aminoglucosides and sulfonamides, giving the bacteria multiresistance patterns [33]. Conjugative plasmids can be easily transferred to other bacteria [34, 35]. This mechanism of gene transfer is particularly important in hospital effluents, because in such environments, the possibility of gene exchange is increased as a result of higher density and greater contact between these microorganisms, which allows for an increased rate of horizontal transfer of these genes.

The incorrect handling and disposal of the hospital sewage into inappropriate locations may compromise public health by contaminating the soil, water supplies or recreational waters, and facilitate dissemination of microorganisms and resistance genes into the environment. In this way, a high prevalence of resistant micro-organisms in hospitals shows a potential risk for disseminating pathogenic agents into the environment, particularly through the wastewaters.

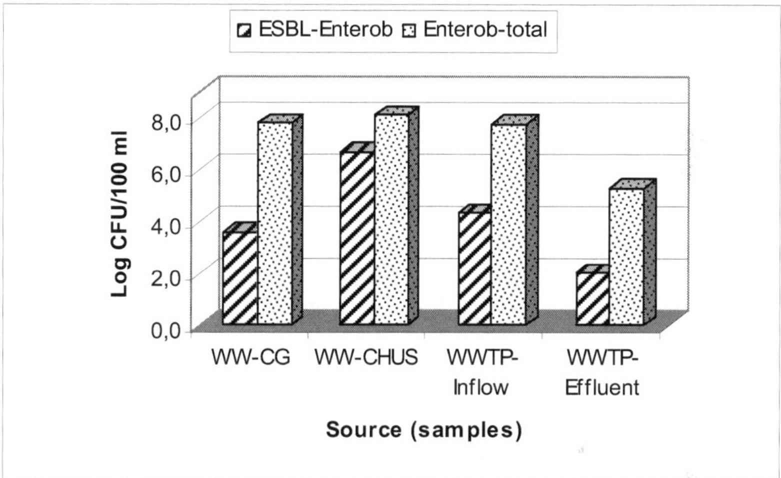

Figure 1: Average counts of *Enterobacteriaceae* and ESBL producing *Enterobacteriaceae* obtained from the water samples. WW-CG, waste water general collector of Santiago de Compostela. WW-CHUS, waste water from University hospital complex of Santiago de Compostela. WWTP, waste water treatment plant.

By the other hand, the sewage treatment plant studied did not perform well in removing pathogenic micro-organisms. Although *Enterobacteriaceae* concentrations were reduced in the final effluent by almost two log units as compared with the crude sewage concentrations, a high number of *Enterobacteriaceae* were detected in the effluent, representing a risk for microbiological pollution of water resources. Thus, as shown in Figure 2, the wastewater discharge from WWTP causes a strong increase in ESBL-producing bacteria in the receiving waters, so that 15 km downstream of discharge point, is still detected a high level of resistant bacteria. In Figure 2, also notes the presence of ESBL-producing bacteria in the water of shellfish cultivation area affected by the river waters at that WWTP discharged wastewater. Thus, the prevalence of ESBL-producing *Enterobacteriaceae* was notable in all the environments studied, suggesting a global expansion of these enzymes, as recorded Mesa *et al* [18].

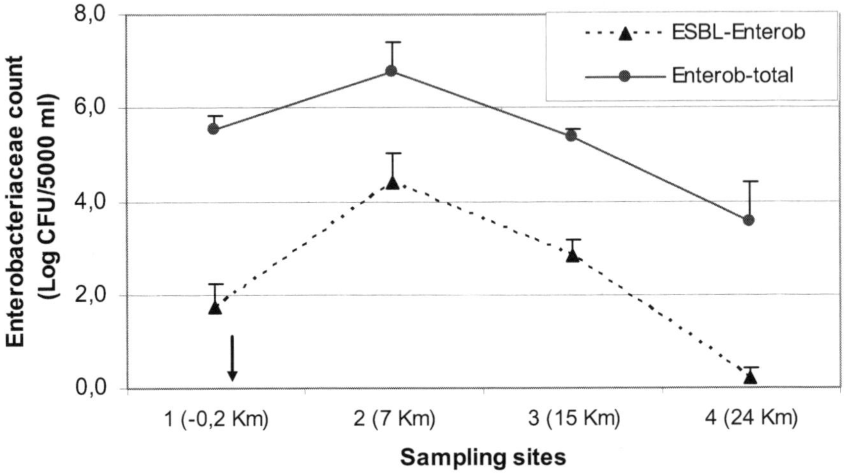

Figure 2: *Enterobacteriaceae* (Enterob-total) and ESBL producing *Enterobacteriaceae* (ESBL-Enterob) counts in four sampling sites along the river Sar. Values shown are geometric means of six determinations. The arrow indicates the dumping site of treatment plant into the river.

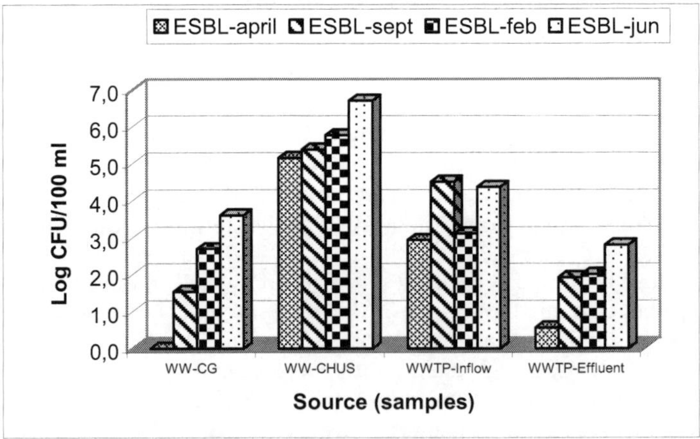

Figure 3: ESBL-producing *Enterobacteriaceae* (ESBL) counts obtained in water samples collected in April, September (sept), February (feb) and June (jun). WW-CG, waste water general collector of Santiago de Compostela. WW-CHUS, waste water from University hospital complex of Santiago de Compostela. WWTP, waste water treatment plant.

When analyzing the results is done taking into account the time of year, Figure 3 shows that the release of ESBL-producing *Enterobacteriaceae* from hospital waste water was very high throughout the year, particularly in the period of time for June. This same behaviour was observed in waste after samples collected in the other three points. A temporal distribution of ESBL-producing *Enterobacteriaceae* was also observed by Mesa *et al* [18] in Spain, who found a higher prevalence between July and November and lowest between February and May.

The results obtained have demonstrated that the hospital sewage also release a high concentration of VRE (> 3.0 x 10^5 CFU/100 ml), which represents 25% of the total population of enterococci presents in these waters (Figure 4). The hospital sewage contains a proportion 10 times higher of VRE in comparison with the one that was detected in the urban sewage. In Figure 4 also may be observed that the proportion of VRE was very similar in influent and effluent of the waste water treatment plant.

The VRE proportion compared to the total population of Enterococci in the hospital wastewater throughout the year ranged between 37.05% in April and 17.6% in February. Nevertheless, the median values were constant along year (Figure 5). The VRE proportion in the urban sewage was very low (2.6%) and constant along the different samplings. In the other hand, sewage discharged from municipal sewage treatment plant effluent increased the VRE proportion in the river Sar water (1.5% of EVR in upstream to 3.8% downstream from the plant). The results obtained shown that the presence of VRE in environment aquatic is clearly influenced by hospital sewage, which are led to municipal treatment plants and mixed with urban sewage (Figure 6).

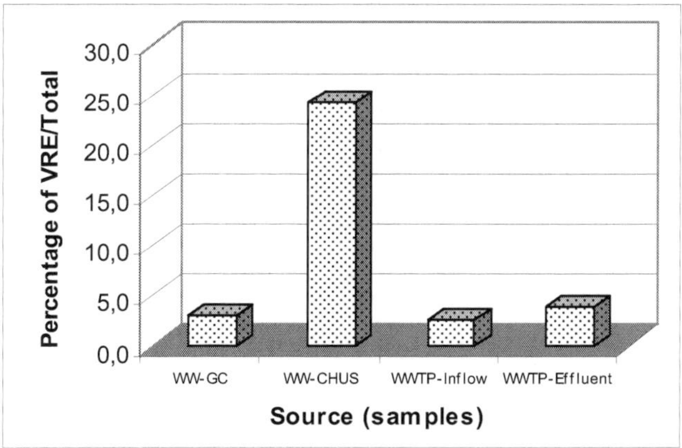

Figure 4: Proportion of vancomycin-resistant Enterococci (VRE) compared to the total population of Enterococci obtained in the water samples. WW-CG, waste water general collector of Santiago de Compostela. WW-CHUS, waste water from University hospital complex of Santiago de Compostela. WWTP, waste water treatment plant.

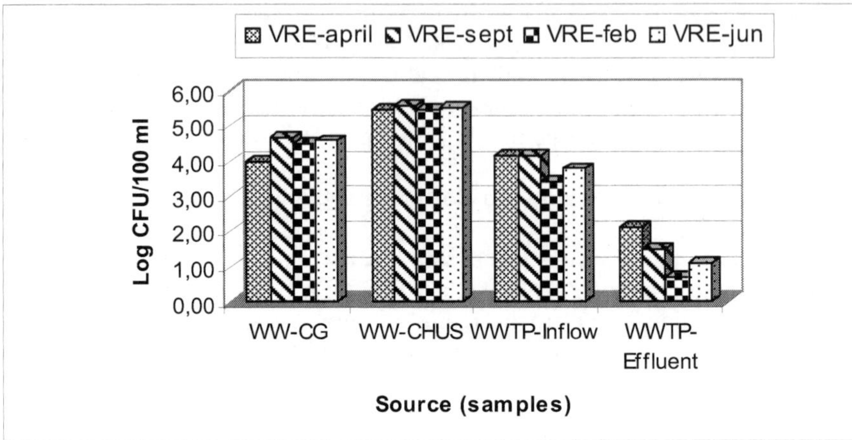

Figure 5: Vancomycin-resistant Enterococci (VRE) counts obtained in water samples collected in April, September (sept), February (feb) and June (jun). WW-CG, waste water general collector of Santiago de Compostela. WW-CHUS, waste water from University Hospital Complex of Santiago de Compostela. WWTP, waste water treatment plant.

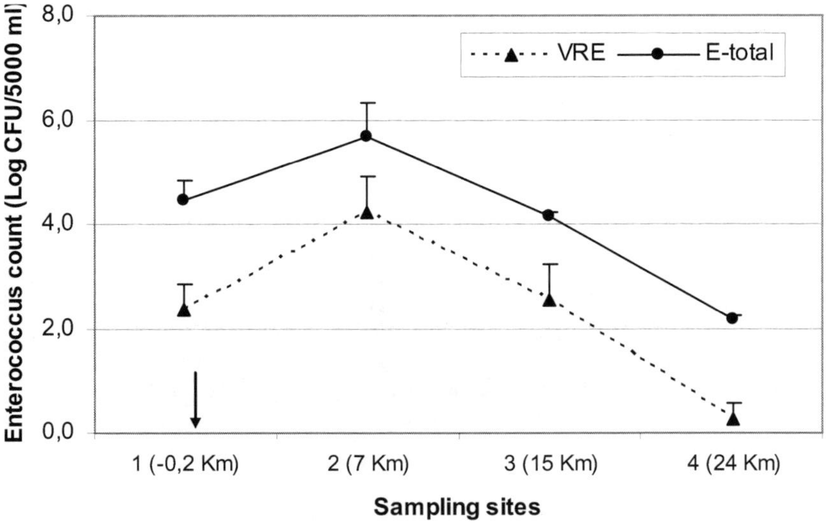

Figure 6: Vancomycin-resistant enterococci (VRE) and total enterococci (E-total) counts in four sampling sites along the river Sar. Values shown are geometric means of six determinations. The arrow indicates the dumping site of treatment plant into the river.

In conclusion, sewage (particularly, sewage hospital) contains a high proportion of ESBL-producing *Enterobacteriaceae* and vancomycin resistant enterococci, implying a potential risk of spreading these resistant bacteria and resistance genes to the environment and possibly to the human and animal population. The high incidence of resistant bacteria in sewage implies a need for high efficient hygienic treatment of sewage (application of final disinfection processes for to reduce the microbial burden), with the intention of minimizing the impact from the release of hospital wastewaters into water resources, preventing dissemination of multiresistant microorganisms and their genes of resistance into the environment, thus promoting prevention measures to protect public health.

Acknowledgement

The work was financially supported by a grant from the Xunta de Galicia (PGIDIT03TAM03E).

References

[1] Kummerer, A. & Henninger, K. Promoting resistance by the emission of antibiotics from hospitals and households into effluent. *Clin. Microbiol. Infect.,* 9, pp. 1203-1214, 2003.

[2] Kümmerer, K. Resistance in the environment. *J. Antimicrob. Chemother.,* 54, pp. 311-320, 2004.

[3] Bartoloni, A., Pallecchi, L., Benedetti, M., Fernandez, C., Vallejos, Y., Guzman, E., Villagran, A.L., Mantella, A., Lucchetti, C., Bartalesi, F., Strohmeyer, M., Bechini, A., Gamboa, H., Rodríguez, H., Falkenberg, T., Kronvall, G., Gotuzzo, E., Paradisi, F. & Rossolini, G.M. Multidrug-resistant commensal *Escherichia coli* in children, Peru and Bolivia. *Emerg. Infect. Dis.,* 12, pp. 907-913, 2006.

[4] Bradford, PA. Extended-spectrum b-lactamases in the 21st century: characterization, epidemiology, and detection of this important resistance threat. *Clin. Microbiol. Rev.,* 14, 933–51, 2001.

[5] Kotapati, S., Kuti, J.L. and Nightingale, C.H. Clinical implications of extended spectrum b-lactamase (ESBL) producing Klebsiella species and Escherichia coli on cefepime effectiveness. *J. Infect.,* 51, pp. 211–217, 2005.

[6] Lopes, A.C.S., Rodrigues, J.F. & Morais Junior, M.A. Molecular typing of *Klebsiella pneumoniae* isolates from public hospitals in Recife, Brazil. *Microbiol Res*., 160, pp. 37–46, 2005.

[7] Grover, S.S., Sharma, M., Chattopadhya, D., Kapoor, H., Pasha, S.T. & Singh, G. Phenotypic and genotypic detection of ESBL mediated cephalosporin resistance in *Klebsiella pneumoniae*: emergence of high resistance against cefepime, the fourth generation cephalosporin. *J. Infect.*, 53, pp. 279–288, 2006.

[8] Sun, H.Y., Chen, S.Y., Chang, S.C., Pan, S.C., Su, C.P. & Chen, Y.C. Community-onset *Escherichia coli* and *Klebsiella pneumoniae* bacteremia: influence of health care exposure on antimicrobial susceptibility. *Diagn Microbiol. Infect. Dis.,* 55, pp. 135–141, 2006.

[9] Briñas, L., Moreno, M.A., Teshager, T., Sáenz, Y., Porrero, M.C., Domínguez, L. & Torres, C. Monitoring and characterization of extended-spectrum b-lactamases in *Escherichia coli* strains from healthy and sick animals in Spain in 2003. *Antimicrob. Agents Chemother.*, 49, pp. 1262–4, 2005.

[10] Mirelis, B., Navarro, F., Miró, E., Mesa, R.J., Coll, P. & Prats, G. Community transmission of extended-spectrum b-lactamase. *Emerg. Infect. Dis.*, 9, pp. 1024–1035, 2003.

[11] Peña, C., Pujol, M., Ricart, A., Ardanuy, C., Ayats, J., Liñares, J., Garrigosa, F., Ariza, J. & Gudiol, F. Risk factors for faecal carriage of Klebsiella pneumoniae producing extended spectrum b-lactamase (ESBL – KP) in the intensive care unit. *J. Hosp. Infect.,* 35, pp. 9–16, 1997.

[12] Araque, M., Nieves, B., Lauretti, L. & Rossolini, G.M. Molecular basis of extended-spectrum b-lactamase production in nosocomial isolates of Klebsiella pneumoniae from Mérida, Venezuela. *Int. J. Antimicrob. Agents* 15, pp. 37–42, 2000.

[13] Seid, J. & Asrat, D. Occurrence of extended spectrum b-lactamase enzymes in clinical isolates of *Klebsiella* species from Harar region, eastern Ethiopia. *Acta Trop.,* 95, pp. 143–148, 2005.

[14] Kolar, M., Latal, T., Cermak, P., Bartonikova, N., Chmelarova, E., Sauer, P. & Kesselova, M. Prevalence of extended-spectrum b-lactamase-positive *Klebsiella pneumoniae* isolates in the Czech Republic. *Int. J. Antimicrob. Agents*, 28, pp. 49–53, 2006.

[15] Lee, H., Yong, D., Yum, J.H., Roh, K.L.H., Lee, K.L., Yamane, K.L., Arakawa, Y. and Chong, Y. Dissemination of 16S rRNA methylas-mediated highly amikacin-resistant of *Klebsiella pneumoniae* and *Acinetobacter baumannii* in Korea. *Diagn. Microbiol. Infect. Dis.,* 56, pp. 305–312, 2006.

[16] Paterson, D.L. Resistance in gram-negative bacteria: Enterobacteriaceae. *Am. J. Med.,* 119, S20–S28, 2006.

[17] Yu, Y., Ji, S., Chen, Y., Zhou, W., Wei, Z., Li, L. & Ma, Y. Resistance of strains producing extended-spectrum b-lactamases and genotype distribution in China. *J. Infect.,* 54, pp. 53–57, 2006.

[18] Mesa, R.J., Blanc, V., Blanch, A.R., Cortés, P., González, J.J., Lavilla, S., Miró, E., Muniesa, M., Saco, M., Tórtola, M.T., Mirelis, B., Coll, P., Llagostera, M., Prats, G. & Navarro, F. Extended-spectrum b-lactamase-producing *Enterobacteriaceae* in different environments (humans, food, animal farms and sewage). *J. Antimicrob. Chemother.,* 58(1), pp. 211-5, 2006.

[19] Prado, T., Pereira, W.C., Silva, D.M., Seki, L.M., Carvalho, A.P. & Asensi, M.D. Detection of extended-spectrum beta-lactamase-producing *Klebsiella*

pneumoniae in effluents and sludge of a hospital sewage treatment plant. *Lett. Appl. Microbiol.*, 46(1), 136-41, 2008.

[20] Werner, G., Coque, T.M., Hammerum, A.M., Hope, R., Hryniewicz, W., Johnson, A., Klare, I., Kristinsson, K.G., Leclercq, R., Lester, C.H., Lillie, M., Novais, C., Olsson-Liljequist, B., Peixe, L.V., Sadowy, E., Simonsen, G.S., Top, J., Vuopio-Varkila, J., Willems, R.J., Witte, W. & Woodford, N. Emergence and spread of vancomycin resistance among enterococci in Europe. *Euro Surveill.*, 13(47), pp. 1-11, 2008.

[21] Mascini, E.M., Bonten, M.J. Vancomycin-resistant enterococci: consequences for therapy and infection control. *Clin. Microbiol. Infect.*, 11(Suppl 4), pp. 43-56, 2005.

[22] Calfee, D.P., Giannetta, E.T., Durbin, L.J., Germanson, T.P. & Farr, B.M. Control of endemic vancomycin-resistant Enterococcus among inpatients at a university hospital. *Clin. Infect. Dis.*, 37, pp. 326–332, 2003.

[23] Bonten, M.J., Willems, R. & Weinstein, R.A. Vancomycin-resistant enterococci: why are they here, and where do they come from? *Lancet Inf. Dis.*, 1, pp. 314-325, 2001.

[24] Coque, T.M., Tomayko, J.F., Ricke, S.C., Okhyusen, P.C. & Murray, B.E. Vancomycin-resistant enterococci from nosocomial, community, and animal sources in the United States. *Antimicrob. Agents Chemother.*, 40(11), pp. 2605-2609, 1996.

[25] Rice, E.W., Messer, J.W., Johnson, C.H. and Reasoner, D.J. Occurrence of high-level aminoglycoside resistance in environmental isolates of enterococci. *Appl. Environ. Microbiol.*, 61, 374–376, 1995.

[26] Iversen, A., Kühn, I., Franklin, A. & Möllby, R. High prevalence of vancomycin-resistant enterococci in Swedish sewage. *Appl. Environ Microbiol.*, 68, pp. 2838–2842, 2002.

[27] Kühn, I., Iversen, A., Finn, M., Greko, C., Burman, L., Blanch, A., Vilanova, X., Manero, A., Taylor, H., Caplin, J., Domínguez, L., Herrero, I., Moreno, M.A. & Möllby, R. Occurrence and Relatedness of Vancomycin-Resistant Enterococci in Animals, Humans, and the Environment in Different European Regions. *Appl. Environ. Microbiol.*, 71(9), pp. 5383–5390, 2005.

[28] Novais, C., Coque, T.M., Ferreira, H., Sousa, J.C. & Peixe, L. Environmental contamination with vancomycinresistant enterococci from hospital sewage in Portugal. *Appl. Environ. Microbiol.*, 71, pp. 3364–3368, 2005.

[29] Poole, T.L., Hume, M.E., Campbell, L.D., Scott, H.M., Alali, W.Q. & Harvey, R.B. Vancomycin-resistant *Enterococcus faecium* strains isolated from community wastewater from a semiclosed agri-food system in Texas. *Antimicrob. Agents Chemother.*, 49, pp. 4382–4385, 2005.

[30] ISO 21528-2:2004. Microbiology of food and animal feeding stuffs -- Horizontal methods for the detection and enumeration of *Enterobacteriaceae* -- Part 2: Colony-count method.

[31] Steward, C.D., Rasheed, J.K., Hubert, S.K., Biddle, J.W., Raney, P.M., Anderson, G.J., Williams, P.P., Brittain, K.L., Oliver, A., McGowan, J.E.

Jr, & Tenover, F.C. Characterization of clinical isolates of *Klebsiella pneumoniae* from 19 laboratories using the National Committee for Clinical Laboratory Standards extended-spectrum β-lactamase detection methods. *J. Clin. Microbiol.*, 39, pp. 2864-2872, 2001.
[32] Norma ISO 7899-2:2000. Water quality -- Detection and enumeration of intestinal enterococci -- Part 2: Membrane filtration method.
[33] Paterson, D.L. (2006) Resistance in gram-negative bacteria: *Enterobacteriaceae. Am. J. Med.* 119, pp. S20–S28, 2006.
[34] Heuer, H., Krogerrecklenfort, E., Wellington, E.M.H., Egan, S., Van Elsas, J.D., Van Overbeek, L., Collard, J.M. & Guillaume, G. Gentamicin resistance genes in environmental bacteria: prevalence and transfer. *FEMS Microbiol. Ecol.*, 42, pp. 289–302, 2002.
[35] Tennstedt, T., Szczepanowski, R., Braun, S., Pühler, A & Schlüter, A. Occurrence of integron-associated resistance gene cassetes located on antibiotic resistance plasmids isolated from a wastewater treatment plant. *FEMS Microbiol. Ecol.*, 45, pp. 239–252, 2003.

Section 3
Remediation

Clearance of explosive ordnance: the answer to all evils?

F. Jaenig
German Scientific-Technical Association for Environmental Remediation and Brownfield Redevelopment, Germany

Abstract

Wars do bring harm and death and, even after the end of war, legacies create continuing danger to human beings. Thus, in former operational zones, every 22 minutes a human being is injured or killed by unexploded ammunition. As a first activity a clearance of explosive ordnance is carried out, in order to eliminate the danger of contact. Depending on expected depth and type of ammunition, various methods of detection and clearance are applied, all of which have in common the detection of metals, because of suiting, such as boxes for energetic materials (EMs). After that, the clearance of the explosive ordnances detected – such as shells, bombs, ammunitions – is stipulated, whereupon a serious intervention in the soil in both area and depth can take place. As an inevitable consequence of this, EMs that have leaked from bins or EMs that are in broken bins remain in the soil. In addition, the soil's primarily punctual contamination with EMs could be dispersed across the complete area of clearance. In the soil the particles of the EM are exposed to elution, thus groundwater-contamination with EM is affected due to the type and duration of exposition. Because of the toxic and carcinogenic properties of these substances, such groundwater- and soil-contaminations can cause hazard to sensitive goods. The anonymous German practical example presented shows that high and persistent concentrations of EMs in the leachate were detected after an extremely cost-intensive clearance of explosive ordnance, resulting from a temporary mobilization of EM-particles within the scope of clearance and from EM-soil contaminations not cleared. Thus, a conjunction of clearance of explosive ordnance and soil remedial actions is mandatory, in order to ensure satisfying results of a remediation or clearance.

Keywords: clearance of explosive ordnance, geophysical detection, remediation.

WIT Transactions on Ecology and the Environment, Vol 135, © 2010 WIT Press
www.witpress.com, ISSN 1743-3541 (on-line)
doi:10.2495/WP100091

1 Introduction

In former operational zones human beings are injured or killed periodically by unexploded ordnance devices (UXOs) or left behind ammunition. Even in Germany explosive ordnances deriving from World Wars I and II and military use of areas are found despite an intensive clearance of explosive ordnance for decades.

The imminent danger for human beings is not as high as in accordant conflict areas discussed in the media though. In Germany many areas are identified, which have the suspicion of existence of explosive ordnances, so that it is possible to take correspondent measures in the run-up to any activities in the area. However, unexpected UXOs or ammunition are encountered, mainly during infrastructural works.

In Germany the clearance of explosive ordnance consists of investigation, sounding, excavation, transport and finally demolition of the explosive ordnances. The demolition (e.g. by detonation) is mostly carried out governmentally. The other activities can also be carried out by private specialty firms with accordant individual-related authorizations.

Corresponding to German federal state law "explosive ordnances" are defined as "bombs and ammunition and their components, which contain energetic materials (EM)" [1]. The nomenclature "explosive ordnance" refers to the bin of the EM being enclosed in it, like TNT, but not to the EMs themselves. Hence, for instance a bulk TNT-lump is not classified as explosive ordnance and will not be incorporated in the clearance of explosive ordnance. For this reason, after the clearance of explosive ordnance has taken place the imminent danger for human beings is eliminated, but often these cleared areas still present a danger for soil and groundwater: EM-lumps remaining in the area are dissolved by precipitation and elute into the topsoil, where they can be adsorbed reversibly, resulting in a soil contamination. By permanent ad- and desorption, caused by environmental changes in the soil followed by rainfalls and changes in the soil, the EMs dissolved in the leachate finally reach the groundwater.

2 Clearance of explosive ordnance

The clearance of explosive ordnance follows a methodical procedure consisting of four phases [1], see Table 1.

In the framework of the historical investigation on the basis of archival materials, the potential explosive ordnance load is reproduced regionally and genetically in order to evaluate the potential impact [1].

If the suspicion of an explosive ordnance load is confirmed by the historical investigation, a technical investigation follows with the objective of a risk assessment. This comprises the representation of type, position and amount of explosive ordnances assumed/confirmed, their actual situation, their facility for detonation and thus, their potential effect on subjects of protection. In the framework of the technical investigation test areas are planned, which shall represent the distribution of explosive ordnances on the site. These test areas are

Table 1: Four phases of the methodical proceeding of the clearance of explosive ordnance.

Phase	
1	Historical Investigation
2	Technical Investigation including risk assessment
3	Clearance concept including tendering and awarding
4	Clearance

investigated geophysically and cleared subsequently, so that an assessment for the whole site can be made on that basis [1].

If a clearance of explosive ordnance is inevitable, a clearance concept is developed, which presents different models of clearance methods for a defined objective. Objectives can be the clearance of explosive ordnances without restriction of any kind (clearance of the whole area under the terms of state of the art), the restricted clearance of explosive ordnance (clearance is limited to depth/area/type) or simply safety measures [1]. The tendering and awarding is carried out on the basis of the clearance concept after that.

Finally, the clearance of explosive ordnance is carried out. This consists of sounding and localizing of the explosive ordnances, their excavation and recovery and their transport for demolition (or their detonation in situ if necessary if safe transport cannot be ensured).

The clearance discerns five methods, depending on the problem, which can also be applied combined [1]: 1. Visual clearance: Here, visible explosive ordnances shallow tipped or protruding from soil are removed from the area. 2. Clearance during the construction phase: Here, during excavation in layers, the horizontal and vertical faces of the pit are sounded and located explosive ordnances are removed. 3. Clearance of UXOs/punctual soil influenced clearance: UXOs are excavated and removed after sounding via special excavation techniques. 4. Holohedral, punctual soil influence clearance: The whole area is sounded, and localized explosive ordnances are removed after excavation. 5. Clearance after excavation of soil and other materials ("volume clearance"/separation): Mutually, the soil is sounded and removed in layers. Major explosive ordnances are recovered during removal. The excavated soil is stored temporarily on cleared areas and is checked visually by sensor. Alternatively, the excavated soil is treated in screening plants, magnetic separators or whirler-type separators.

Because "explosive ordnance" signifies bins containing EMs, the sounding and locating provides for searching for metals, which serves for bins. The relevant bins can be magnetic (mostly ferrous) or antimagnetic (e.g. brazen bins). The detection can be carried out laminarily and in the borehole. Methods of detection are geomagnetics, electromagnetic-2-inductor-systems and radar [2].

Figure 1: Geomagnetic detection.

Table 2: Different ranges depending on the bore diameter.

bore diameter [cm]	2	6	10	16	20	30	34
range [m]	0.6	1.5	2.1	2.75	3.1	3.7	4.0

Figure 2: Detection with an electromagnetic-2-inductor-system.

Geomagnetics utilize the earth's natural magnetic field, which is overlain by local magnetic fields of ferromagnetic compounds in the subsoil, see figure 1. By detecting the variation of the magnetic field, or of the gradient, the compound is located. Depending on the dimension of the compound the location is made in different depths, see table 2. Magnetometers only detect magnetic compounds. In addition, this method is not useful in areas containing magnetic compounds, such as cables, rails, cars etc., because the small subsurface fields of the explosive ordnances are overlain by the influence of the major components [2].

The electromagnetic-2-inductor-system induces a primary magnetic field by an emitting inductor, which excites a secondary magnetic field in the conductive subsoil by induction, recorded with a receiving inductor, see figure 2. With this method the conductivity of the subsoil is examined, so that antimagnetic compounds can also be detected. However, the method is susceptible to metallic compounds, such as cables or metallic objects on the surface, whose signals can overlay the ones of the explosive ordnances [2].

Radar emits electromagnetic waves into the subsoil by a transmitting aerial, see figure 3. These waves are leaked and reflected by compounds or geologic layers, resulting in changes of the conductivity. A receiving aerial records these waves, so that changes of the conductivity in the subsoil can be detected. The range of the waves depends highly on the geology: sandy subsoil is applicable; clayey soil is barely applicable, because the range is strongly reduced. In additionally, humidity has an effect on range. In dry, sandy soils ranges of ca. 5.0 m are obtained. The significant advantage of radar is the applicability within splitting distance to cables, rails or other compounds. However, radar also localizes cavities, groundwater or geological compounds, so that the cause of an anomaly cannot be determined definitely [2].

Figure 3: Detection with radar.

Figure 4: Screening plant in situ.

After locating, the excavation takes place, which is usually carried out by an excavator. The subsoil is excavated and stocked in piles. From the piles the explosive ordnances are separated manually or mechanically. By manual separation the single explosive ordnances are localized by geophysical methods and then picked out. By mechanical separation the excavated soil is sieved. In doing so, the mesh size is attuned to the smallest explosive ordnance estimated. In addition, the separation can be carried out with screening plants (see figure 4) that have different methods of separation (sieve, magnetic separator, whirler-type separator). The soil cleared from the explosive ordnance is backfilled again in situ.

3 Soil and groundwater contamination

The clearance of explosive ordnances can result in harmful impacts to the environment, due to the significant changes in the soil: The environment is highly affected by the use of forest clearance, the destruction of the plant canopies and the soil layers and the adherent loosening of the soil [1]. In addition, a hazard to soil and groundwater can be highly enhanced by the clearance of explosive ordnances: On the one hand, explosive ordnances can be highly corroded, because of long-lasting exposure time, so that the metallic bins burst with moderate mechanical strain, resulting in a discharge of the EM contained, see figure 5. In particular, the application of separation equipment causes soil contamination. On the other hand, exposed remnants of EMs (e.g. TNT, RDX) are existent, particularly in areas where explosive ordnances were launched or blasted (e.g. former artillery ranges, demolition areas constructed after the World Wars).

Figure 5: Corroded bin with scattered EMs.

Due to the soil changes during clearance, the EM-rests will be displaced deeply and extensively, because they are not included in the separation. Thus, the EMs remain in the soil and will be refilled after clearance in combination with the "cleaned" soil, i.e. the sieved piles.

In consequence of the high loosening of the soil, the extensive dispersion of the EMs and the potential crushing by the mechanical strain, the mobilization of the EMs is promoted: Leachate can penetrate the soil more easily and dissolve finely dispersed EMs. Thus, EMs such as TNT and RDX – after the filling of the ordnance – can infiltrate into the groundwater by elution. Due to their carcinogenicity and toxicity, various EMs and their metabolites represent a relevant hazard to sensitive goods, not only the microflora and fauna in the soil and the aquifer, but also for human beings in the case of untreated water being consumed.

In Germany various former ammunition production facilities and former demolition areas (e.g. artillery ranges) show groundwater contamination plumes downstream, which result from not only remaining EM-Hot-Spots, but also from former mine craters, which are cleared of explosive ordnances but not cleared of EMs.

4 Anonymous practical example

In World War I an ammunition production facility for "the manufacture of high explosives for projectiles' and ammunitions' filling" was constructed in Germany. The facility consisted of a pressing plant for compacts made of picric acid, a filler point of artillery shells and a plant for delaboration.

The production and manufacture was limited to the wartime of World War I; during this time several thousand tons of TNT and DNB subsidiary were produced, several million compacts of picric acid were pressed and several million shells/demolition charges made of casted charges of TNT, DNB ammonium-/sodium-nitrate and their mixtures were filled. The plant for delaboration temporarily had the capacity of several thousand units per day, and after the end of the war an interim storage of several thousand shells was documented.

After the war a spate of explosions happened during delaboration, resulting in a nearly complete demolition of the facility and an extensive dispersion of ammunition in the circumference. Because of the lack of environmental awareness at that time, both small-area contaminations by production/delaboration and extensive contamination by the explosion accidents of the soil with miscellaneous EMs occurred. The aftermath debris of the explosions and the ammunition dispersed were collected and the remained inventory of ammunition was demolished by detonation or burning out. The activities were discontinued in 1929 and the area of the ammunition plant was declared "cleared of explosive ordnance".

During clearance of the explosive ordnance huge masses of soil were excavated and cleared by screening plants that have mesh openings of some centimetres and backfilled after clearance. Thus, the cleared area was free of explosive ordnances but not free of EMs: EM-lumps with diameters of several centimetres remained in the soil. In addition, major lumps were shred mechanically by sieving and heterogeneously dispersed in the whole excavation/sievings. EMs eventually enclosed in damaged bins got into the soil due to the break-up of the bins while sieving. The remaining EMs reached the ground- and surfacewater by erosion and elution. Due to an average annual rainfall of 1,200 mm and a presumed infiltration of 10%, the ground- and surfacewater has per m^2 an input of 120 l p.a. of EM-contaminated water, which means 36,000 m^3 p.a. for the whole area of ca. 300,000 m^2.

Thereby the EM-input depends on the particle size of the EM particles and on the substrate: major EM-lumps have per se a smaller surface for elution than smaller lumps. That is why they remain longer, particularly in the soil, than small lumps or powdery dispersed EMs. However, major EM-lumps are a long-lasting source of emission if they remain in the soil. Depending on the dimension of the contaminated area and the amount of the existent EMs in it, significant ground- and surfacewater contaminations can be generated.

The transport and the spreading of the dissolved EMs in the saturated and unsaturated zone depends on the characteristics of the EMs and the soil: EMs are able to adsorb temporarily on clay minerals, organic substances and ferric (hydr)oxides present in the soil, so that their transport can be retarded by reversible adsorption.

Thus, a decreasing of the spreading velocity is indicated when the amount of adsorbents in the soil increases. In the process, the adsorption of EMs is co-determined by the concentration equilibrium, which exists between the dissolved phase (concentration of the pore water) and the adsorbed phase (concentration of the adsorbent/the soil). Thus, the real spreading velocity in situ is overlain by the changing of sorption and release processes resulting from changes in the solution phase concentration of the EMs caused, e.g., by the discontinuity of the source. Therefore, changes in the soil resulting from the clearance of explosive ordnance produce an alteration of the source of contamination, which influences the concentration equilibrium seriously. Hence, after changes, increased mobilization of EMs arises, which becomes manifest in more or less long-lasting increased ground- and surfacewater concentrations of EMs, until a new concentration equilibrium is recalibrated, adapted to the new conditions.

In the case of the practical example presented, a mobilization of EMs also occurred because of a temporary leakage of red water coloured by EMs.

However, the described leakage of EMs was taken into account initially in a soil remediation and soil protection measurement carried out up to now. In the framework of the physical safeguarding of the site, an infiltration ditch was passed around the core area of the former facility, with which contaminated leachate is collected and fed into an activated carbon filter. Concerning the emission of the EMs in the deep subsurface, initially only monitoring was carried out, because the bedrock is relatively low-permeable and the transport of groundwater happens via fissures. Here, at the identified emersion points, the water is chemically analyzed regularly.

Despite eliminating the imminent danger for human beings by explosive ordnances, the former ammunition production facility still represents a danger for human beings and the environment, because the soil in the area of the former facility will be a source of emission for hundreds of years, caused by finely dispersed EMs resulting from the production, delaboration, accidental explosions and clearance of explosive ordnances.

5 Conclusions and perspectives

The practical example shows that the site was cleared with extraordinary expense, but the clearance of explosive ordnance resulted in an extensive, heterogenic, fine dispersion of the accordant EMs. Because easily accessible contamination-Hot-Spots were not eliminated and EM-contaminated soil was backfilled, high EM-concentrations of the leachate are expected for an indefinite period.

To protect both the environment and human beings and in order to save expenses in the same way, a conjunction of the clearance of explosive ordnance and soil remedial actions is reasonable in the run-up to any activities. With a conjunction of both procedures it can be ensured that both soil free of explosive ordnance and of EMs is backfilled, Hot-Spots of EMs (and other contaminants) are excavated separately, so that the contaminants in the framework of the clearance of explosive ordnances will not be dispersed extensively or remain in the soil and temporary mobilizations of EMs by changes in the soil are considered in the run-up and their inhibition will be assured by appropriate attended techniques.

Because the progression of the methodical procedure of a soil remediation is similar to that of a clearance of explosive ordnance (see tables 1 and 3), a conjunction means no extravagant additional expenditure per se. Data-recall facilities, appropriate investigations and interpretations can be carried out hand in hand. In addition, the clearance and remedial activities can be technically carried out at the same time.

Thus, it is not for technical reasons but on legal grounds by the mentioned definition of “explosive ordnance”, that the clearance of explosive ordnances solely eliminates explosive ordnances and does not clean the area of EMs as well. In addition, one has to deal with the competences of diverse authorities for

Table 3: Four phases of the methodical proceedings of a soil remediation.

Phase	
1	Historical Investigation
2	Technical Investigation including risk assessment
3	Concept of remediation (including tendering and awarding)
4	Remediation

the clearance of explosive ordnances and the enforcement of the "German Federal Protection Act and Ordinance" (BBodSchG), an environmental law to be applied in Germany for the investigation, assessment and remediation of soil contaminations.

Thus, the aim is that, via associations, occupational unions, professional organisations and similar institutions, which are engaged in both problems, attention is paid to formulating appropriate articles for the developing policy, suitable settlements in the form of guidelines and to ensuring future adoption into standardized guidelines, rules and regulations.

Acknowledgements

I thank the German Scientific-Technical Association for Environmental remediation and Brownfield Redevelopment, namely Mrs. Sabine Gier, for permission for publication and my colleagues Mr. Uwe Kaufmann and Mr. Tobias Bausinger, PhD, for the critical review of the manuscript.

Thanks as well go to my colleagues Mr. Alexander Schwendner and Mr. Jürgen Klatt for providing photos.

Furthermore I thank the Tauber company, namely Mr. Norbert Schmidt and Mr. Simon Gremmler for providing photos and for the elaboration of the geophysical investigation methods.

References

[1] *Arbeitshilfen Kampfmittelräumung – Arbeitshilfen zur wirtschaftlichen Erkundung, Planung und Räumung von Kampfmitteln auf Liegenschaften des Bundes (AH KMR)*, Bundesministerium für Verkehr, Bau und Stadtentwicklung & Bundesministerium der Verteidigung, Berlin, 2007.

[2] Gremmler, S., *Erkundungstechniken in der Kampfmittelsuche*, Tauber DeDeComp GmbH, Hannover, 2009.

Removal of volatile organic compounds from soil

R. S. Amano
University of Wisconsin-Milwaukee, USA

Abstract

Pollution by petroleum products is a source of volatile organic compounds in soil. Therefore, laboratory column venting experiments were completed in order to investigate the removal of a pure compound (toluene) and a mixture of two (toluene and n-heptane) and five (toluene, n-heptane, ethylbenzene, m-xylene and p-xylene) compounds. The choice of the compounds, as well as their proportion in the mixture, was made on the basis of the real fuel composition. The objective of this study is a comparison between the experimental volatile organic compounds removal results and the computational fluid dynamics (CFDs) analysis. The proposed model for the contaminants transport describes the removal of organic compounds from soil, the contaminants being distributed among four phases: vapor, nonaqueous liquid phase, aqueous and "solid"; the local phase's equilibrium and the ideal behavior of all four phases was found to be accurate enough to describe the interphase mass transfer. The process developed in the lab consists of a heater/boiler that pumps and circulates hot oil through a pipeline that is enclosed in a larger-diameter pipe. This extraction pipe is vertically installed within the contaminated soil up to a certain depth and is welded at the bottom and capped at the top. The number of heat source pipes and extraction wells depends on the type of soil, the type of pollutants, and the moisture content of the soil and the size of the area to be cleaned. The heat source heats the soil, which is transported in the interior part of the soil by means of conduction and convection. This heating of the soil results in vaporization of the gases, which are then driven out of the soil by the extraction well. The extraction well consists of the blower, which sucks the vaporized gases out of the system. Our previous studies have removed higher boiling compounds, such as naphthalene, etc., to a non-detectable level. Thus, the current technology is very promising for removing most of the chemical compounds; and can also remove these high boiling compounds from the saturated zone. The reasonable

WIT Transactions on Ecology and the Environment, Vol 135, © 2010 WIT Press
www.witpress.com, ISSN 1743-3541 (on-line)
doi:10.2495/WP100101

agreement between the predicted and the experimental results as well as the values of the performance criteria prove that the mathematical model is suitable to describe the removal of volatile organic compound pollutants by venting in the range of experimental conditions used in the pilot plant.

Keywords: soil remediation, vapor extraction, volatile organic compound removal.

1 Introduction

With environmental safety being of prime importance, the remediation systems for cleaning soil of polluting chemicals have gained importance in recent years. Different remediation processes, such as venting [3] for soil contaminants like diesel, the solvent washing of soil [4] for contaminants like pentachlorophenol (PCP), which is a wood preserving agent, have been studied and used for this purpose. Soil vapor extraction (SVE) has been used at many sites to remove volatile organic compounds (VOCs) from soil in the vadose zone. The effectiveness of SVE, however, is limited at sites with complex geology or by the distribution of contaminants in the subsurface and saturated soils. In recent years, research and field demonstrations have been conducted using innovative technologies and procedures to enhance the treatment effectiveness and removal rates of VOCs from vadose zone soil and of VOCs dissolved in groundwater and adsorbed to saturation zone soils. Various pollutants, such as fuels and other chemicals leaking into soil, pose a high environmental threat. The research analysis focuses on heated soil vapor extraction with air sparging, which is a remediation process primarily used to clean the soil of pollutants, such as organic compounds that include solvents, fuels and para nuclear aromatics (PNAs). Figure 1 shows the schematic sketch of this process [2].

The system consists of heating and suction wells embedded into the soil. The heating pipes are connected to the main pipeline, which is connected to the boiler

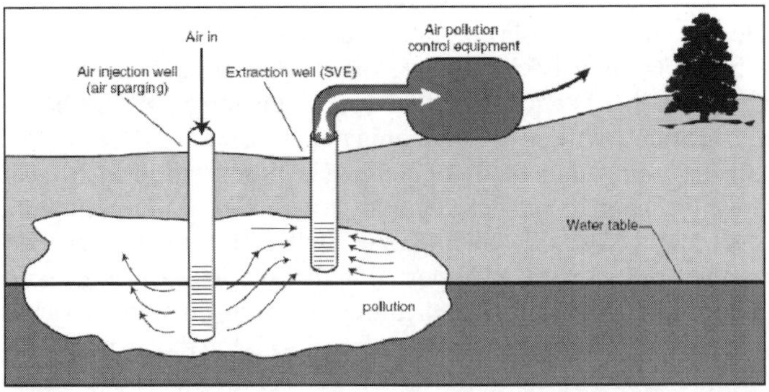

Figure 1: Schematic of a HSVE and air sparging system [2].

and re-circulating pump. Through this heating pipe heated oil at approximately 500°F is supplied to the heating wells. Heat is diffused out of the heating well into the soil and vaporizes the contaminants. These contaminants are removed from the soil by the suction well, which is connected to a storage tank where the contaminants are stored for temporary purposes. Finally the contaminants are disposed of from time to time as air is blown through the air-sparging well to drive out the contaminants from the saturated zone to the unsaturated zone.

Previous research work on this process [5-7] has proved the feasibility and effectiveness of the process. Application of this process in real-life cases has shown that, in soil contaminated with volatile organic compounds (VOCs), the contaminant concentrations can be brought down to non-detectable levels or below acceptable government standards.

A two-month experiment study was conducted to demonstrate the effectiveness in degradation of hydrocarbons and to remove volatile constituents in the saturated soil. The soil samples (approximately 15 lbs) were obtained from two different depths at 10 feet (saturated) that represent typically contaminated MGP soil without light nonaqueous-phase liquid (LNAPL) or dense nonaqueous-phase liquid (DNPL) layers.

Laboratory experiments were conducted at the University of Wisconsin-Milwaukee. Soil placed in the experimental setup was subjected to a heat of about 125°F and a vacuum of nearly 13in-Hg. These values were chosen in order to simulate actual conditions. The samples have been periodically analyzed for VOCs and PAHs using Soxhlet extraction and Gas Chromatography (GC). (See below)

2 Experimental set-up

The experimental setup is constructed to simulate the actual conditions in the closest possible way. It consists of two round bottom (RB) flasks, a super condenser, carbon cells, and vacuum pump. A picture is shown in Figure 2.

The larger of the two round bottom flask contains the soil and is heated from the bottom by a heating mantle. A thermo couple is inserted to monitor the temperature. An air spurge is also provided. An air flow meter is used to control the amount of airflow and is kept at 50cc/min. The air and gases (from the soil) at the beginning of the flow path are led through a super condenser, cooled by tap water, to a smaller RB flask receiver, which is placed in coolant (dry ice), the flow path continues though activated carbon cells, to the vacuum pump. This elaborate procedure ensures that there are no gases escaping to the atmosphere.

Cold tap water and not coolant is used in the condenser to avoid the situation of freezing the water vapor from the soil in the condenser and clogging the flow path. The smaller flask in the dry ice bath acts as collecting flask and collects the condensed liquid. Due to the air spurge, the flow velocities of the gases are high and thus the condensation in both the condenser and the trap would not be very effective. Hence, the carbon cells have been used to further trap any remaining vapors from the soil.

Figure 2: Experiment set up. 1. Flask and soil. 2. Heating mantle. 3. Thermometer. 4. Heat controller (Variac). 5. Airflow meter. 6. Super-condenser. 7. Coolant 8. Receiver flask. 9. Activated carbon cells (2 nos.) 10. Vacuum pump (not showing).

2.1 Running of the experiment

Saturated soil was heated for 600 hours real time (25 days) with a constant temperature of 125°F, airflow of 50cc/min and a vacuum of 13in-Hg. The soil was tested for remaining compounds at 3, 19, and 25 days of heating. A ¼ lbs sample of treated soil was randomly sampled at each interval and Soxhlet extraction was performed for 24hrs using 200ml of dichloromethane (CH_2Cl_2) as solvent. The solvent dissolves, extracts and concentrates the compounds (VOCs) in the soil and the GC-MS is performed to separate and analyze the compounds in the concentrated solvent.

2.2 Soxhlet extraction

The Soxhlet extractor is a piece of laboratory apparatus invented in 1879 by Franz von Soxhlet [9]. It was originally designed for the extraction of a lipid from a solid material.

The contaminated soil is placed inside a "thimble" made from thick filter paper, which is loaded into the main chamber of the Soxhlet extractor. The Soxhlet extractor is placed onto a flask containing the extraction solvent. The Soxhlet is then equipped with a condenser.

The solvent is heated to reflux by heating mantle. The solvent vapor travels up a distillation arm, and floods into the chamber housing the thimble of solid. The condenser ensures that any solvent vapor cools and drips back down into the chamber housing the soil.

(a) Picture of Soxhlet extraction

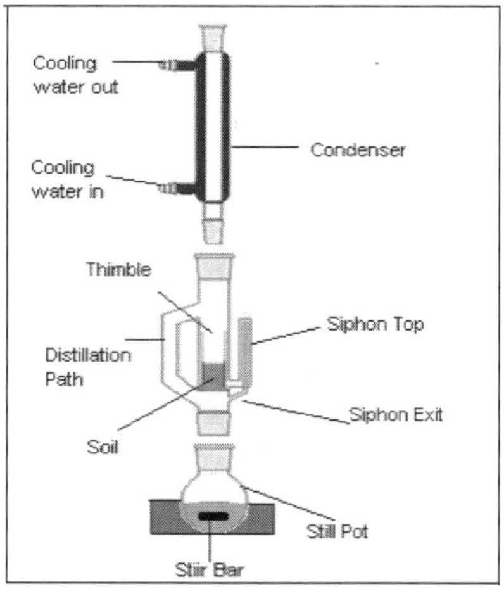

(b) Schematic sketch of Soxhlet extraction

Figure 3: Soxhlet extraction [9].

The chamber containing the soil slowly fills with solvent. Some of the desired compound will then dissolve in the solvent. When the Soxhlet chamber is almost full, the chamber is automatically emptied by a siphon side arm, with the solvent running back down to the distillation flask. This cycle may be allowed to repeat many times, over hours or days; in this study it was repeated for 24hrs.

During each cycle, a portion of the non-volatile compound dissolves in the redistilled solvent. After many cycles the desired compound is concentrated in the distillation flask. To illustrate this apparatus a picture and schematic sketch have been placed in Figure 3(a) and 3(b), respectively.

2.3 Gas chromatography

The GC is a type of chromatography [9] in which the mobile phase is a carrier gas, usually an inert gas such as helium, nitrogen or hydrogen and the stationary phase is a microscopic layer of liquid or polymer on an inert solid support, inside glass or metal tubing, called a column. The instrument to perform gas chromatographic separations is called a gas chromatograph. A gas chromatograph is a chemical analysis instrument for separating chemicals in a complex sample. The gas chromatograph uses a flow-through narrow tube known as the column, through which different chemical constituents of a sample pass in a gas stream (carrier gas, mobile phase) at different rates depending on their various chemical and physical properties and their interaction with a specific column filling, called the stationary phase. As the chemicals exit the end of the column, they are detected and identified electronically. The function of the stationary phase in the column is to separate different components, causing each one to exit the column at a different time (retention time). Other parameters that can be used to alter the order or time of retention are the carrier gas flow rate, and the temperature.

In a GC analysis, a known volume of gaseous or liquid analyte (a substance or chemical constituent that is determined in an analytical procedure) is injected into the "entrance" (head) of the column, usually using a micro-syringe (or, solid phase micro-extraction fibers, or a gas source switching system). Although the carrier gas sweeps the analyte molecules through the column, this motion is inhibited by the adsorption of the analyte molecules either onto the column walls or onto packing materials in the column.

The rate at which the molecules progress along the column depends on the strength of adsorption, which in turn depends on the type of molecule and on the stationary phase materials. Since each type of molecule has a different rate of progression, the various components of the analyte mixture are separated as they progress along the column and reach the end of the column at different times (retention time). A detector is used to monitor the outlet stream from the column; thus, the time at which each component reaches the outlet and the amount of that component can be determined. Generally, substances are identified by the order in which they emerge (elute) from the column and by the retention time of the analyte in the column.

2.4 Physical components [8]

Inlet: inlet or injector provides the means to introduce a sample into a continuous flow of carrier gas.

Columns: There are two types of columns (packed and capillary) that are used in GC: packed columns which are 1.5-10 m in length and have an internal diameter of 2-4 mm. Capillary columns have a very small internal diameter, on the order of a few tenths of millimeters, and lengths between 25-60 meters are common.

Detectors: A number of detectors are used in gas chromatography. The most common are the flame ionization detector (FID) and the thermal conductivity detector (TCD). Both are sensitive to a wide range of components, and both work over a wide range of concentrations. Some gas chromatographs are connected to a mass spectrometer, which acts as the detector; the combination of both is known as GC-MS. A GC-MS can identify many of the separated compounds from the GC, since the mass spectrometer will identify the component's molecular weight but this still takes time and skill to do properly.

Methods: The method is the collection of conditions in which the GC operates for a given analysis. Method development is the process of determining what conditions are adequate and/or ideal for the analysis required (see Figure 4). These experiments utilized a Hewlett Packard GC equipped with a J&W Scientific DB-1,30 meter, wide bore capillary column with a 1 micron film thickness and an FID detector. The column was temperature programmed from 35C (8 min) to 130C at 4C/min. hold 1 min., then to 250C at 20C/min. for 40 min. The injector and detector were held at 250C. A Hewlett Packard GC/MS 5971 mass spectrometer was utilized in the identification of soil contaminants using the NBS/Wiley probability based library search.

Figure 4: Picture of oven gas chromatography [9].

3 Results and discussion

Saturated Soil: Based on the concentrations of the VOCs and sulfur; the initial concentration in untreated soil was 117 part per million (ppm). After 7 cumulative days of heating, the concentrations came down by 10% of the initial, to 106ppm. After 15 days the concentrations were reduced to 103ppm that is, 12% of the volatiles were removed. After 26 cumulative days 20.51% of the compounds were removed with the concentration being 93ppm. These concentrations are plotted against time and shown in Figure 5 and the values are in Table 1. It was observed that the sulfur concentrations remain almost the same at 80ppm, throughout the heating period (0 –26 days).

Considering only VOCs, the concentrations came down to 26ppm from the initial 37ppm after 7 cumulative days of heating, showing 29% reduction. The concentrations were 23ppm and 13ppm after 15 cumulative days and 26 cumulative days of heating respectively. These values show a removal of 37% and 65% respectively. Projecting this data by a third degree polynomial it is

Table 1: Concentration of VOCs and sulfur.

Cumulative days of heating	Total ppm
0	117
7	106
15	103
26	93.5

Table 2: Concentration of VOCs only.

Cumulative days of heating	Total ppm
0	37.25
7	26.00
15	23.00
26	13.47

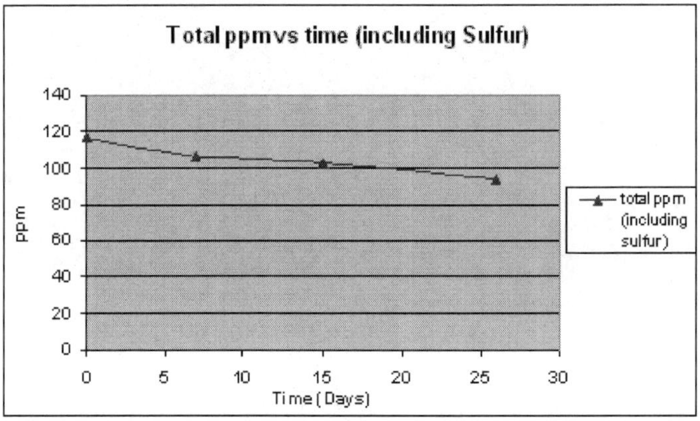

Figure 5: Graph of total ppm v/s time (including sulfur).

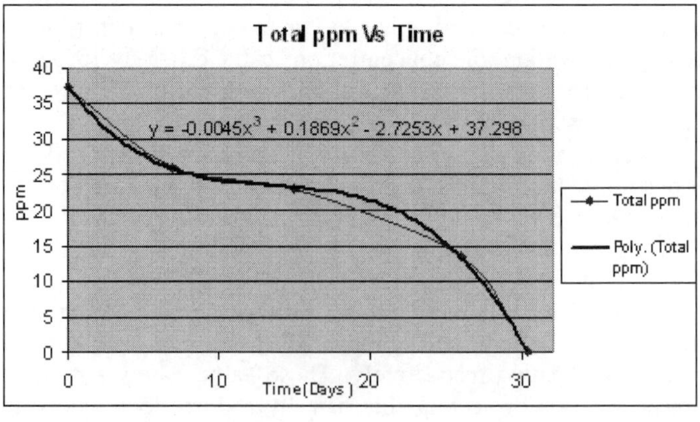

Figure 6: Graph of total ppm v/s time.

observed that, the concentrations fall below 0.37ppm (99% removal) in approximately 30 cumulative days of heating. The plot of ppm v/s time is shown in Figure 6 and the values in Table 2.

4 Conclusions

Approximately 65% of VOCs were removed in 625 hours (26 cumulative days) of heating and over 99% of VOCs are expected to be removed after 720 hours (30 cumulative days) of heating, as is shown in Figure 6. HSVE is not an effective method to remove sulfur, after heating for 26 cumulative days the concentration of sulfur remains at the same initial value. Apart from testing the effectiveness of the method, this also provides data for CFD modeling and validation.

Enhancement technologies should be considered when contaminant or soil characteristics limit the effectiveness of SVE or when contaminants are present in saturated soil. The five enhancement technologies are as follows [10]:

- Air Sparging
- Dual-phase Extraction
- Directional Drilling
- Pneumatic and Hydraulic Fracturing
- Thermal Enhancement

One of the limitations of SVE alone is that it does not effectively address contaminated soils within the capillary fringe and below the groundwater table. Among the methods listed above, air sparging can enhance the remediation capabilities of SVE in the capillary fringe zone to include remediation of chemicals with lower volatilities and/or chemicals that are tightly sorbed. This technique also enhances biodegradation of aerobically-degradable contaminants and can significantly reduce the remediation time for contaminated sites. These are the studies to be conducted as a next step.

References

[1] Maureen C. Leahy, Anthony M. Fiorentine, Richard J. Schmitz "Biosparging for In Situ Treatment of Manufactured Gas Plant Residuals".

[2] P.M. Mohan Das, R.S. Amano, "Numerical Modeling of Heated Soil Vapor Extraction Process With Air Sparging", 2006 ASME International Mechanical Engineering Congress & Exposition November 5-10, 2006, Chicago, Illinois.

[3] A. Fotonich, V.K. Dhir and S. Lingineni, "Experimental Investigation of Remediation of Soils Contaminated With Diesel Fuel", Journal of Environmental Engineering, 1999.

[4] Amid P. Khodadoust, George A. Sorial, Gregory J. Wilson, Makaram T. Suidan, Richard A Griffiths, and Richard C. Brenner, "Integrated System

for Remediation of Contaminated Soils", Journal of Environmental Engineering, 1999.
[5] Jadhav, R.S. Amano, R. Lind, "Numerical Analysis of Heated Soil Vapor Extraction System", ASME IMECE2002- 32092, November 2002, New Orleans, USA.
[6] T. Roy, R.S. Amano, E.K. Lee and R.S. Jadhav, "Computer Simulation of Heated Soil Vapor Extraction System for Effective Remediation".
[7] T. Roy and R.S. Amano, "A Transient Simulation of Heated Soil-Vapor Extraction System", ASME Heat Transfer/Fluid Engineering Summer Conference, 2004.
[8] En-Wikipedia GNU Free Documentation License Copyright (C) 2000, 2001, 2002 Free Software Foundation, Inc. 51 Franklin St, Fifth Floor, Boston, MA 02110-1301 USA Everyone is permitted to copy and distribute verbatim copies of this license document, but changing it is not allowed.
[9] http://en.wikipedia.org/wiki/Soxhlet_extractor
[10] Analysis of Selected Enhancement for Soil Vapor Extraction, EPA-542-R-97-007, Agency (5102G) September 1997

Kinetic and thermodynamic aspects of arsenic (III) adsorption onto iron oxide obtained from iron oxalate

A. Negrea, M. Ciopec, L. Lupa, C. Muntean, R. Lazau & P. Negrea
University Politehnica Timisoara, Faculty of Industrial Chemistry and Environmental Engineering, Romania

Abstract

Drinking water has been identified as one of the major sources of arsenic exposure by the general population. A variety of treatment processes has been studied for arsenic removal from water. One of the most used methods is adsorption. The most efficient adsorbent materials are those with iron content considering the affinity of arsenic towards iron. Relatively limited information is available regarding the impact of temperature on the arsenic (As) (III) adsorption kinetic and equilibrium capacities on iron oxide. In this paper we studied the possibility of using iron oxide as an adsorbent for the removal of As (III) from aqueous solutions. The iron oxide was obtained through annealing of iron oxalate at 550°C. The effects of contact time, initial concentration of the solutions and temperature on the removal of As (III) were investigated in order to explain the adsorption mechanism. The kinetic of As (III) adsorption can be described well by the pseudo-second-order models. The equilibrium adsorption data were fitted using Freundlich and Langmuir isotherm equations and the corresponding parameters were calculated and discussed in detail. An increase in temperature increases the values of the overall adsorption reaction rate constant. An examination of thermodynamic parameters shows that the adsorption of As (III) by iron oxide is an endothermic process and is spontaneous at the specific temperature investigated.
Keywords: arsenic, adsorption, iron oxide, kinetic, thermodynamic.

WIT Transactions on Ecology and the Environment, Vol 135, © 2010 WIT Press
www.witpress.com, ISSN 1743-3541 (on-line)
doi:10.2495/WP100111

1 Introduction

Water is essential for sustaining human, animal and plant life. Drinking water contains many trace contaminants, both cationic and anionic [1]. Among common inorganic pollutants, the arsenic (As) has been largely studied because of its potential harmfulness to human health [2]. Background arsenic concentrations in natural water are low, but elevated arsenic concentration are common in groundwater as a result of natural conditions (volcanic emission, minerals) or anthropogenic impacts (mining activities, combustion of fossil fuels and the use of arsenic pesticides) [3–5].

Literature survey reveals that there are a good number of approaches for arsenic remediation from drinking water. Amongst the various methods like oxidation-reduction, precipitation, co precipitation, sorption, electrolysis and cementation, solvent extraction, ion exchange, ion flotation and bioremediation [5-8] proposed to negotiate the problem of arsenic contamination in drinking water, sorption technique is, however, the most common and is considered to be an effective method.

Sorbents of different types like biomaterials [2, 7, 9], metal oxide/hydroxide [3, 6, 10–14], zeolite [5], activated carbon [5, 16], laterite [17] etc have been used by workers to achieve the goal. Iron oxides are present in the environment as a wide range of minerals. These are the most important adsorbents for arsenic removal from drinking water because of their abundance and strong affinity, and both arsenite and arsenate show a high affinity for the iron oxide surface [10–14, 16–20].

The objective of our research were to gain an understanding of the adsorption process kinetics and to describe and explain some important thermodynamic parameters for the arsenic removal process from water when we used as adsorbent an iron oxide obtained from iron oxalate.

2 Experimental

2.1 Sorbent characterization

The sorbent for this study was obtained trough annealing of the iron oxalate ($Fe(COO)_2 \cdot 2H_2O$) at 550°C. For the resulted product, iron oxide (Fe_2O_3) was determined the specific surface area and the acid base properties.

The specific surface of the sample was measured using a Micrometrics ASAP 2020 BET surface area analyzer, by cold nitrogen adsorption.

Colloidal chemical properties of aqueous dispersions of various solids represents a subject of intensive research. The dispersion behavior is dictated by interfacial chemistry of solids and suspending media. The particles of these solids in aqueous solutions develop their surface charge through acid-base surface reactions. The point of zero charge, pH_{pzc}, and the number of acidic and basic groups on these materials surfaces are important guides to interfacial properties [5, 13, 15, 21].

The point of zero charge of the iron oxide was determined by the batch equilibration technique. Samples of iron oxide (0.2 g) have been treated with 100 mL NaCl solution (0.01, 0.005 and 0.001 mol L^{-1}) of various pH values, in the pH range from 2 to 12. The suspensions have been stirred with 400 rpm and left for equilibration 24 h at room temperature. The initial pH values (pH_i) have been adjusted by adding a small amount of HNO_3 or NaOH solution (concentration 0.1 or 2 mol L^{-1}, so that the volume variation was minimal). The role of the background electrolyte NaCl was to keep the ionic strength constant. All used chemicals were in analytical reagent grade. The pH values have been measured using a "CRISON MultiMeter MM41" pH-meter.

2.2 Batch experiments

In the first instance the effect of contact time was studied at different temperature (291, 295, 313 K). In each experiment, 0.1 g of sorbent was suspended in 100 mL of 100 µg/L solution of As (III). The samples were stirred using agitating device with glass rod (stirring speed 200 rpm) at different stirring times (15, 30, 45, 60, 90, 120 and 150 minutes). The temperature of the sample was kept using a CRISON MultiMeter MM41 apparatus. After stirring the samples were centrifuged at 1200 rpm for 30 min using a ROTINA 420 centrifuge. Arsenic concentration in the supernatants was measured trough atomic adsorption spectrometry with hydride generation using a spectrophotometer VARIAN SpectrAA 110 with hydride generation VGA 77.

Similar batch experiments were performed to study the influence of the initial As (III) concentration (100, 200, 300, 400, 500, 600 and 700 µg/L). The initial pH of the solutions was kept around the value 7, which was at the middle of the plateau corresponding to the plot of pH_f versus pH_i who gives the pH_{pzc} and which also represent the most common pH value found in the natural waters. The pH of the solutions was measured using a multi parameter instrument WTW Multi 197i.

The equilibrium adsorption capacity was calculated using the following equation:

$$q_e = (C_0 - C_e)\frac{v}{m}$$

where: C_0 - initial concentration of arsenic in solution, µg/L;
C_e - equilibrium concentration of arsenic in solution, µg/L;
v - solution volume, L;
m - amount of the absorbent, g.

3 Results and discussion

3.1 Sorbent properties

Figure 1 illustrates the representation for the determination of pH_{pzc}. The plateau corresponding to the plot of final pH values of the suspensions (pH_f) against the initial values (pH_i) gives the value of pH_{pzc}. The presence of this plateau shows that the material presents amphoteric properties, and acts as a buffer. The plateau of the pH_i–pH_f plot corresponds to the pH range where the buffering effect of the iron oxide surface takes place, i.e. where for all pH_i in this range the pH_f is almost the same and corresponds to pH_{pzc}.

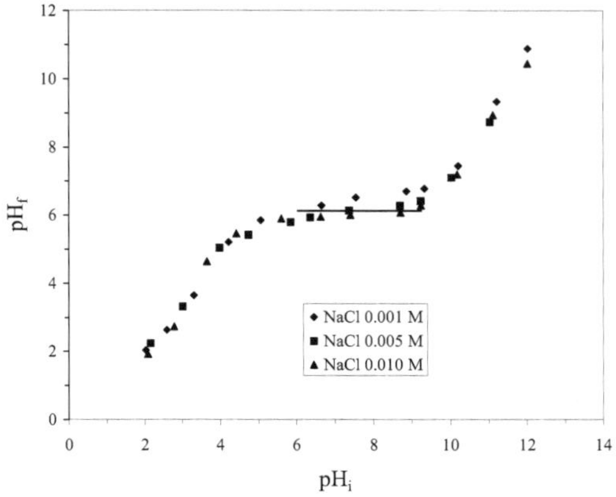

Figure 1: The plot of final solution pH (pH_f) versus initial solution pH (pH_f).

The pH region where the iron oxide buffers the solution is from ~6 to ~9.5. It means that for all values of pH_i, in this range, the pH_f is the same and equal to pH_{pzc}. The pH_{pzc} determined in 0.005 and 0.01 M NaCl solutions is practically the same. This leads to the conclusion that pH_{pzc} is independent of the ionic strength of NaCl solutions, i.e. that NaCl is an indifferent electrolyte. In the case of 0.001 M NaCl solution one may notice that in the pH range where the plateau should be, the pH_f values are somehow higher than the values for the other two NaCl solutions and are also slightly increasing. This may be due to the fact that the 0.001 M NaCl solution is too diluted to ensure a constant ionic strength. Therefore we considered as pH_{pzc} the value ~6.1 given by the other two solutions (table 1). One may also notice that in the pH range ~4.5 to ~6 the iron oxide tends to act as a buffer but the buffering effect is incomplete and the pH_f values remain under the plateau. The pH_{pzc} value suggests that the surface of the material should be predominantly positive at pH values lower than 6.1 and negative at pH values higher than 6.1. The surface charge density of the material should increase or decrease with the decrease or increase of the pH value from the pH_{pzc}.

Table 1: Some characteristic properties of iron oxide.

Parameter	pH_{pzc}	S_{BET}, m^2/g
Value	6.1	14.6

3.2 Kinetic studies

To attain the adsorption equilibrium time the results of the effect of stirring time on the adsorption capacity of the iron oxide in the process of As (III) removal from water for all the studied temperatures are presented in figure 2.

As can be seen the equilibrium times lie in the range of 90 – 120 minutes at all of studied temperature. The absorbed amount of As (III) ions slightly increase when increasing temperature from 291 to 313 K. For subsequent experiments, an equilibration time of 90 minutes was chosen for the sake of convenience.

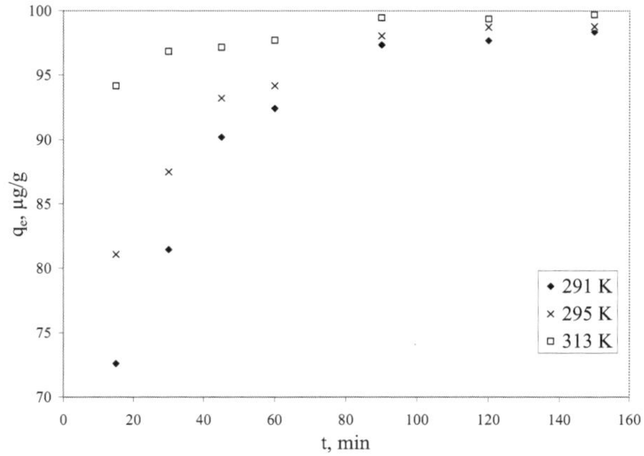

Figure 2: The effect of stirring time on arsenic adsorption capacity of iron oxide at different temperatures.

In order to express the kinetics of arsenic adsorption onto iron oxide the results were analysed using the following models.

The pseudo-first-order kinetic model based on the solid capacity and proposed by Lagergren can be used to determine the rate constant for the adsorption process and the integrated form is expressed by the following equation:

$$\ln(q_e - q_t) = \ln q_e - k_1 t$$

where: q_t and q_e represent the amounts of the arsenic adsorbed on the iron oxide at time t and at equilibrium time, respectively, μg/g;
k_1 is the specific adsorption rate constant, min^{-1}.

Figure 3 display the plots of $\ln(q_e\text{-}q_t)$ versus t for As (III) adsorption onto iron oxide for all studied temperatures. The slopes and intercept of the plots are used to estimate the pseudo-first-order rate constant (k_1) and the equilibrium adsorption capacity (q_e), respectively.

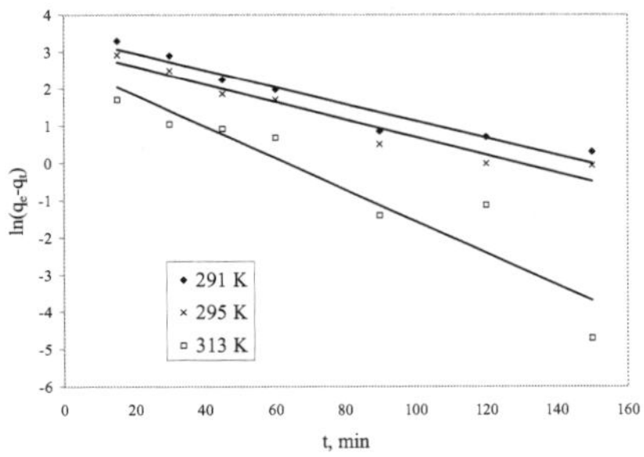

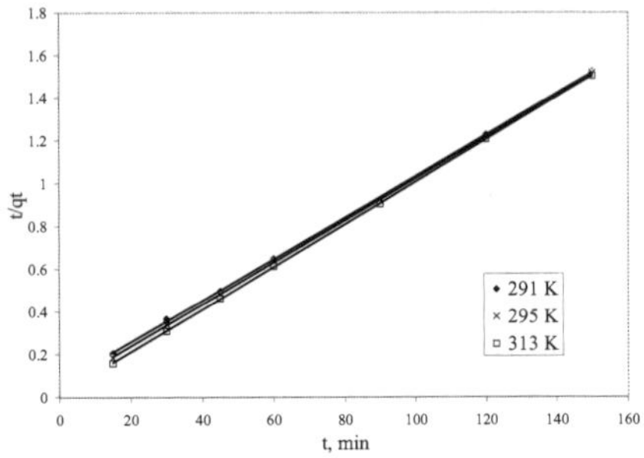

Figure 3: Pseudo-first-order kinetic plot.

Figure 4: Pseudo-second-order kinetic plot.

The linear form of the pseudo-second-order rate expression of Ho and Mckay, based on the solid phase sorption, is given by:

$$\frac{t}{q_t} = \frac{1}{h} + \frac{t}{q_e}$$

where: $h=k_{s2}\cdot q_e^2$; k_2 is the pseudo-second-order constant, $min^{-1}(\mu g/g)^{-1}$. Other terms have their usual meanings.

A plot of t/q_t versus t should yield a straight line. From the intercept and slop (figure 4) are calculated the second-order rate constant (k_2) and the equilibrium adsorption capacity (q_e) for all the studied temperatures.

The values of the constants, together with the regression coefficients (R^2) obtained in both cases are summarized in table 2.

Table 2: Kinetic parameters for As (III) sorption onto iron oxide.

Parameter	q_e, exp, µg/g	Pseudo-first-order model			Pseudo-second-order model		
Temp.		q_e, calc, µg/g	k_1 min^{-1}	R^2	q_e, calc, µg/g	k_2 $min^{-1}(\mu g/g)^{-1}$	R^2
291 K	98.4	30.5	0.0228	0.9433	103	0.00144	0.9997
295 K	98.8	21.8	0.0237	0.9925	102	0.00227	0.9999
313 K	99.7	14.6	0.0424	0.887	100	0.00794	1

It was observed that correlation coefficient for the pseudo-first-order model was much lower than for the pseudo-second-order rate. Furthermore, the calculated equilibrium sorption capacity for the first order model, at all temperatures, $q_{e,calc}$, values are not close to the experimental values $q_{e,\,exp}$, while, the theoretically predicted equilibrium sorption in the case of the pseudo-second-order model is close to that determined experimentally, at all temperatures. The pseudo-second-order rate constant (k_2) increase with increasing temperature on the reactions. This shows that the kinetics of As (III) removal by iron oxide is described by a pseudo-second-order expression instead of a pseudo-first-order.

3.3 Thermodynamic studies

The study of the temperature effect on As (III) removal by iron oxide enabled us to determine the thermodynamic parameters (ΔG°, ΔH° and ΔS°) by using the following equation:

$$\ln K_d = \frac{\Delta S^\circ}{R} - \frac{\Delta H^\circ}{RT}$$

where: R is the ideal gas constant,
T is the temperature, K
K_d is the distribution coefficient (amount of removed arsenic per gram of material divided by its concentration in the liquid phase)

The changes in free energy (ΔG°) have been calculated using the Gibbs-Helmholtz relation as:

$$\Delta G^\circ = \Delta H^\circ - T\Delta S^\circ$$

where: the symbols have their usual meanings.

The plot of $\ln K_d$ versus 1/T gives a straight line (figure 5) and the obtained thermodynamic parameters are shown in table 3.

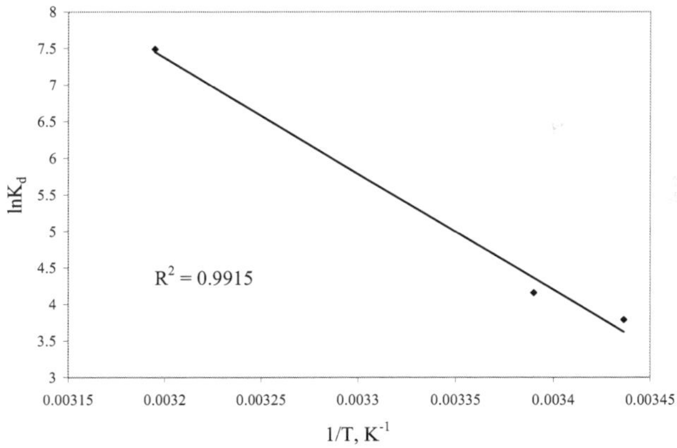

Figure 5: The plot of $\ln K_d$ versus 1/T for As (III) removal with iron oxide.

Table 3: Thermodynamic parameters evaluated for As (III) sorption on iron oxide.

Temperature, K	ΔG°, kJ/mol	ΔH°, kJ/mol	ΔS°, kJ/mol·K
291	- 8.79	131.7	0.483
295	-10.7		
313	-19.4		

Results showed the increase in magnitude of - $\Delta G°$ with rising temperature of the sorption, which indicate the increase in spontaneity of the sorption with increasing temperature. Based on the $\Delta H°$, it can be conclude that the sorption of As (III) on iron oxide is endothermic ($\Delta H°$ positive) and take place with increasing entropy ($\Delta S°$ positive). This suggests the increase in number of species during sorption process at the solid-liquid interface when hydrated solutes are sorbed by the solid phase.

3.4 Adsorption isotherms

The adsorption isotherm of As (III) is presented in figure 6.

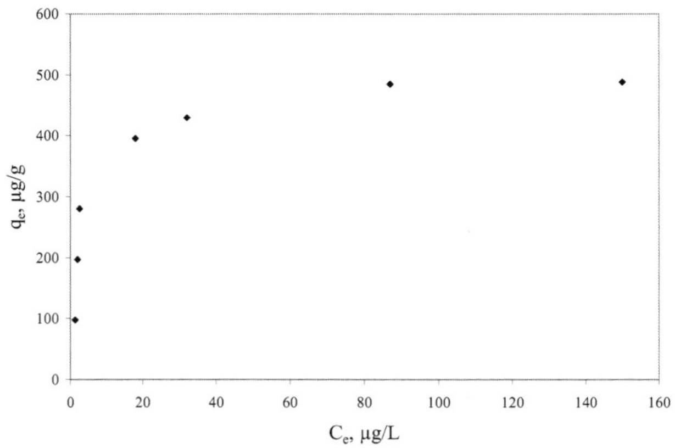

Figure 6: The adsorption isotherm of As (III).

The adsorption capacity increased with increasing equilibrium concentration of arsenic. Then, they approached a constant value at the high equilibrium concentration. The maximum adsorption capacity of iron oxide determined experimental is 490 µg/g.

Several models have been published in the literature to describe experimental data of adsorption isotherms. The Langmuir and Freundlich models are the most frequently employed models. In this work, both models were used to describe the relationship between the amount of As (III) ions adsorbed by iron oxide and its equilibrium concentration in solution for 90 minutes at the temperature of 295 K.

The linear form of the Freundlich isotherm equation can be written as:

$$\ln q_e = \ln K_F + \frac{1}{n}\ln C_e$$

and the Langmuir isotherm as the following equation:

$$\frac{C_e}{q_e} = \frac{1}{K_L q_m} + \frac{C_e}{q_m}$$

where: q_e is the amount of arsenic adsorbed per gram of sorbent, μg/g;
C_e is the equilibrium concentration of arsenic, μg/L;
K_f and 1/n are characteristic constants that can be related to the relative adsorption capacity of the adsorbent and the intensity of adsorption, respectively;
q_m is a measure of monolayer adsorption capacity [μg/g] and K_L is a constant related to the free energy of adsorption.

The curves and parameters, as well as the correlation coefficients (R^2), for As (III) removal trough adsorption onto iron oxide are presented in figures 7, 8 and table 4.

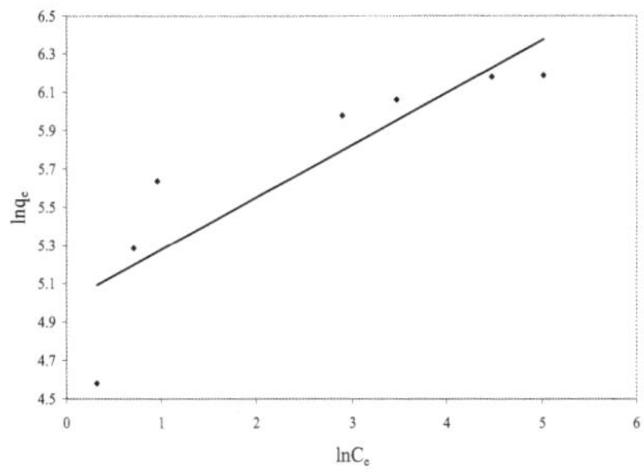

Figure 7: Freundlich plot of As (III) adsorption on iron oxide.

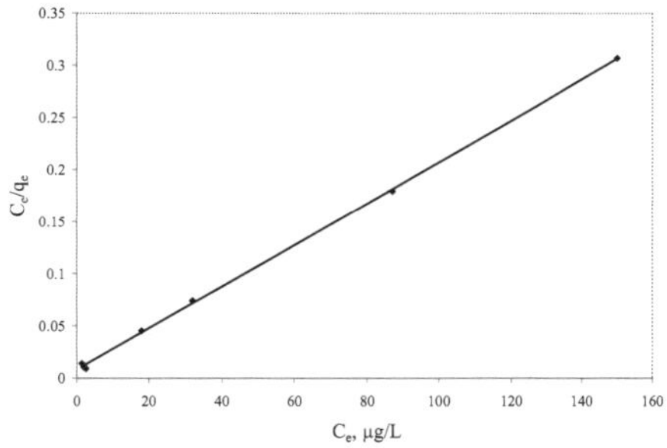

Figure 8: Langmuir plot of As (III) adsorption on iron oxide.

Table 4: Parameters of different isotherm for the As (III) adsorption on iron oxide.

Freundlich isotherm			Langmuir isotherm		
K_F, μg/g	1/n	R^2	K_L, L/μg	q_m, μg/g	R^2
149	0.2744	0.7689	0.247	500	0.9994

The Freundlich plot has a correlation coefficient very low; this suggests a restriction on the use of Freundlich isotherms. The numerical value of 1/n< 1, which provides information about surface heterogeneity and surface affinity for the solute, indicates a favourable sorption of As (III) and a very high affinity of the iron oxide for As (III).

The Langmuir model effectively describes the sorption data with a correlation coefficient of 0.9994. Thus the isotherm follows the sorption process in the entire concentration range studied. Furthermore, the maximum adsorption capacity obtained from the Langmuir plot is almost the same with the maximum adsorption capacity obtained experimentally.

The dimensional constant, called separation factor (R_L), was used to describe the essential characteristics of Langmuir isotherm (R_L >1, unfavourable; R_L = 1, linear; 0<R_L<1, favorable; and R_L=0, irreversible),

$$R_L = \frac{1}{1 + K_L \cdot C_0}$$

In fact, the separation factor is a measure of the adsorbent capacity used. R_L values were calculated for the entire concentration range studied and found to be greater than 0 and less than 1, showing favourable adsorption.

4 Conclusions

The sorption performance of a synthetic iron oxide for the removal of As (III) removal was evaluated. The pH_{pzc} of the iron oxide was found to be 6.1, close to the neutral pH.

A detailed laboratory study was conducted to evaluate the impact of temperature during the adsorption of arsenic by the synthetic iron oxide obtained from the annealing of the iron oxalate at 550°C.

The initial rate of adsorption is fast, followed by a slower rate that gradually approaches a plateau. The equilibrium is reached in 90 minutes for all the studied temperatures. Adsorption rate data are best described by a pseudo-second-order kinetic model.

Temperature dependence of sorption reveals the increase in sorption performance of the sorbent with temperature. The values of Gibbs free energy ($\Delta G°$) show the spontaneous nature of arsenite sorption and the positive values of the standard enthalpy ($\Delta H°$) show the endothermic nature of arsenite sorption on iron oxide. Positive entropy ($\Delta S°$) suggest the affinity of iron oxide for the arsenite in solution.

The equilibrium sorption data were modelled using Freundlich and Langmuir isotherms. Comparisons of the correlation coefficient demonstrate that the sorption process is best described by the Langmuir model. The dimensional separation factor used to predict the essential characteristics of Langmuir isotherm indicated favourable sorption in the entire range concentration. The maximum adsorption capacity of iron oxide, in the process of arsenite removal from water, determined experimentally and also predicted from the Langmuir isotherm is 500 μg/g.

Acknowledgements

The authors gratefully acknowledge the financial support provided by the UEFISCSU, under grant no. 694/19.01.2009, Code 927, "Integrated concept about depollution of waters with arsenic content, through adsorption on oxide materials, followed by immobilization of the resulted waste in crystalline matrices".

References

[1] Streat, M., Hellgardt, K., Newton & N.L.R., Hydrous ferric oxide as an adsorbent in water treatment. Part 2. Adsorption studies, *Process Safety and Environmental Protection,* 86, pp. 11-20, 2008.

[2] Jonsson, J. & Sherman, D.M., Sorption of As (III) and As(V) to siderite, green rust (fougerite) and magnetite; Implications for arsenic release in anoxic groundwaters, *Chemical Geology*, 255, pp. 173-181, 2008.

[3] Mamindy-Pajany, Y., Hurel, C., Marmier, N. & Romeo, M., Arsenic adsorption onto hematite and goethite, *Comptes Rendys Chimie,* 12(8), pp. 876-881, 2009.

[4] Hartley, W. & Lepp, N.W., Remediation of arsenic contaminated soils by iron-oxide application, evaluated in terms of plant productivity arsenic and phytotoxic metal uptake, *Science of the Total Environment,* 390, pp. 35-44, 2008.

[5] Borah, D., Satokawa, S., Kato, S. & Kojima, T., Sorption of As(V) from aqueous solution using acid modified carbon black, *Journal of Hazardous Materials*, 162, pp. 1269-1277, 2009.

[6] Camacho, J., Wee, H.Y., Kramer, T.A. & Autenrieth, R., Arsenic stabilization on water treatment residuals by calcium addition, *Journal of Hazardous Materials,* 165, pp. 599-603, 2009.

[7] Chen, Y.N., Chai, L.Y. & Shu, Y.D., Study of arsenic(V) adsorption on bone char from aqueous solution, *Journal of Hazardous Materials*, 160, pp. 168-172, 2008.

[8] Nguyen, V.T., Vigneswaran, S., Ngo, H.H., Shon, H.K. & Kandasamy, J., Arsenic removal by a membrane hybrid filtration system, *Desalination*, 236, pp. 363-369, 2009.

[9] So, H.U., Postma, D., Jakobsen, R. & Larsen, F., Sorption and desorption of arsenate and arsenite on calcite, *Geochimica et Cosmochimica Acta,* 72, pp. 5871-5884, 2008.

[10] Jeong, Y., Fan, M., Singh, S., Chuang, C.L., Saha, B. & van Leeuwen, J.H., Evaluation of iron oxide and aluminium oxide as potential arsenic(V) adsorbents, *Chemical Engineering and Processing*, 46, pp. 1030-1039, 2007.

[11] Zeng, L., A method for preparing silica-containing iron(III) oxide adsorbents for arsenic removal, *Water Research*, 37, pp. 4351-4358, 2003.

[12] Thirunavukkarasu, O. S., Viraraghavan, T. & Subramanian, K.S., Arsenic removal from drinking water using granular ferric hydroxide, *Water SA*, 29, pp. 161-170, 2003.

[13] Gupta, K. & Ghosh, U.C., Arsenic removal using hydrous nanostructure iron(III)-titanium(IV) binary mixed oxide from aqueous solution, *Journal of Hazardous Materials,* 161, pp. 884-892, 2009.

[14] Banerjee, K., Amy, G.L., Prevost, M., Nour, S., Jekel, M., Gallagher, P.M. & Blumenschein, C.D., Kinetic and thermodynamic aspects of adsorption of arsenic onto granular ferric hydroxide (GFH), *Water Research*, 42, pp. 3371-3378, 2008.

[15] Chutia, P., Kato, S., Kojima, T. & Satokawa, S., Adsorption of As(V) on surfactant-modified natural zeolites, *Journal of Hazardous Materials,* 162, pp. 204-211, 2009.
[16] Mondal, P., Majumder, C.B. & Mohanty, B., Effects of adsorbent dose, its particle size and initial arsenic concentration on the removal of arsenic, iron and manganese from simulated ground water by Fe^{3+} impregnated activated carbon, *Journal of Hazardous Materials,* 150, pp. 695-702, 2008.
[17] Partey, F., Norman, D., Ndur, S. & Nartey, R., Arsenic sorption onto laterite iron concretions: Temperature effect, *Journal of Colloid and Interface Science*, 321, pp. 493-500, 2008.
[18] Pedersen, H.D., Postma, D. & Jakobsen, R., Release of arsenic associated with the reduction and transformation of iron oxides, *Geochimica et Cosmochimica Acta,* 70, pp. 4116-4129, 2006.
[19] Ohe, K., Tagai, Y., Nakamura, S., Oshima, T. & Baba, Y., Adsorption behaviour of arsenic(III) and arsenic(V) using magnetite, *Journal of Chemical Engineering of Japan*, 38, pp. 671-676, 2005.
[20] Hsu, J.C., Lin, C.J., Liao, C.H. & Chen, S.T., Removal of As(V) and As (III) by reclaimed iron-oxide coated sands *Journal of Hazardous Materials,* 153, pp. 817-826, 2008.
[21] Cerovic, Lj. S., Milonjic, S.K., Todorovic, M.B., Trtanj, M.I., Pogozhev, Y.S., Blagoveschenskii, Y. & Levashov, E.A., Point of zero charge of different carbides, *Colloids and Surfaces A: Physicochem. Eng. Aspects,* 297, pp. 1–6, 2007.

Section 4
Lakes and rivers

Alkylphenol ethoxylates and bisphenol A in surface water within a heavily urbanized area, such as Paris

M. Cladière, J. Gasperi, S. Gilbert, C. Lorgeoux & B. Tassin
Université Paris-Est, LEESU, UMR-MA102 – AgroParisTech, France

Abstract

This work focused on alkylphenol ethoxylates (APEOs) and bisphenol A (BPA) in surface water at the scale of the upper part of the Seine basin. Although this basin is vulnerable, since it undergoes a combination of strong human pressures (Paris conurbation) with very limited dilution via the Seine River, no comprehensive data was available for these compounds. In this context, the evolution of the pollutant concentrations and loads was assessed during two sampling campaigns (April and July 2009) along a 300 km-transect. Hence, 18 stations (10 sites on the Seine River, four tributaries and four wastewater treatment plants – WWTPs) were considered. Additional campaigns were also conducted over 2009 on four sites on the Seine River. The BPA concentrations range from 7.2 to 127 ng.l^{-1} in surface water, while higher concentrations of APEOs (324 ± 153 ng.l^{-1}) were observed. Based on our first results, no spatial variability of concentrations between up- and downstream sites, or any seasonal variability, were highlighted. During the longitudinal campaigns, an increase of the pollutant loads between up- and downstream sites was clearly noticed. Contrary to BPA, for which a limited seasonal variation of loads was observed at the basin outlet (1,400 and 1,500 g.d^{-1}), the APEO loads significantly vary (10,000 and 5,500 g.d^{-1}). Mass balances between the inputs (tributaries + WWTPs) and the exported loads reveal that the BPA inputs are lower than the exported loads, suggesting that other inputs have to be considered along the transect studied. For APEOs, the inputs appear higher than the exported loads, underscoring the in-stream removal of APEOs. Globally, the results reveal that the removal of APEOs and the number of ethoxylate units are correlated – the longer the ethoxylate chain, the higher the removal.
Keywords: alkylphenol ethoxylates, bisphenol A, endocrine-disrupting compounds, mass load, nonylphenol, octylphenol, surface water.

WIT Transactions on Ecology and the Environment, Vol 135, © 2010 WIT Press
www.witpress.com, ISSN 1743-3541 (on-line)
doi:10.2495/WP100121

1 Introduction

Among the endocrine-disrupting compounds commonly pointed out in the literature, alkylphenol ethoxylates (APEOs) and bisphenol A (2,2-bis(4-hydroxyphenyl)propane, BPA) deserve special attention, not only because of their disrupting properties but also for their widespread distribution in the environment [1, 2].

APEOs, mainly composed of nonylphenol ethoxylates (NPEOs, 80%) and octylphenol ethoxylates (OPEOs, 20%), are widely used for industrial, domestic and commercial applications, such as lubricating, oil additives, detergents and antistatic agents. Similarly, BPA is mainly used as a monomer in the synthesis of epoxy resins and polycarbonate plastics [3, 4]. Due to these applications, the worldwide annual production is approximately 500,000 tons for APEOs (in 2002, according to Ying *et al.* [5]) and 450,000 tons for BPA (in 2007, according to Vandenberg *et al.* [4]).

The occurrence of APEOs and BPA in the environment is closely correlated with anthropogenic activities. Many studies suggested that these chemicals preferentially enter into environment through the effluents of wastewater treatment plants (WWTPs) [5, 6].

Contrary to BPA, for which the estrogenic activity was clearly observed [1, 2], APEOs are of concern because their degradation products are found to be more toxic than their precursors [7, 8]. Biotransformation pathways of alkylphenol ethoxylates are well documented [9, 10]. This biodegradation can occur through the oxidative hydrolytic (aerobic) or non-oxidative hydrolytic pathways. This last pathway leads to the reduction of the number of ethoxy units generating APEOs with shorter ethoxylate chains, such as alkylphenol diethoxylate (NP_2EO and OP_2EO) and alkylphenol monoethoxylate (NP_1EO and OP_1EO). Ultimately, the last products of biodegradation are 4-nonylphenol (NP) and 4-tert-octylphenol (OP), which are identified as endocrine-disrupting compounds [1, 11].

The Seine Basin, located in north-western France and responsible for draining approximately 32,000 km² from its headwaters to Paris, can be considered as representative of river basins exposed to the impacts of intense human activity (Meybeck *et al.* [12]). This basin accommodates a combination of strong human pressures with very limited dilution via the Seine River, due to its low flow rate. However, no comprehensive data on the occurrence of APEOs and BPA is presently available for the upper part of the Seine basin. As regards the environmental issue of these chemicals, an accurate knowledge at the scale of the Seine Basin, and more especially within the Paris Region, corresponding to the most heavily urbanised area (with a density of roughly 3,540 inhabitants.km^{-2}), has definitely proved necessary.

As a consequence, this study was launched to provide the first relevant information on the occurrence and significance of the concentrations of APEOs and BPA in surface water at the scale of the upper part of the Seine basin. In this paper, we particularly examine (i) the concentrations in surface water and

WWTP effluents, (ii) the spatio-temporal variability of loads and finally (iii) the environmental behaviour and fate of APEOs and BPA along the Seine River.

2 Materials and methods

2.1 Sampling sites

Two sampling campaigns, corresponding to the median-water period (14/04/09, median flow at 154 $m^3.s^{-1}$) and the low-water period (07/07/09, median flow at 83 $m^3.s^{-1}$), were carried out along a 300-km transect of the Seine River. For each campaign 14 sites, including 10 stations on the Seine River (S_{1-10}) and four main tributaries (T_1-$_4$), were considered (Figure 1). In parallel, the effluents of the four major WWTPs along this section were investigated ($WWTP_{1-4}$). The main characteristics of these WWTPs (type of treatment, population equivalent, capacity, etc.) are reported in Figure 1. Two additional campaigns were also performed on 11/05/09 and 28/09/09 on four sites (S_1, S_3, S_5 and S_8 stations).

2.2 Analytical procedure

For each site, a 2L-sample was manually collected and stored in glass amber bottles until analysis. After filtration on glass fibre filters (GF/D 2.7 µm and GF/F 0.7 µm), dissolved and particulate phases were individually analyzed.

2.2.1 Dissolved phase

Each water sample was spiked with internal standards (BPA-d6; n-OP-d17 and NP_1EO-d2) for extraction follow-up and stored overnight at 4 °C to reach an equilibrium state. Volumes of 250 ml were extracted by solid phase extraction

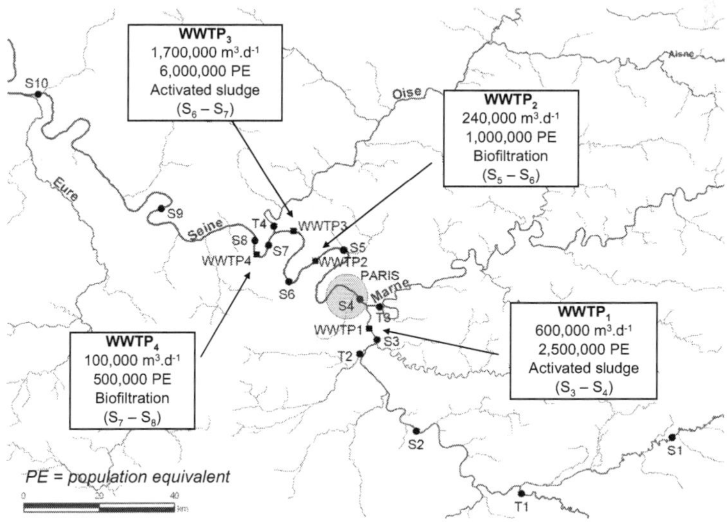

Figure 1: Sampling sites along the Seine River.

(Autotrace SPE Workstation, Caliper LifeScience) using Oasis HLB (200 mg, 6 ml) cartridges. After conditioning with 10 ml of methanol and 10 ml of ultrapure water, samples were extracted at neutral pH. After drying under vacuum, elutions were performed with 12 ml of a mixture methanol/dichloromethane/ethylacetate (40:40:20, v/v).

2.2.2 Particulate phase

After freeze-drying, the filters spiked with internal standards are extracted by microwave assisted extraction (MAE) with 20 ml of dichloromethane/methanol (90:10, v/v) (Multiwave 3000, Anton Paar) during a 30 min-cycle. Extracts are then filtered and evaporated until dryness under nitrogen for subsequent clean up using silica gel cartridges. These cartridges are conditioned with 5 ml of heptane before deposing the sample. Elution is done sequentially in three steps: 18 ml of heptane/dichloromethane (98:2, v/v), 10 ml of heptane/dichloromethane (80:20, v/v) and 10 ml of heptane/dichloromethane/ ethylacetate (10:30:60, v/v). Only the last fraction is kept for BPA and APEOs analysis.

2.3 LC-MS-MS conditions

Before analysis, both extracts are spiked with internal standard for quantification (BPA-d16, n-NP and n-NP_1EO). The analysis is performed by liquid chromatography coupled to a tandem mass spectrometry, LC-MS-MS (AQUITY UPLC / TQD, Waters). APEOs and BPA are separated on an AQUITY UPLC / BEH C_{18} column, heated at 40 °C, with ultrapure water containing 4.5 mM NH_4OH (A) and methanol containing 4.5 mM NH_4OH (B) as mobile phase. Equilibration takes place with 50% B at 0.4 ml.min^{-1} and 10 µl are injected. The mass spectrometer is equipped with an electrospray interface used in positive ionization mode (ESI+) for $NP_{1\text{-}2}EO$ and $OP_{1\text{-}2}EO$ and negative ionization mode (ESI-) for all other analytes. The compounds were detected in multiple reactions monitoring mode (MRM) with two mass transitions for each analyte. Cone voltage and collision energy were optimized for all analytes (Table 1).

Table 1: Optimized parameters for LC-MS-MS analysis.

Compound	M (g.mol^{-1})	ESI Mode	Retention time (min)	Quantification Transition (m/z)	Quantification Collision energy (eV)	Qualification Transition (m/z)	Qualification Collision energy (eV)	Cone voltage (V)
BPA	228	ESI-	1.82	226.9 → 133.0	25	226.9 → 212.2	25	30
OP	206	ESI-	3.25	205.5 → 134.2	17	205.2 → 133.2	30	45
OP_1EO	250	ESI+	3.34	268.2 → 113.4	10	268.2 → 139.2	12	20
OP_2EO	294	ESI+	3.38	312.3 → 183.3	10	312.3 → 121.2	18	20
NP	220	ESI-	3.52	219.2 → 147.2	26	219.2 → 133.2	26	40
NP_1EO	264	ESI+	3.58	282.4 → 127.3	8	282.4 → 85.4	15	20
NP_2EO	308	ESI+	3.61	326.4 → 183.3	10	326.4 → 71.2	20	40

2.4 Quantification and quality control

The quantification was performed using internal calibration based on areas relative to the internal standards. Internal standard n-NP was used for the quantification of NP, OP and n-OP-d17. Similarly, BPA-d16 was used for BPA and BPA-d6 and n-NP_1EO for OP_1EO, OP_2EO, NP_1EO, NP_2EO and n-NP_1EO-d2. For each compound, calibration curves were drawn and satisfactory determination coefficients were obtained ($r^2 > 0.99$). Experimental blanks for SPE and MAE extractions, as well as mobile phase blanks were processed. No quantifiable contamination was observed. During a sequence, two calibration controls (lower and higher concentrations) were injected every 10 samples. A deviation between the control and the calibration curve lower than 20% validates the quantification.

3 Results and discussion

3.1 APEO and BPA in surface water and WWTP effluents

3.1.1 APEO and BPA concentrations

Total concentrations (median ± standard deviation, SD) measured for the Seine River, its tributaries and the WWTP effluents are shown in Figure 2.

As illustrated in Figure 2, BPA concentrations are quite similar in surface water (from 7.2 to 127 $ng.l^{-1}$, median concentration being 50 $ng.l^{-1}$) and in WWTP effluents (from 24 to 163 $ng.l^{-1}$, median concentration being 36 $ng.l^{-1}$). No significant difference was also observed between the Seine River and its tributaries. For APEOs, the distributions within surface water are featured by the predominance of NP, followed by NP_2EO and NP_1EO while OPEOs exhibit far lower concentrations (< 1 $ng.l^{-1}$ for OP_1EO to 20 $ng.l^{-1}$ for OP). The relative concentrations of NPEOs and OPEOs reflect their consumption pattern and are consistent with ratio reported in literature [6] but are slightly different from

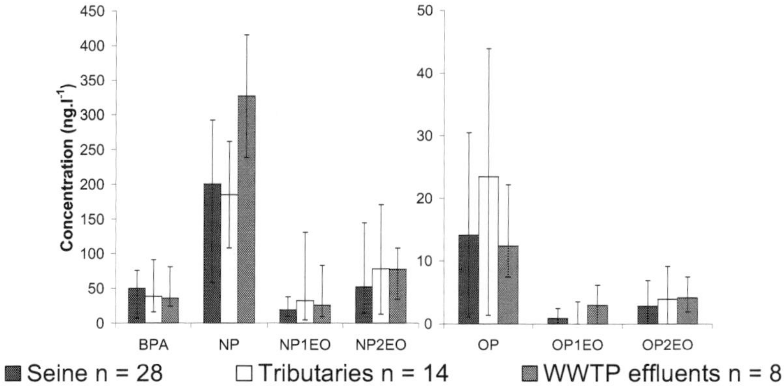

Figure 2: Total concentrations in surface water (Seine River and tributaries) and in WWTP effluents.

Table 2: Concentrations (ng.l^{-1}) in surface water, this study *vs.* literature.

Location	BPA	NP	NP_1EO	NP_2EO	OP	OP_1EO	OP_2EO	Reference
Glatt River	9.4	64	18	9.4	2.4	-	-	[15]
Masan Bay	-	18.9	24.2	115.4	-	-	-	[18]
Seine River downstream	-	168	82	59	-	-	-	[13]
Jamaica bay	-	201	157	320	3.3	26	4.9	[19]
Tiber River	80	240	360	270	-	-	-	[20]
Seine River upstream*	50 7-127	200 58-426	19 10-79	52 15-414	14 1-81	1 < 1-4	3 < 1-14	This study

* *Median, min – max values.*

values given by Ying *et al.* [5], i.e. 80% for NPEOs and 20% for OPEOs. As regard the Water Framework Directive, the total concentrations for NP (200 ± 92 ng.l^{-1}) and OP (14 ± 16 ng.l^{-1}) do not exceed the environmental quality standards (300 and 100 ng.l^{-1} for NP and OP, respectively). Concentrations found in this study on the upper part of the Seine basin are in good agreement with concentrations reported in the lower part by Cailleaud *et al.* [13]. Moreover, and in spite of the strong human pressures on the Seine basin, the BPA and APEO concentrations measured in the Seine River appear in the same order of magnitude than concentrations reported in literature (Table 2). Contrary to some studies emphasising higher levels of APEOs and BPA in WWTP effluents comparatively to surface water [14-16], comparable concentrations were found in WWTP effluents and in surface water, except for NP which is more concentrated in WWTP effluents. Comparatively to other studies, the WWTP effluent concentrations found in this study are similar to the results reported by Jonkers *et al.* [15] in Switzerland but far lower than those reported in China and Austria [16, 17]. The low concentrations in WWTP effluents probably reflect the decrease of APEO consumption for both industrial and domestic applications due to the European Directive 2003/53/CE.

In spite of their hydrophobicity (log K_{ow} ranging from 3.2 for BPA to 4.6 for NP, Ahel and Giger [21]), APEOs and BPA are weakly associated with suspended solids, except for OP. Indeed, particulate phase only represent 20% of total concentration of APEOs and 5% of BPA for each surface water and WWTP effluents. Similar distributions between dissolved and particulate phase were found in the downstream part of the Seine basin by Cailleaud *et al.* [13]. The higher particulate proportion of OP (between 40 and 50%) could probably be explained by the sorption of this compound on suspended solids. Indeed, Johnson *et al.* [22] demonstrated that up to 40% of OP can be sorbed onto suspended solids under low flow condition.

3.1.2 Evolution of the pollutant concentrations along the Seine River

The individual concentrations along the 300-km transect of the Seine River (S_{1-10}) are illustrated in Figure 3. For stations downstream of $WWTP_{1,\ 2,\ 4}$, no significant impact on the concentrations was noticed. This is mainly related to the low concentrations of WWTP effluents and low volumes discharged (1 - 4

$m^3.s^{-1}$) comparatively to the Seine River (about 150 $m^3.s^{-1}$). To the opposite, a significant increase of concentrations is observed at S_7 downstream of $WWTP_3$ (from 324 at S_6 to 751 $ng.l^{-1}$ at S_7 in April and from 304 to 526 $ng.l^{-1}$ in July). This impact is related to higher volume discharged (17 $m^3.s^{-1}$ against 150 $m^3.s^{-1}$ for the Seine River) and to a lesser extent to the higher concentrations measured in the effluents. Actually, the $WWTP_3$ is the biggest plant of the Parisian conurbation treating approximately 75% of the Parisian wastewater, i.e. 1.7 million m^3 (6.0 millions of population-equivalents). This impact is clearly visible at S_7 but becomes less perceptible at S_8 due to the Seine River dilution by the Oise tributary (T_4, Figure 1), accounting for 40% of total river flow.

3.2 Alkylphenol ethoxylate and bisphenol A mass loads

For both sampling campaigns, daily mass loads ($g.d^{-1}$) were calculated according to the total concentrations ($ng.l^{-1}$) and the river flows ($m^3.s^{-1}$). These values were obtained from different river monitoring stations.

3.2.1 Temporal variation of the pollutant loads

In order to assess the temporal variation of the pollutant loads, the loads ($g.d^{-1}$) at S_8 and S_{10} for all sampling campaigns available were examined (Figure 4). The distance between S_8 and S_{10} is about 150 km and the hydraulic time is evaluated between 3 and 5 days.

As underlined by Figure 4, a limited temporal variation is observed for BPA between median- and low-water periods. While the BPA loads in April (S_8 and S_{10}) or in May (S_8) ranged from 1,360 to 1,776 $g.d^{-1}$, those observed during low-water period are comprised between 611 and 1,370 $g.d^{-1}$. In contrast, the APEO loads significantly vary between spring and summer. While the APEO loads (sum of 6 congeners) in April reach 10,741 and 10,108 $g.d^{-1}$ at S_8 and S_{10}, far lower loads are indeed observed in July (3,822 and 3,931 $g.d^{-1}$, Figure 4). This temporal variability is mainly explained by the important decrease of Seine River

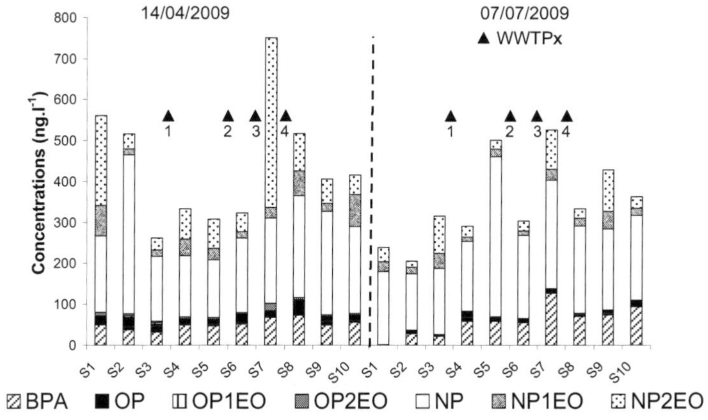

Figure 3: APEO and BPA concentrations in the Seine River.

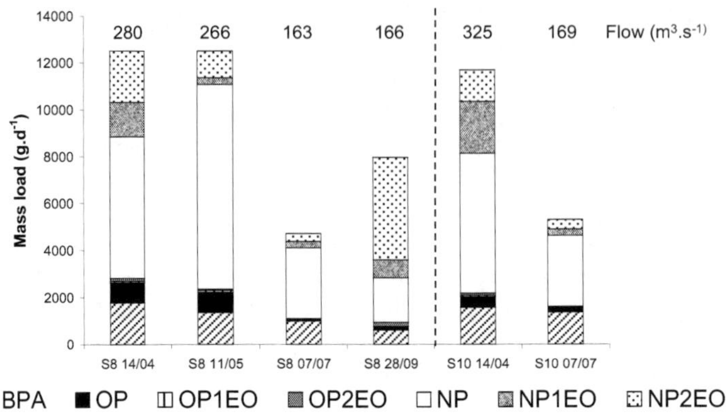

Figure 4: Temporal variation of the APEO and BPA loads.

flow between both dates (280 *vs.* 163 $m^3.s^{-1}$, i.e. 58%) and to a lesser extent by the difference of concentrations at S_8. Whether no clear seasonal variation of APEO concentrations was highlighted in our study, the lower concentration at S_8 was however noted in July (263 $ng.l^{-1}$) comparatively to April (444 $ng.l^{-1}$).

Figure 4 also underlines an evolution of the $NP_{1\text{-}2}EO$ and $OP_{1\text{-}2}EO$ distributions between S_8 and S_{10}. This change probably results from *in-situ* processes, more likely through biodegradation. It is actually known that the biodegradation leading to the formation of APEOs with shorter ethoxylate chain affect the distributions of the different congeners (John and White [9]). As attested by the change of $NP_{1\text{-}2}EO$ and $OP_{1\text{-}2}EO$ proportions between both stations, biodegradation seems to be more important in April than in July.

3.2.2 Evolution of pollutant loads along the Seine River

Figure 5 illustrates the evolution of the pollutant loads along the Seine River (from S_1 to S_{10}) for both sampling campaigns. Whatever the campaign considered, results show an increase of the pollutant loads from up- (S_1) to down-stream (S_{10}) stations. Typically, the BPA loads in April rise from 154 $g.d^{-1}$ at S_1 to 1,576 at S_{10} while the APEO loads increase from 1,580 to 10,108 $g.d^{-1}$ in the same time (Figure 5).

At the Seine basin outlet (S_{10}), the exported loads in April and July to the lower part of the Seine basin were respectively evaluated at 1,576 and 1,373 $g.d^{-1}$ for BPA and at 10,108 and 3,931 $g.d^{-1}$ for APEOs. The $WWTP_{1,2,4}$ seem to have a moderate impact on the pollutant loads conveyed by the Seine River. As previously mentioned, this can be mainly related to their low effluent discharges comparatively to the Seine River and their low concentrations.

To the opposite, the $WWTP_3$ has a significant impact on the Seine River pollutant loads. Due to the $WWTP_3$ discharge, the APEO and BPA loads between S_6 and S_7 increase from 5,173 to 11,681 $g.d^{-1}$ in April and from 2,576 to 5,240 $g.d^{-1}$ in July. As previously explained, this impact is related to the high flow of the $WWTP_3$ effluents (17 $m^3.s^{-1}$) representing up to 15% of Seine River flow (S_7). Contrary to the results found for concentrations, the $WWTP_3$ impact

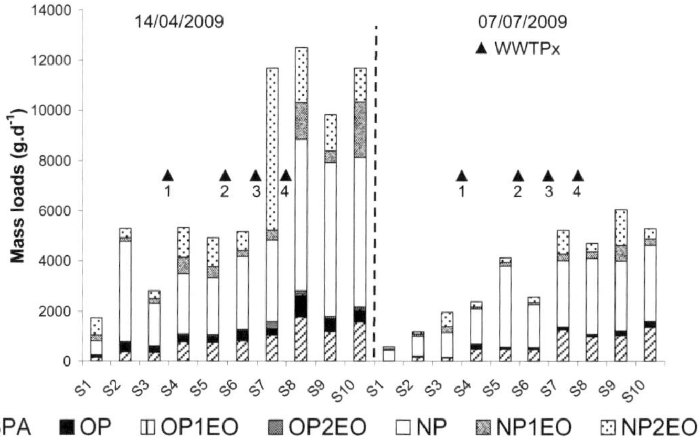

Figure 5: Evolution of pollutant loads along the Seine River.

on the pollutant loads is still visible on the downstream stations. A comparison between the WWTP loads (sum of 4 WWTPs) and the total inputs to the Seine River system (S_1, tributaries and WWTPs) attests to the higher contribution of the WWTPs in July comparatively to April. For instance, the contribution of WWTP effluents to total inputs rises from 13% to 36% for BPA and from 10% to 40% for APEOs. These results are mainly related to the decrease of tributaries and S_1 flows between spring and summer whereas the WWTP discharge flows are constant.

3.2.3 Inputs and exported pollutant loads at the scale of the Seine basin

To assess the dynamic of the pollutants within the Seine River, mass balances between the inputs (S_1, tributaries and WWTPs) and the pollutant loads measured at the basin outlet (S_{10}) were carried out. Thus, the ratio between the inputs and the exported loads were calculated for both campaigns (Table 3).

The comparison between the inputs and the exported loads highlights two different cases. For BPA, the inputs for both campaigns appear lower than the exported loads (ratio inputs *vs.* exported loads respectively equal to 0.7 and 0.5). Considering that a production of BPA cannot occur within the Seine River [4], these ratios indicate that not all sources were account for. To asses more accurate mass balances, the other WWTPs present in the Seine basin (about 250 plants with capacity below 2,000 population-equivalents) have to be considered along the Seine transect.

To the opposite, APEOs exhibit ratios above 1 suggesting that the inputs are higher than the exported loads (Table 3). These values point out the in-stream removal of APEOs along the Seine River, probably through *in-situ* processes such as adsorption on particles and sedimentation [22] and/or biodegradation [9, 10]. Out of these different processes, the biodegradation is likely the predominant one [5]. Results found in April and to a lesser extent in July reveal that the removal of APEOs and the number of ethoxylate units are correlated – the longer ethoxylate chain, the higher the removal (3.5 for NP_2EO > 1.7 for

Table 3: Inputs and exported loads ($g.d^{-1}$) in the Seine River.

compound	April		July		Inputs / exported	
	Inputs	exported	Inputs	exported	14/04/09	07/07/09
BPA	1,164	1,576	728	1,373	0.7	0.5
OP	544	447	135	235	1.2	0.6
OP_1EO	132	68	10	NC	2.0	-
OP_2EO	282	87	20	NC	3.2	-
NP	6,138	5,959	3,466	3,019	1.0	1.1
NP_1EO	3,282	2,205	467	268	1.7	1.7
NP_2EO	5,175	1,342	676	409	3.5	1.7

NC = not calculated since concentrations are below the detection limit.

NP_1EO > 1.0 for NP in April, Table 3). Considering the views of transformation reactions of nonylphenol ethoxylates provided by Giger *et al.* [23], results implied that the transformation of NP_2EO into NP_1EO or nonylphenolethoxyacetic acid (NP_2EC) is faster than its formation from NPnEO. Similarly, our results suggest that the transformation of NP_1EO into NP or nonylphenolacetic acid (NP_1EC) is faster than its formation from longer homolog.

4 Conclusion

This study provides first relevant information on the occurrence and fate of APEOs and BPA at the scale of the upper part of the Seine basin. BPA concentrations range from 7.2 to 127 $ng.l^{-1}$ in surface water. For APEOs, higher concentrations for surface water (324 ± 153 $ng.l^{-1}$) were observed. Based on our first results, no spatial variability of concentrations between up- and downstream sites as well as any seasonal variability were highlighted. To the opposite, an increase of the pollutant loads between up- and down-stream sites was noticed. Contrary to the limited seasonal variation of BPA, the APEO loads significantly vary (10,108 and 3,931 $g.d^{-1}$). At present, this seasonal variation has to be better examined in relation to the hydrological dynamic conditions. Therefore, further investigations will be soon carried out.

The assessment of the pollutant mass balances between the inputs and the exported loads reveal that the BPA inputs are lower than the exported loads suggesting that not all sources were considered. For APEOs, the inputs appear higher than the exported loads, underscoring their in-stream removal during their transport within the Seine River. Results reveal that this removal and the number of ethoxylate units are correlated – the longer ethoxylate chain, the higher the removal. Whether our result do not allow the identification of the transformation reactions involved in the aquatic system, our sampling approach nevertheless enables a better understanding of the APEO behaviour. In the future, the analyses of $NP_{1\text{-}2}EC$ intermediates will improve this understanding.

Acknowledgement

The authors would thank the Interdepartmental Association for Sewage Disposal in the Paris Conurbation (SIAAP) for their technical help.

References

[1] Brian, J.V., Harris, C.A., Scholze, M., Kortenkamp, A., Booy, P., Lamoree, M., Pojana, G., Jonkers, N., Marcomini, A., & Sumpter, J.P., Evidence of estrogenic mixture effects on the reproductive performance of fish. *Environmental Science & Technology*, **41(1)**, pp. 337-344, 2007.

[2] Wetherill, Y.B., Akingbemi, B.T., Kanno, J., McLachlan, J.A., Nadal, A., Sonnenschein, C., Watson, C.S., Zoeller, R.T., & Belcher, S.M., In vitro molecular mechanisms of bisphenol A action. *Reproductive Toxicology*, **24(2)**, pp. 178-198, 2007.

[3] Soares, A., Guieysse, B., Jefferson, B., Cartmell, E., & Lester, J.N., Nonylphenol in the environment: A critical review on occurrence, fate, toxicity and treatment in wastewaters. *Environment International*, **34(7)**, pp. 1033-1049, 2008.

[4] Vandenberg, L.N., Hauser, R., Marcus, M., Olea, N., & Welshons, W.V., Human exposure to bisphenol A (BPA). *Reproductive Toxicology*, **24(2)**, pp. 139-177, 2007.

[5] Ying, G.-G., Williams, B., & Kookana, R., Environmental fate of alkylphenols and alkylphenol ethoxylates--a review. *Environment International*, **28(3)**, pp. 215-226, 2002.

[6] Sharma, V.K., Anquandah, G.A.K., Yngard, R.A., Kim, H., Fekete, J., Bouzek, K., Ray, A.K., & Golovko, D., Nonylphenol, octylphenol, and bisphenol-A in the aquatic environment: A review on occurrence, fate, and treatment. *Journal of Environmental Science and Health Part a-Toxic/Hazardous Substances & Environmental Engineering*, **44(5)**, pp. 423-442, 2009.

[7] Bhattacharya, H., Xiao, Q., & Lun, L., Toxicity studies of nonylphenol on rosy barb (Puntius conchonious): A biochemical and histopathological evaluation. *Tissue and Cell*, **40(4)**, pp. 243-249, 2008.

[8] Uguz, C., Iscan, M., Ergüven, A., Isgor, B., & Togan, I., The bioaccumulation of nonyphenol and its adverse effect on the liver of rainbow trout (Onchorynchus mykiss). *Environmental Research*, **92(3)**, pp. 262-270, 2003.

[9] John, D.M. & White, G.F., Mechanism for Biotransformation of Nonylphenol polyethoxylates to xenoestrogens in Pseudomonas putida. *Journal of Bacteriology*, **180(17)**, pp. 4332-4338, 1998.

[10] Jonkers, N., Knepper, T.P., & De Voogt, P., Aerobic biodegradation studies of nonylphenol ethoxylates in river water using liquid chromatography-electrospray tandem mass spectrometry. *Environmental Science & Technology*, **35(2)**, pp. 335-340, 2001.

[11] Zhang, L., Gibble, R., & Baer, K.N., The effects of 4-nonylphenol and ethanol on acute toxicity, embryo development, and reproduction in Daphnia magna. *Ecotoxicology and Environmental Safety*, **55(3)**, pp. 330-337, 2003.

[12] Meybeck, M., Lestel, L., Bonte, P., Moilleron, R., Colin, J.L., Rousselot, O., Herve, D., de Ponteves, C., Grosbois, C., & Thevenot, D.R., Historical

perspective of heavy metals contamination (Cd, Cr, Cu, Hg, Pb, Zn) in the Seine River basin (France) following a DPSIR approach (1950-2005). *Science of the Total Environment*, **375(1-3)**, pp. 204-231, 2007.

[13] Cailleaud, K., Forget-Leray, J., Souissi, S., Lardy, S., Augagneur, S., & Budzinski, H., Seasonal variation of hydrophobic organic contaminant concentrations in the water-column of the Seine Estuary and their transfer to a planktonic species Eurytemora affinis (Calanoïd, copepod). Part 2: Alkylphenol-polyethoxylates. *Chemosphere*, **70(2)**, pp. 281-287, 2007.

[14] González, S., Petrovic, M., & Barceló, D., Simultaneous extraction and fate of linear alkylbenzene sulfonates, coconut diethanol amides, nonylphenol ethoxylates and their degradation products in wastewater treatment plants, receiving coastal waters and sediments in the Catalonian area (NE Spain). *Journal of Chromatography A*, **1052(1-2)**, pp. 111-120, 2004.

[15] Jonkers, N., Kohler, H.P.E., Dammshauser, A., & Giger, W., Mass flows of endocrine disruptors in the Glatt River during varying weather conditions. *Environmental Pollution*, **157(3)**, pp. 714-723, 2009.

[16] Yu, Y., Zhai, H., Hou, S., & Sun, H., Nonylphenol ethoxylates and their metabolites in sewage treatment plants and rivers of Tianjin, China. *Chemosphere*, **77(1)**, pp. 1-7, 2009.

[17] Clara, M., Scharf, S., Scheffknecht, C., & Gans, O., Occurrence of selected surfactants in untreated and treated sewage. *Water Research*, **41(19)**, pp. 4339-4348, 2007.

[18] Li, D., Dong, M., Shim, W.J., Yim, U.H., Hong, S.H., & Kannan, N., Distribution characteristics of nonylphenolic chemicals in Masan Bay environments, Korea. *Chemosphere*, **71(6)**, pp. 1162-1172, 2008.

[19] Ferguson, P.L., Iden, C.R., & Brownawell, B.J., Analysis of alkylphenol ethoxylate metabolites in the aquatic environment using liquid chromatography-electrospray mass spectrometry. *Analytical Chemistry*, **72(18)**, pp. 4322-4330, 2000.

[20] Patrolecco, L., Capri, S., De Angelis, S., Pagnotta, R., Polesello, S., & Valsecchi, S., Partition of nonylphenol and related compounds among different aquatic compartments in Tiber River (Central Italy). *Water Air and Soil Pollution*, **172(1-4)**, pp. 151-166, 2006.

[21] Ahel, M. & Giger, W., Aqueous solubility of alkylphenols and alkylphenol polyethoxylates. *Chemosphere*, **26(8)**, pp. 1461-1470, 1993.

[22] Johnson, A.C., White, C., Besien, T.J., & Jürgens, M.D., The sorption potential of octylphenol, a xenobiotic oestrogen, to suspended and bed-sediments collected from industrial and rural reaches of three English rivers. *Science of the Total Environment*, **210-211**, pp. 271-282, 1998.

[23] Giger, W., Gabriel, F.d.r.L.P., Jonkers, N., Wettstein, F.E., & Kohler, H.-P.E., Environmental fate of phenolic endocrine disruptors: field and laboratory studies. *Philosophical Transactions of the Royal Society A: Mathematical, Physical and Engineering Sciences*, **367(1904)**, pp. 3941-3963, 2009.

Trophic state evaluation for selected lakes in Yellowstone National Park, USA

A. W. Miller
Department of Civil & Environmental Engineering, Brigham Young University, USA

Abstract

The purpose of this study is to classify the trophic state for selected lakes in Yellowstone National Park, USA. This paper also documents that monitoring methods and perspectives used in this study meet current acceptable practices. For selected lakes in Yellowstone National Park, phosphorus, nitrogen, chlorophyll-a, and other lake characteristics are studied to identify lake behavior and to classify the annual average trophic state of the lakes. The four main trophic states are oligotrophic, mesotrophic, eutrophic and hyper-eutrophic. The greater the trophic state, the greater the level of eutrophication that has taken place. Eutrophication is the natural aging process of a lake as it progresses from clear and pristine deep water to more shallow, turbid, and nutrient rich water where plant life and algae are more abundant. The trophic state of a lake is a measurement of where the lake is along the eutrophication process. Human interaction tends to speed up eutrophication by introducing accelerated loadings of nitrogen and phosphorus into aquatic systems. As lakes advance in the eutrophication process, water quality generally decreases. Four models are used in this study to classify the trophic state of the lakes: the Carlson Trophic State Index, the Vollenweider Model, the Larsen-Mercier Model, and the Nitrogen-Phosphorus Ratio Model. Simple models are commonly used where steady-state conditions and lake homogeneity are assumed. There is concern that natural processes and human activity on and around the Yellowstone Lakes are causing the water quality to decline. The objectives of this study are to identify possible areas of concern and develop a baseline to which future evaluations can be compared. This paper presents results for some 20 lakes in Yellowstone Park, which have been studied over the past 12 years.
Keywords: trophic state, lake water quality, eutrophication, phosphorus, chlorophyll-a, Secchi depth, modeling.

WIT Transactions on Ecology and the Environment, Vol 135, © 2010 WIT Press
www.witpress.com, ISSN 1743-3541 (on-line)
doi:10.2495/WP100131

1 Introduction

The purpose of this study was to classify the trophic state for selected lakes in Yellowstone National Park, USA. This paper also documents the study monitoring methods and perspectives as meeting current acceptable practice.

For selected lakes in Yellowstone National Park, phosphorus, nitrogen, chlorophyll-a, and other lake characteristics were studied to identify short-term lake behavior and to classify the annual average trophic state of the lakes. By studying short-term lake behavior, there may be a greater understanding of how the trophic state of a lake can be averaged annually.

There is concern that natural processes and human activity on and around the lakes are causing the water quality to decline. We were unable to find any previous studies that defined the trophic state of the lakes. Therefore, an objective of this study is to evaluate the current trophic state of the lakes and develop a preliminary baseline to which future evaluations can be compared.

This study has provided a benchmark trophic state survey not only for comparison to future evaluations, but to identify possible areas of concern. The greater the trophic state, the greater the eutrophication that has taken place. As lakes advance in the eutrophication process, water quality generally decreases.

2 Eutrophication

Eutrophication is the natural aging process of a lake as it progresses from clear and pristine deep water, that does not support much plant or animal life, to more shallow, turbid, and nutrient rich water where plant life and algae is abundant. This process occurs in all lakes happening over thousands of years, and Chapra [1] says the eventual end is the filling of the lake with sediments and formation of a meadow. Human interaction tends to speed up the eutrophication process by adding accelerated loadings of nitrogen and phosphorus into aquatic systems. The trophic state of a lake is a measurement of where the lake is in the eutrophication process. Tchobanoglous and Schroeder [2] give the four states explained below.

2.1 Trophic states

<u>Oligotrophic:</u> The water is clear and devoid of most plants and algae because there are not many nutrients such as nitrogen and phosphorus found in the water.
<u>Mesotrophic</u>: Plant and animal life is abundant and diverse in all forms and in all levels of the food chain.
<u>Eutrophic:</u> The water quality is poor but there still exists an abundance of life. However, there are fewer species and these species are usually of lower quality. In addition, there are more algal blooms and slime as well as algal turbidity.
<u>Hyper-eutrophic</u>: The water quality is exceptionally poor with an overabundance of algae. Animal life is similar to eutrophic state in that there are fewer species.

2.2 Nutrients and algae

Nutrients found in the water are the fuel of the eutrophication process.

Phosphorus and nitrogen are particularly important since they stimulate plant and algae growth. Phosphorus and nitrogen enter a wilderness area lake in various ways. One is by leaching from the subsurface bedrock into groundwater that eventually flows into the lake. Another is by tributaries eroding soil and carrying the sediment as well as material from the forest floor into the lake. Animal waste in or around the lake can contribute to the inflow of phosphorus and nitrogen as can human activity. Trash or other waste can enter the lake directly, while hiking or other activities can cause more erosion of the soil which enters the tributaries and ultimately the lake. The human acceleration of the eutrophication process is commonly called cultural eutrophication [1].

Phosphorus concentrations and algae levels in a lake can fluctuate markedly over time. Comparison of short-term results to annual averages may give insight as to a lake's response to nutrients and how longer-term eutrophication may be progressing. During the growing season or summer, growth is nurtured by a rise in water temperature. Phosphorus is used by algae and is typically depleted significantly during summer months. Phosphorus concentrations and algae may be increased by storms, which cause mixing and contribute phosphorus loads from eroded soils in the watershed. NALMS [3] states some of this phosphorus and nitrogen is available to algae in the photic zone and triggers algae growth.

For this study, one focus was the correlation of trophic state and algae growth. Algae in lakes can be seen and therefore is often of concern by lake users. A national survey in 1998 showed that 44% of all U.S. lakes contained excessive nutrients that contribute to aquatic growth, especially algae. Data and information on algae growth can help to explain short-term variations in lake water nutrient levels and hence trophic state [3].

3 Trophic state indicators

Four widely used trophic state indicators are the Secchi disk transparency, chlorophyll-a levels, total phosphorus and nitrogen. Typically these parameters are all measured and compared to provide valuable information about the relationship among water nutrients and algal growth. The parameters together provide a more complete picture, than if one or more is not included. US EPA [4] guidelines encourage using all four of these indicators.

3.1 Transparency

Secchi disk transparency measures water clarity. Transparency reduces as algal populations grow. The Secchi disk is the most commonly used parameter in monitoring programs. Microorganisms other than algae, suspended sediments, and other dissolved materials that color the water can effect Secchi disk measurements. In some shallow lakes, a Secchi disk reading will be inaccurate if the disk reaches the bottom before disappearing from site [4].

3.2 Chlorophyll-a

Chlorophyll-a, contained in most algae, is a photosynthetic pigment in the algal cells. Different algal species produce differing amounts of chlorophyll-a. Two lakes can have the same amount of chlorophyll-a, but one have more algae because of a difference in algae species. Chlorophyll-a increases during the day and decreases at night, and is changed significantly by seasons. The exact amount of algae cannot be determined from the amount of chlorophyll-a, but it is still one of the best "simple" indicators of algal biomass [4].

3.3 Phosphorus

Phosphorus is a measurement of water fertility, is rapidly used by algae, and hence is often found in low concentrations. Other forms of phosphorus are more dynamic and convert over time to orthophosphate. Lakes are tested for total phosphorus [4]. For eutrophication analysis, total phosphorus is generally the single most important nutrient to determine when lakes are phosphorus limited [3]. Because phosphorus tends to remain bound to soil particles, drainage from eroded soils carries higher amounts of phosphorus than from vegetated soils. Similarly, surface runoff yields higher concentrations of total phosphorus than infiltration water according to Harper [5]. Particulate phosphorus may be deposited in bottom sediments, where much of it is acted upon by micro-bacteria and ultimately converted into dissolved orthophosphate according to Correll [6].

4 Sampling methods

Samples were taken in 1 liter bottles and fixed with sulfuric acid. Inlet and in-lake samples were analyzed for total phosphorus, and in-lake samples were also analyzed for chlorophyll-a and nitrogen. Sampling was in one or more locations at each lake, and also in the inlets and outlets, and was consistent from lake to lake for comparison. Sampling was every one or two months over the spring to fall seasons when lakes were accessible. Information on field forms included human and animal activity and estimates of the Secchi depth and stream flow.

5 Trophic state models

Three models were used in this study to classify the trophic state of the lakes; the Carlson Trophic State Index Model, the Vollenweider Model, and the Larsen-Mercier Model. Simple models are used where steady-state conditions and lake homogeneity are assumed. Furthermore, the models are spatially and temporally averaged. These assumptions are reasonable for the lakes in this study because these lakes are generally small and well mixed. Models with these assumptions are commonly used because they still make useful predictions [3].

5.1 Carlson trophic state index

Carlson's Model utilizes the measurements of three water quality characteristics to graphically estimate an average trophic state. See figure 1 as an example. The

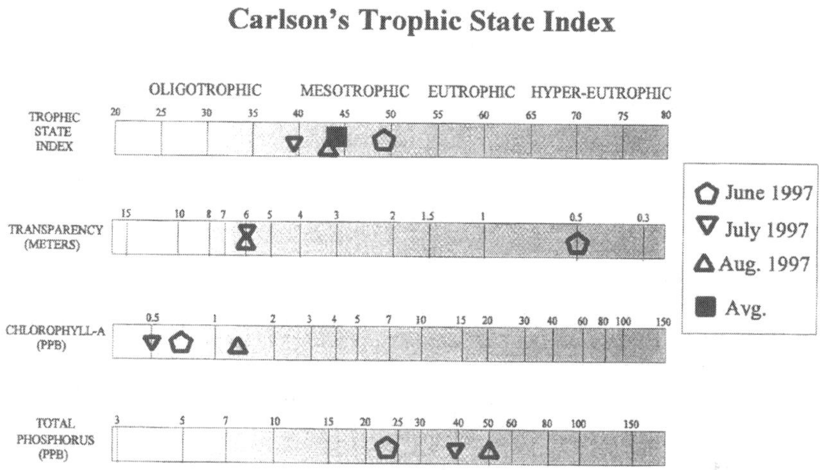

Figure 1: Example of the Carlson model.

Table 1: General trophic states and their sub-states.

Oligotrophic			Mesotrophic			Eutrophic		
Strong Oligo	Oligo	Slight Oligo	Slight Meso	Meso	Strong Meso	Slight Eutro	Eutro	Strong Eutro
< 26	26-33	33-38	38-43	43-49	49-54	54-58	58-62	> 62

three measurements are transparency, chlorophyll-a, and total phosphorus concentration in the lake. Each measured value is plotted on a logarithmic scale specific to the measured parameter. Carlson [7] says that based upon the value's location, an average can be found that describes the trophic state. Carlson Trophic State Index (TSI) is the most widely used trophic state indicator and is commonly used in lake protection and restoration [3]. Sampling measurements are applied to the following TSI equations to determine TSI values [3]:

TSI (Secchi disk) = 60 - 14.41*ln (depth in meters)

TSI (chlorophyll-a) = 30.6 + 9.81*ln (concentration in ug/L)

TSI (total phosphorus) = 4.15 + 14.42*ln (concentration in ug/L)

Usually TSI values are classified as follows: mesotrophic at 38 <= TSI <= 54, eutrophic at TSI > 54 and oligotrophic at TSI < 38 [4]. We subdivided the TSI values for a finer classification of the trophic states, as shown in table 1. The Carlson phosphorus TSI is generally valid for phosphorus limited lakes. TSI results for the individual parameters may not be close to the same values if unique conditions exist. For example, a growth of zooplankton may suppress the algae, lowering the chlorophyll-a, while the TSI for phosphorus remains high.

5.2 Vollenweider model

The Vollenweider Model utilizes measurements of total inflowing phosphorus concentration plotted against the hydraulic residence time of the lake. The

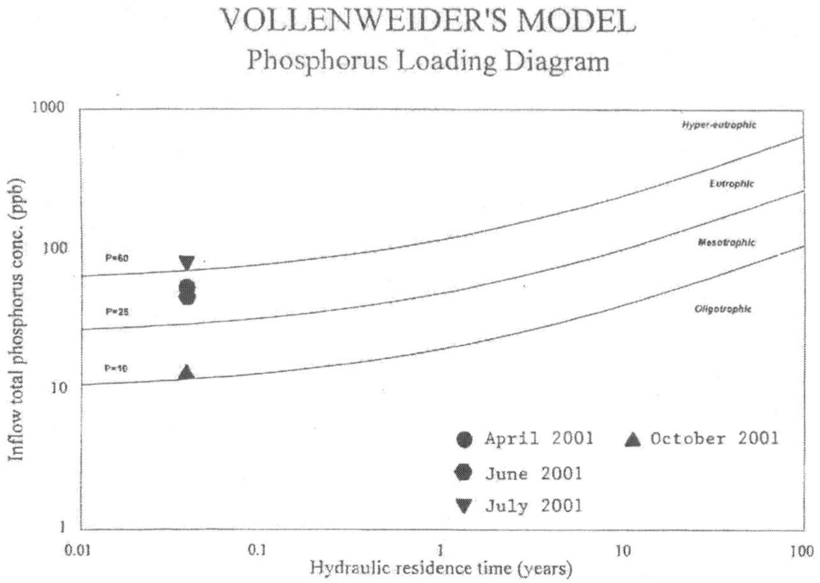

Figure 2: Example of the Vollenweider model.

hydraulic residence time is the average outflow divided into the volume of the lake. Vollenweider [8] states that the location of the values on the graph determines the trophic state of the lake. See Figure 2 as an example.

5.3 Larsen-Mercier model

The Larsen-Mercier Model [9] uses measurements of the total inflowing phosphorus concentration plotted against the phosphorus retention coefficient.

See figure 3 as an example. This phosphorus retention coefficient is calculated by the following equation:

$$\text{Phosphorus Retention Coefficient} = \frac{P_{IN}Q_{IN} - P_{OUT}Q_{OUT}}{P_{IN}Q_{IN}}$$

where: P_{IN} = total phosphorus concentration in the inflow
Q_{IN} = flow rate of the incoming water
P_{OUT} = total phosphorus concentration in the outflow
Q_{OUT} = flow rate of the outgoing water

In this model, sometimes the average phosphorus concentration in the lake water was used for the outflow phosphorus concentration. This is an appropriate assumption for small, well-mixed lakes that typically have small hydraulic residence times, but may not be for larger lakes with longer residence times.

6 Large lakes in south Yellowstone Park

This paper presents the trophic state results for large lakes in south Yellowstone Park, namely, Shoshone Lake, Lewis Lake and Heart Lake as shown on figure 4.

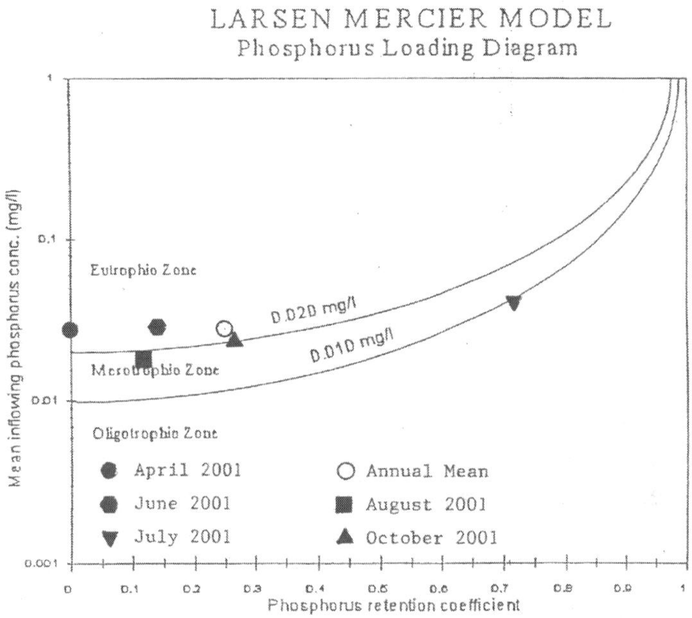

Figure 3: Example of the Larsen-Mercier model.

6.1 Shoshone Lake

Shoshone Lake is approximately 6 miles (10 km) southwest of Yellowstone Lake (see figure 4). The elevation is 7791 ft (2375 m) and the depth is 200 ft (61 m). The surface area is approximately 8100 acres (32.8 km^2) and the approximate volume is 778,000 acre-ft (0.96 km^3). Shoshone Lake has three main tributaries; DeLacy, Shoshone and Moose Creeks. Lewis River is the only outlet of the lake.

Water samples were taken from the lake at three different locations; from the northeast side of the lake near DeLacy Creek, from the west side near Shoshone Creek, and from the southeast side near the Lewis River outlet. The results of the model analyses over the summer months show that Shoshone Lake is mesotrophic at DeLacy Creek and slightly oligotrophic at both Shoshone Creek and near the outlet. The lake as a whole is classified as slightly mesotrophic (see table 2). DeLacy Creek and the lake near DeLacy Creek are a concern because of their high phosphorus concentration and trophic state, respectively.

6.2 Lewis Lake

Lewis Lake, at elevation of 7,779 ft (2371 m), is about 3 miles (5 km) southeast of Shoshone Lake (see figure 4). It has a surface area of 2,700 acres (10.9 km^2) and an approximate volume of 121,000 acre-ft (0.15 km^3). The depth is 100 ft (30 m) at the center, but in the western quarter the depth is no more than 10 ft (3 m).

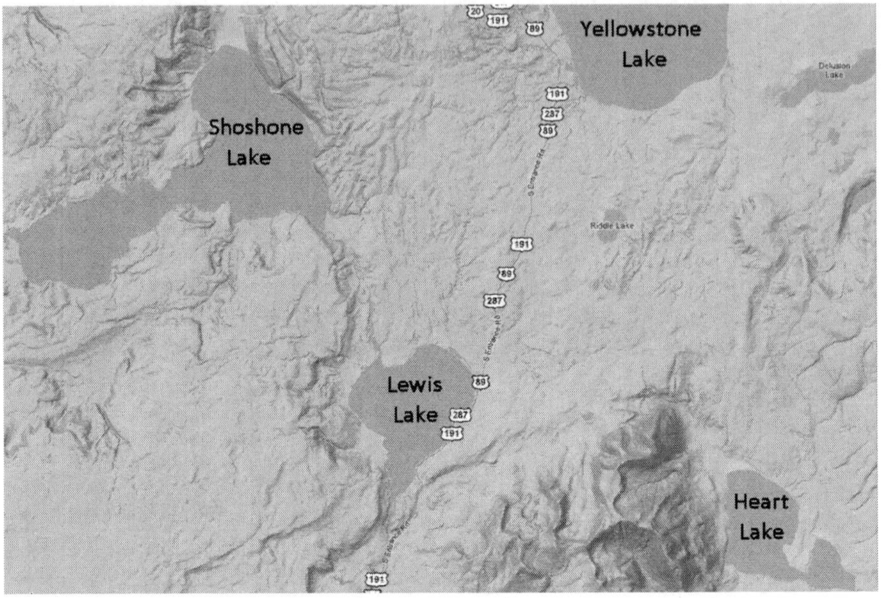

Figure 4: Map of large lakes in south Yellowstone Park.

Table 2: Large lakes' trophic states according to the three models.

Model	Classification
Shoshone Lake	
Carlson	Slightly Mesotrophic
Vollenweider	Slightly Mesotrophic
Larsen-Mercier	Mesotrophic
Shoshone Lake Average	*Slightly Mesotrophic*
Lewis Lake	
Carlson	Slightly Mesotrophic
Vollenweider	Slightly Mesotrophic
Larsen-Mercier	Slightly Mesotrophic
Lewis Lake Average	*Slightly Mesotrophic*
Heart Lake	
Carlson	Slightly Oligotrophic
Vollenweider	Slightly Oligotrophic
Larsen-Mercier	Mesotrophic
Heart Lake Average	*Slightly Mesotrophic*

The main inlets to the lake are Lewis River and Dogshead Creek. The only outlet is the Lewis River at the south end of the lake.

Sampling results show that Lewis Lake is mesotrophic in the northeast at Dogshead Creek and slightly mesotrophic in both the south and in the northwest near the inlet. The lake as a whole is classified as slightly mesotrophic (table 2).

6.3 Heart Lake

Heart Lake is located about 6 miles (10 km) to the southeast of Lewis Lake (see figure 4). The lake is at elevation 7,450 ft (2271 m), has a surface area of about 2,500 acres (10.1 km^2), has an approximate volume of 125,000 acre-ft (0.15 km^3) and a maximum depth of 180 ft (55 m). Heart Lake's eastern bay is very shallow with depths less than 10 ft (3 m). The major inflowing streams are Witch and Beaver Creeks and the only outlet is the Heart River.

The results of sampling near Witch Creek and on the southwest side show Heart Lake as slightly mesotrophic, but slightly oligotrophic on the northwest side (see table 2). Although the lake is classified as slightly mesotrophic, if samples were taken on the east side, there may be a different classification.

7 Mid-sized lakes in central Yellowstone Park

Water samples were also taken at 4 lakes in central Yellowstone Park, namely, Cascade, Grebe, Ice and Wolf Lakes. Model evaluation led to the trophic classification or trophic state of each lake. The trophic results from each model and the overall averages are given in table 3. The lakes are shown on figure 5.

Table 3: Mid-sized lakes' trophic states according to the three models.

Model	Classification
Cascade Lake	
Carlson	Strongly Mesotrophic
Vollenweider	Mesotrophic
Larsen-Mercier	Strongly Mesotrophic
Cascade Lake Average	*Strongly Mesotrophic*
Grebe Lake	
Carlson	Mesotrophic
Vollenweider	Slightly Mesotrophic
Larsen-Mercier	Strongly Mesotrophic
Grebe Lake Average	*Mesotrophic*
Ice Lake	
Carlson	Slightly Mesotrophic
Vollenweider	Mesotrophic
Larsen-Mercier	Slightly Eutrophic
Ice Lake Average	*Mesotrophic*
Wolf Lake	
Carlson	Mesotrophic
Vollenweider	Mesotrophic
Larsen-Mercier	Slightly Eutrophic
Wolf Lake Average	*Strongly Mesotrophic*

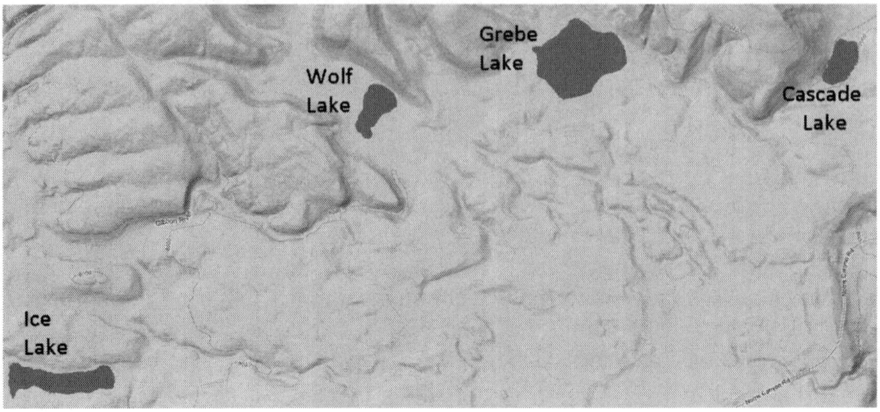

Figure 5: Map of mid-sized lakes in central Yellowstone Park.

7.1 Summary

Grebe Lake, as mesotrophic, appears to be in good trophic condition. Ice Lake also appears to be in good condition, based mainly on the Carlson Model and minimizing the Larsen-Mercier results. There is some concern for Cascade and Wolf Lakes as strongly mesotrophic. While trophic states varied from month to month, the water quality of the lakes, as measured by the trophic state, is generally good. Total nitrogen was also measured in these lakes and while the nitrogen concentrations were near or below the detection limit, the lakes appear to be on the border of being nitrogen limited or phosphorus limited. It will be very useful to compare these results with future studies in order to determine the effects of the significant human interaction on these lakes.

8 Small lakes in northwestern Yellowstone Park

Finally, this paper presents the results for other small lakes in northwestern Yellowstone Park, namely, South Nymph Lake, Nymph Lake, South Twin Lake, North Twin Lake, Lake of the Woods, Beaver Lake, Swan Lake, Blacktail Pond, Harlequin Lake and Goose Lake (see figures 6 and 7). The trophic state of each lake from each model and the overall average classifications are given in table 4.

8.1 Summary

Nymph Lake, Lake of the Woods and Beaver Lake were classified strongly mesotrophic because of high in-lake total phosphorus, and should be of concern. The eutrophication of these three lakes is less severe if they are not phosphorus-limited. However, Beaver Lake has almost gone away. Classifications of Lake of the Woods and Goose Lake were determined from only one sample, so they may need further study to determine if there is concern. The trophic states of the remainder of the lakes are based on several samples and are not of concern.

Table 4: Small west lakes' trophic states according to each of the three models.

Model	Classification
South Nymph Lake	
Carlson	Slightly Mesotrophic
Vollenweider	Slightly Eutrophic
Larsen-Mercier	Slightly Eutrophic
South Nymph Lake Average	*Mesotrophic*
Nymph Lake	
Carlson	Slightly Eutrophic
Vollenweider	Mesotrophic
Larsen-Mercier	Strongly Mesotrophic
Nymph Lake Average	*Strongly Mesotrophic*
South Twin Lake	
Carlson	Slightly Oligotrophic
Vollenweider	Slightly Oligotrophic)
Larsen-Mercier	Slightly Mesotrophic
South Twin Lake Average	*Slightly Mesotrophic*
North Twin Lake	
Carlson	Slightly Mesotrophic
Vollenweider	Strongly Eutrophic
Larsen-Mercier	Slightly Mesotrophic
North Twin Lake Average	*Mesotrophic*
Lake of the Woods	
Carlson	Strongly Mesotrophic
Lake of the Woods Average	*Strongly Mesotrophic*
Beaver Lake	
Carlson	Slightly Mesotrophic
Vollenweider	Eutrophic
Larsen-Mercier	Eutrophic
Beaver Lake Average	*Strongly Mesotrophic*
Swan Lake	
Carlson	Mesotrophic
Swan Lake Average	*Mesotrophic*
Blacktail Pond	
Carlson	Slightly Mesotrophic
Vollenweider	Strongly Mesotrophic
Larsen-Mercier	Slightly Eutrophic
Blacktail Pond Average	*Mesotrophic*
Harlequin Lake (south of maps)	
Carlson	Slightly Mesotrophic
Harlequin Lake Average	*Slightly Mesotrophic*
Goose Lake (south of maps)	
Carlson	Oligotrophic
Vollenweider	Strongly Mesotrophic
Larsen-Mercier	Strongly Mesotrophic
Goose Lake Average	*Slightly Mesotrophic*

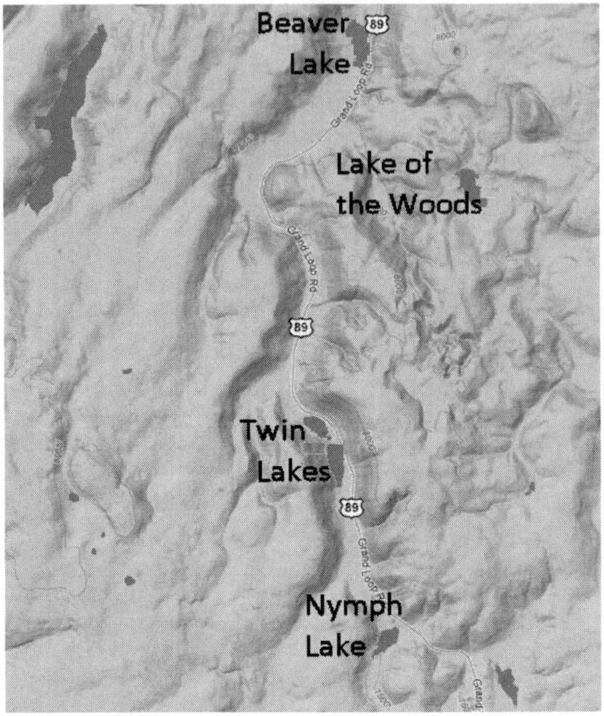

Figure 6: Map of small lakes in west Yellowstone Park.

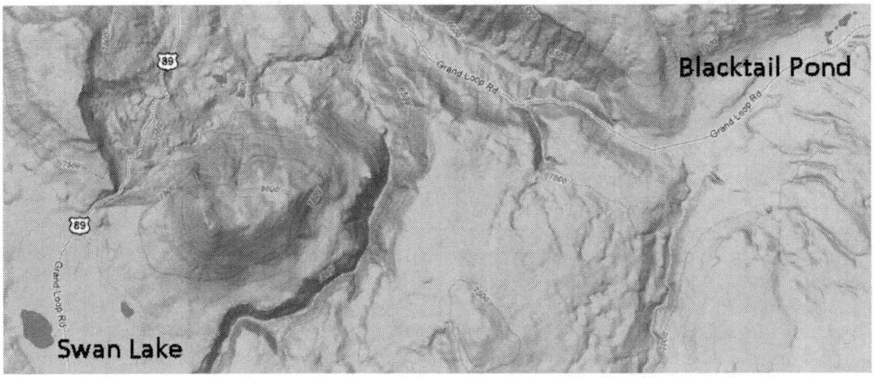

Figure 7: Map of small lakes in north Yellowstone Park.

References

[1] Chapra, S.C., *Surface Water-Quality Modeling*, McGraw-Hill, New York, 1997.
[2] Tchobanoglous, G. & Schroeder, E., *Water Quality*, Addison-Wesley, 1985.

[3] North American Lake Management Society (NALMS), Managing lakes and reservoirs, *Terrene Institute*, 2001.
[4] Environmental Protection Agency, Volunteer lake monitoring, JT&A, 1991.
[5] Harper, D., *Eutrophication of Freshwater*, Chapman and Hall, New York, pp. 36-53, 1992.
[6] Correll, D.L., The role of phosphorus in the eutrophication of receiving waters, *Journal of Environmental Quality,* 27(2), pp. 261-266, 1998.
[7] Carlson, R.E., Trophic state index for lakes, *Limnology and Oceanography,* 22(2), pp. 361-369, 1977.
[8] Vollenweider, R.A., Fundamentals of the eutrophication of lakes and flowing water, with reference to N and P as factors in eutrophication, *Technical Report OECH*, Paris, France, DAS/CSI/68.27, 1968.
[9] Larsen, D.P. & Mercier, H.T. Mercier, Lake phosphorus loading graphs: an alternative, *Working Paper No. 174*, Pacific Northwest Environmental Research Lab, EPA, Corvallis, Oregon, 1975.

The tale of two lakes: managing lake degradation, Rotorua lakes, New Zealand

P. Scholes[1] & J. McIntosh[2]
[1]*Environment Bay of Plenty, New Zealand*
[2]*Lochmoigh Limited, New Zealand*

Abstract

Lakes Rotorua and Rotoiti, two co-joined lakes of the Rotorua Group, North Island, New Zealand, have had an increasing prevalence of cyanobacterial blooms due to lake eutrophication. Targets for water quality restoration are set in statutory planning objectives expressed quantitatively by trophic level indices. Quantification of external and internal nutrient loads underpins the restoration targets. Comprehensive stream and groundwater monitoring, together with a monthly lake monitoring programme, provide data for models that delineate nitrogen inputs from sub-catchments to lake ecological modelling. Lake Rotorua catchment is dominated by pastoral land use and receives significant nitrogen (N) and phosphorous (P) load from groundwater, as well as geothermal springs. Increasing groundwater nitrogen (predominantly as nitrate) and internal phosphorus loads need to be reduced to meet water quality objectives. Reduction of the external nutrient loads to Lake Rotorua has begun by addressing several sources. These include: the Rotorua City wastewater treatment plant upgrade; stream dosing with aluminium sulphate; areas with on-site wastewater treatment reticulated to the central wastewater treatment plant; and stormwater upgrades. Hydrological and limnological studies have found nutrient inputs to Lake Rotoiti to be dominated by the inflow from Lake Rotorua, which accounts for approximately 73% N and 76% P. A diversion wall has been installed to interrupt the underflow current from Lake Rotorua to Lake Rotoiti with Lake Rotorua waters deflected straight to the Lake Rotoiti outlet. The diversion wall has resulted in significant water quality gains for Lake Rotoiti. In this paper we explore strategies to improve lake water quality as part of an integrated land and water management programme, and how this led to the building of a diversion wall. We examine the steps taken to implement restoration works and explore

WIT Transactions on Ecology and the Environment, Vol 135, © 2010 WIT Press
www.witpress.com, ISSN 1743-3541 (on-line)
doi:10.2495/WP100141

research into sediment capping, dredging and chemical treatments as some of the remediation options to address Lake Rotorua's internal nutrients.

Keywords: action plans, trophic level index, eutrophication.

1 Introduction

Lakes Rotorua and Rotoiti form the second largest lake complex in the North Island of New Zealand. Lying in an area known as the Taupo Volcanic Zone, the lakes were formed from two separate eruption events. Lake Rotorua was formed from an ignimbrite caldera over 200,000 years ago [1] and is a shallow lake with a maximum depth of 45 metres, an average depth of 11 meters, and a surface area of 79.8 km^2. The more recently formed Lake Rotoiti was created in the Haroharo caldera by a lava dam around 8,500 years ago. Lake Rotoiti has a surface area of 33.5 km^2 with a maximum depth of 125 metres, and an average depth of 31.5 metres.

The 508 km^2 catchment of Lake Rotorua is dominated by pastoral land use (52%) followed by indigenous vegetation (25%) and exotic forest (14%). Lake Rotoiti's 124 km^2 catchment has less pastoral land use (16%) and is predominantly exotic forest (46%) and indigenous forest (36%) [13]. Up until the 1870s most of the catchments were covered by indigenous forest, which was subsequently clear felled. Some areas were established with *Pinus radiata* plantation forests, with sheep and beef and dairy farming remaining small scale until the 1940s, when mineral deficiencies were corrected with mineral amended fertilizers. Urban areas expanded, particularly around the lake fringes, and until 1991 treated wastewater was discharged to Lake Rotorua, but is now irrigated in forest plantations.

Lake Rotorua is polymictic with an annual mean total nitrogen (TN) and total phosphorus (TP) concentrations of 0.408 mg L^{-1} and 0.034 mg L^{-1} (2008/2009) [13], respectively. There are nine major inflows contributing over 288 t yr^{-1} TN and 21.6 t yr^{-1} TP as well as 131 t yr^{-1} TN and 8. t yr^{-1} TP from minor streams, springs and groundwater [3].

Stratifying from October to May, monomictic Lake Rotoiti's major inflow is from the Ohau Channel (fig 1) carrying around 259 t yr^{-1} TN and 31.1 t yr^{-1} TP. An estimated 73% TN and 76% TP was entering the lake, the surplus carried down the Kaituna River at the lake outlet (fig. 1). Minor inflows contribute approximately 110 t yr^{-1} TN and 4 t yr^{-1} TP, although recent monitoring shows this may be higher. Much of the nitrogen comes from elevated ammonium (NH_4-N) concentrations in geothermal inflows.

Degradation of Lakes Rotorua and Rotoiti were recognised as occurring since the early 1970s. Intensification of pastoral farming systems and increasing urban populations have exacerbated eutrophication problems amplifying calls for further action to arrest lake water quality deterioration.

Local government bodies together with the local Maori indigenous tribe (iwi), Te Arawa, and local stakeholders have instigated an integrated programme of land and water management, the Rotorua Lakes Restoration Programme. This paper examines the processes and results the Restoration Programme has had on

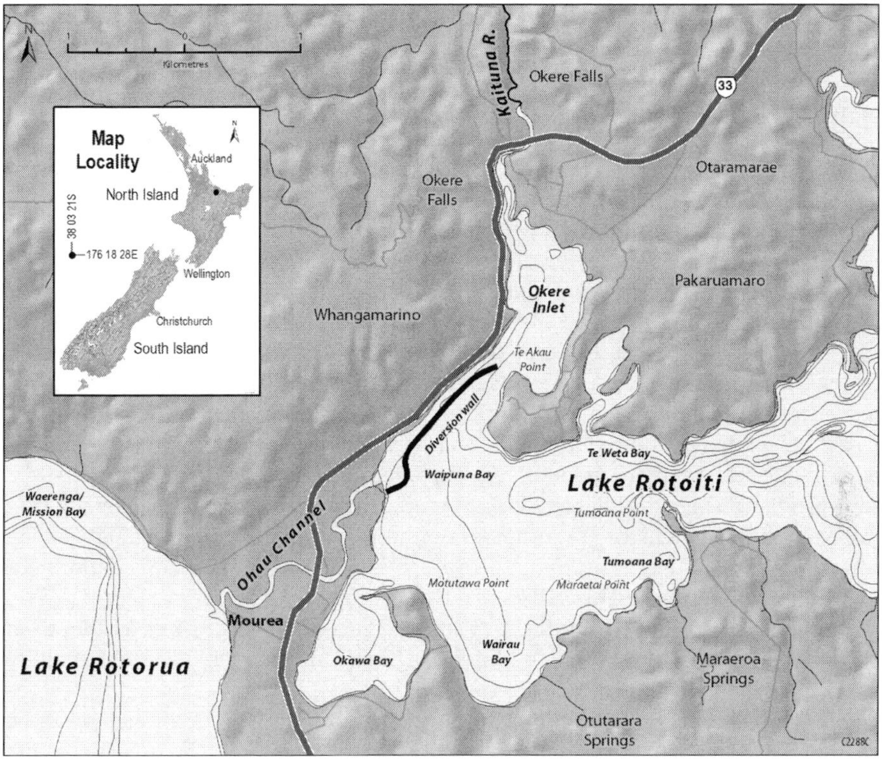

Figure 1: Map of Lakes Rotorua and Rotoiti showing the location of the Ohau Channel, Kaituna River and the diversion wall.

Lakes Rotorua and Rotoiti and focuses on recent remediation works for the lakes.

2 A plan of action

Environment Bay of Plenty Regional Council (EnvBOP) and Rotorua District Council (RDC) are the two local government agencies charged with implementing the principles of the Resource Management Act (1991, RMA), the overarching legislation guiding the sustainable management of natural and physical resources in New Zealand. The RMA charges Regional Councils with developing regional plans which describe issues, and set objectives, methods and rules for sustainable water and land management. The Regional Water and Land Plan (W&LP) has its objectives set by the community in a consultative process and these objectives provide the structure for the Rotorua Lakes Restoration Programme.

Each of the twelve Rotorua lakes has a trophic level index (TLI) set as an objective in the W&LP which forms the basis for developing nutrient reduction targets to protect and restore water quality. EnvBOP calculates a TLI for each lake annually using the annual average of four key parameters: TN, TP, chlorophyll a (Chla), and Secchi depth (SD) using the protocols described by Burns *et al.* (2000) [2].

Lakes that don't meet their TLI objective trigger a public process for development of a strategy to arrest nutrient pollution and return the lake to its objective TLI. Strategies formulated in consultation with the wider community and stakeholders, help the community to gain an understanding of basic limnology and issues contributing to lake eutrophication.

Focus groups have been developed with wide community representation and have been challenged to find solutions to the many issues giving rise to lake eutrophication. These groups identify gaps in the science and decision making and feedback to an Action Plan Working Party made up of key stakeholders which is overseen by a Rotorua Lakes Strategy Group. The Working Party develops recommended actions to achieve the long term water quality objectives based on guidance and recommendations of focus groups and under the advice from scientists brought together as Technical Advisory Groups (TAGs). An Action Plan is then formulated which synthesises the information gained from the councils, iwi, scientific and local communities; identifies and quantifies the source and magnitude of nutrient inputs; and specifies key recommendations for improving lake water quality.

In defining nutrient reduction targets at sustainable levels of nutrient input to maintain the lake's W&LP TLI objective, N and P loads have been obtained by various methods. Initially N and P inputs are estimated based on land use area and land use loss coefficients (kg N,P ha^{-1} yr^{-1}). Nutrient loss coefficients can be used to develop annual mass balance models which in turn can be used to predict the effect of external load changes on lake nutrient concentrations [12]. Hoare [9] used the Vollenweider [15] model related to the outflow concentration, C (also assumed to be the lake annual average concentration, mg m^{-3}), the annual nutrient load, M (mg yr^{-1}), and the annual outflow volume, Q (m^3 yr^{-1}); where the retention coefficient, R (dimensionless), is the proportion of the external load retained within the lake; which is rearranged to estimate nutrient load:

$$M = CQ/ (1\text{-}R) \tag{1}$$

Nurnberg [11] has shown R to be inversely related to hydraulic loading Q/A, where A is the lake surface area. For Lake Rotorua 'R' is between 0.6 and 0.7 [12].

Defining nutrient targets and strategic actions in an Action Plan has paved the way for funding to be allocated from central government to boost local government funding for the implementation of actions.

3 Implementing actions

3.1 Lake Rotoiti diversion wall

The Rotorua-Rotoiti Action Plan states that to meet the Lake Rotoiti TLI objective of 3.5 TLI units a nitrogen reduction of 130 t yr^{-1} and 19 t yr^{-1} of phosphorous is required [3]. Throughout the Action Plan development phase several options were examined to help meet this nutrient reduction target concentrating on the Ohau Channel as the single largest source of nutrients and water to Lake Rotoiti.

Piping the flow directly to the Kaituna River (the natural outflow of Lake Rotoiti) was initially looked at but was discounted because of cost and cultural reasons.

Based on a study of particulate nutrient levels on the shallow shelf of Lake Rotorua before entry to the Ohau Channel by Gibbs [4], it was postulated that extending groynes from either side of the channel entrance out into the lake could reduce the amount of particulate material carried down the Ohau Channel. EnvBOP commissioned NIWA to study the potential nutrient reduction to be gained by erection of such a structure and impacts on lake morphology, circulation and sediment. Numerical modelling of hydrodynamics, waves and sediment transport, water sampling and analysis was part of a multi-faceted approach taken to studying particulate nutrient pathways to the Ohau Channel and the processes that cause their mobilisation [6].

Wave gauges and current meters as well as concentration profiles were measured perpendicular to the shore across the shallow shelf offshore of the Ohau Channel entrance. An analytical sediment resuspension model coupled with a wind-wave generation model (SWAN) and a hydrodynamic flow model (MIKE21) were used to relate wind impacts on wave height and water movement.

A number of sediment transport scenarios were performed using variable wind conditions with and without 200m and 400m long groynes extending into Lake Rotorua. The best modelling scenario estimated a reduction of about 15% of the total volatile (organic) suspended load with the 200m groyne, which relates to a similar reduction in nitrogen and phosphorus. Inorganic sediment was found to be reduced by around 70%, but the groynes would have no impact on the dissolved nutrient load.

Overall the modelling of groynes showed that the potential gains in nutrient reduction may not be large enough to be economically practicable and other options should be explored.

The nature of the connection between Ohau Channel and Lake Rotoiti had been established by intensive monitoring in 1985-86. Warmer summer inflows are less dense than lake water and remain on the surface as discrete overflow, mostly short-circuiting to the Kaituna River. Winter and spring inflows are colder and denser than lake water. These flow eastward initially as an underflow current travelling along the bottom of the lake to the deeper eastern basin. Due to seasonal temperature differences, the underflow can insert as an intrusion layer driven by the hydraulic inertia of the density current [6]. Diurnal variation in temperature can also result in the density current switching from an underflow to an interflow.

Limnological and water quality surveys indicated that greater than 70% of the N and P flowing into Lake Rotoiti comes via the Ohau Channel [6]. Inflow from the Ohau Channel enters at an average 15 $m^3 s^{-1}$ but reaches up to 70 $m^3 s^{-1}$ as it entrains lake water distributing nutrients throughout the lake with the winter mixing event.

The proximity of the Kaituna River outlet to the Ohau Channel inflow lends itself to installation of a diversion structure to eliminate nutrient input to Lake Rotoiti. Gibbs *et al.* [6] estimated a diversion could reduce annual N and P loads by 181 tonnes and 25 tonnes respectively. A diversion while reducing nutrients and algal inoculum would reduce oxygen delivery and the lakes flushing rate. Average residence time is predicted to increase from 2.4 years to 6.3 years which could benefit water quality with increased sedimentation and denitrification.

Based on the limnological information gathered and boosted by biofish (a depth changing conductivity-temperature datalogger) monitoring of Lakes Rotorua and Rotoiti, a collaborative effort by NIWA and the University of Waikato, modelled potential diversion scenarios to assess favourable options and ecological impacts. 3-dimensional hydrodynamic modelling showed the longest diversion option would be required to minimise any losses from a plunging underflow current [14]. Hydrodynamic-ecological modelling (DYRESM-CAEDYM model) indicated limited change in temperature, dissolved oxygen, nutrients and diatom biomass, but up to a 40% reduction in summer cyanobacterial chla within four summers [8].

The diversion wall was completed in July 2008 using sheet-piling over a distance of 1275 metres. Lake water quality had been showing some improvements before completion of the diversion wall partly due to a shift in cyanobacteria species from *Annabena planktonica* and *Microcytis aeruginosa* to *Aphanocapsula holistica* a pico-algae.

Since the completion of the wall water clarity has improved slightly as have nitrogen concentrations, leading to an improved TLI resulting in a change to the lake's classification from eutrophic to mesotrophic. Chla and soluble reactive phosphorous (SRP) concentrations did not improve although hypolimnetic volumetric deoxygenation (HVOD) were at levels below those experienced during intensive cyanobacteria blooms in 2002 to 2006 (fig. 2). Although the

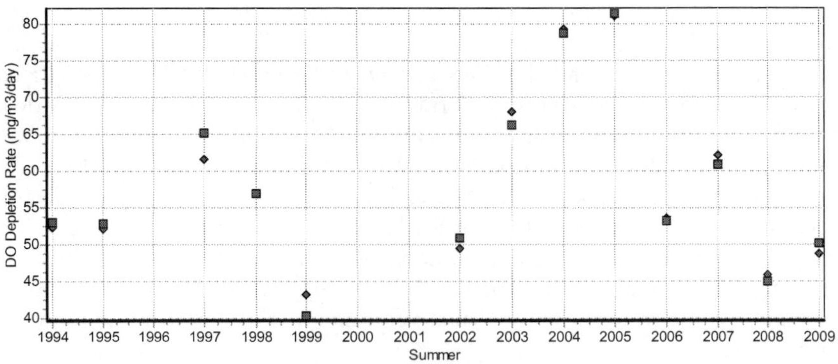

Figure 2: HVOD rates for Lake Rotoiti, 1993 to 2009, squares are HVOD adjusted for standard temperature of 12.3 °C.

HVOD rates are low it remains to be seen if sediment nutrient release during periods of anoxia will lessen.

3.2 Other Lake Rotoiti actions

Upgrading communities from on-site wastewater systems to reticulated centralised sewage treatment has been undertaken for two communities. The community at Okawa Bay has seen the benefits of this action. Okawa Bay is a small shallow embayment on the western side of the lake which underwent anoxic conditions in summer and due to the low exchange rate with the main lake, suffered from severe cyanobacterial blooms.

Removal of an estimated 5.9 tonnes of N and 0.21 tonnes P per annum has within two years resulted in a 25% increase in water clarity. The increase in water clarity has led to a proliferation in the exotic macrophyte *Ceratophyllum demersum*, with the surface reaching macrophyte being harvested for the first time further reducing the nutrient content of the bay.

Other actions to be taken in the Lake Rotoiti catchment will be dependent on the success of the Ohau Channel diversion wall, which meets the bulk of the catchment nutrient reduction targets. Other potential actions include land management, lake oxygenation, and sediment capping.

3.3 Lake Rotorua actions

3.3.1 Inflows

Rutherford [12] estimated the nutrient inputs from baseflow data on the nine major inflows. Particulate P and particulate organic N loads were shown to increase with increasing flow rates, and Hoare [9] showed that baseflow accounted for greater than 90% of the nutrient load. With most of the inflow load attributed to soluble nutrients, Rutherford found that N delivery was increasing with time. Isotopic dating of groundwater inflows [10] showed groundwater ages varying from 15 to 135 years. Increasing nitrate levels can be attributed to increasing agricultural activity following clearance of surrounding lands, and while some catchments will reach equilibrium in 10 to 20 years, it may take up to 100 years for others.

One inflow, the Waiohewa Stream, has a geothermal input at the top of its catchment that has a high concentration of ammonium. A trial nitrification-denitrification plant is underway to test the efficacy of removing the geothermally sourced ammonium. Nitrification will be undertaken using conventional wastewater technology and denitrification will be via a wood bark or maize cob bed, potentially removing 30 t N yr^{-1}.

Addressing the increasing nitrate concentrations is not easily achieved by direct intervention of inflows, hence treatment of inflows has predominantly focused on P. Iron, calcium or aluminium salts have been used effectively in flocculation or precipitation of P. Testing of a range of these salts resulted in aluminium sulphate (alum) being applied to one of the major inflows, the Utuhina Stream, binding an estimated 2 t P yr^{-1} which settles out in the lake.

Many of the inflows to the lake are valued trout fisheries, thus treatment of the stream is closely monitored with fish being examined for potential toxicity effects and macroinvertebrate species measured for abundance and diversity. The success of treatment in this stream has led to the possibility of dosing the Puarenga Stream and one other stream potentially locking a total of 6 t P yr^{-1}.

The greatest P load enters via the Hamurana Stream (6.3 t yr^{-1}) which also delivers 53 t yr^{-1} N, predicted to increase to 92 t yr^{-1} in 50 years time [3]. Treatment is not economically viable or practicable at this time due to the volume of the stream, but diversion to the Ohau Channel has been considered. However, there are potential cultural, aesthetic and ecological problems that would be encountered by a six kilometre diversion wall. Modelling also indicates that the impact of diverting this inflow would have little impact until the internal nutrient releases are addressed (*pers comm.* Professor D. Hamilton, University Of Waikato).

3.3.2 Internal nutrient load

The legacy of land clearance, the discharge of Rotorua's sewage effluent, rising nutrient levels in inflows, and sedimentation of algal biomass, result in an increased sediment oxygen demand. Lake Rotorua stratifies intermittently in the summer during periods of calm weather leading to enhanced conditions for P and N releases. Nutrient releases from the lake bed are the main driver of summer algal productivity in Lake Rotorua.

Burger [1] using benthic chambers and laboratory incubations found high release rates from the lake sediments under variable oxygen conditions, with seasonal means up to 44 mg m^{-2} d^{-1} SRP and 136 mg m^{-2} d^{-1} NH_4-N. Load calculations estimated sediment release rates of SRP and NH_4-N at approximately 98 t N yr^{-1} and 7.4 t P yr^{-1} [3], making it apparent that mitigation of eutrophication will require reducing internal nutrient loads as well as external.

As part of on-going research into remediation of the Rotorua Lakes several sediment capping trials have occurred in two other Rotorua Lakes. Phoslock™, a bentonite clay with the active ingredient lanthium, has been trialled in Lake Okareka. Greater focus has been on sediment capping trials at the smaller Lake Okaro (0.33 km^2, mean depth 12m), which has had successive treatments, the first being alum (0.6 g m^{-3}) followed by repeat applications of a modified zeolite, Z2G1.

Application of the capping medium while the hypolimnion is well oxygenated, precise application techniques, and the best grain size of material to obtain suitable coverage, are some of the lessons learnt [5]. Sequential treatments of Lake Okaro, which include the construction of a wetland to treat the inflowing streams, have resulted in a reduction in the rate of phosphorous release into the hypolimnion over the summer (fig. 3). The continued success of these trials will help determine if use of a sediment capping agent will be applied to Lake Rotorua. Further deliberations and trials of the use of capping agents include economic and ecological aspects of such treatments, and further understanding of their longevity.

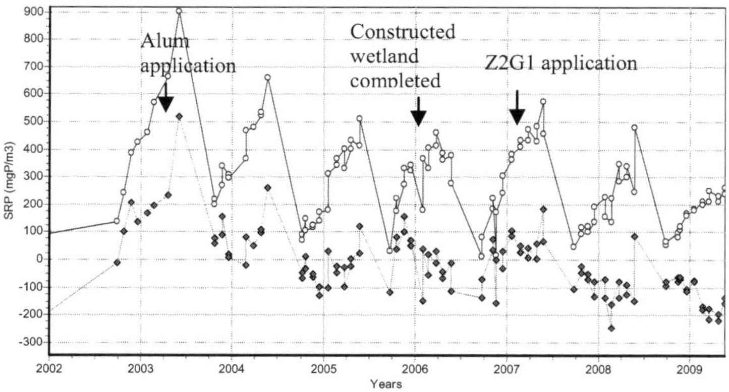

Figure 3: SRP concentration (mg m^{-3}) and deseasonalised residuals (dotted line) in Lake Okaro hypolimnion, 2002 to 2009.

Sediment cores from Lake Rotorua were tested by continuous-flow incubation with four capping agents (alum, Phoslock™, Z2G1, and allophane), at a range concentrations [7]. Results indicated capping agents if left undisturbed should continue to adsorb phosphorous for four years. The eutrophic polymictic Lake Rotoehu is being considered for a trial of a capping agent as it has similarities with Lake Rotorua.

Also under consideration are dredging, aeration and ploughing the lakebed. An assessment of dredging to remove the nutrient rich sediments indicates it to be economically unviable. One problem of this remediation option revealed by the analysis of the lakes sediments, found that they are laden with heavy metals from geothermal inputs. Hence disposal or reuse the organic rich sediment is problematic.

Aeration and ploughing are currently being investigated and may provide an alternative to sediment capping or as part of a mixed remediation solution.

3.3.3 Sewage sources and stormwater

Treated sewage was diverted from the lake to spray irrigation in a forested section of the Puarenga catchment in the early 1990s. Monitoring of the local streams in this catchment displayed increasing nitrate levels and this finding resulted in improvements being made to the irrigation scheme. Nitrate levels have improved but there are indications that soils may be becoming saturated with respect to phosphorous which will in part be addressed by alum treatment.

An upgrade to the Rotorua Wastewater Treatment Plant has taken place in the form of an additional Bardenpho reactor and methanol injection to increase the carbon to nitrogen ratio to enhance denitrification. These modifications will remove approximately 15 t N yr^{-1} from the ultimate nitrogen load leached from the irrigation site.

RDC management initiatives to reduce nutrient contamination of stormwater could reduce N and P inputs to the lake by 3 t yr^{-1} and 0.5 t yr^{-1} respectively. Initiatives include increased network and gully trap maintenance, street sweeping, comprehensive stormwater consents (constraints on permitted discharge standards), increased industrial regulation, and upgrading of trade waste bylaws.

3.3.4 Land management

Catchment land use has passed a sustainable limit. This statement is based on the TLI target set by the community in the W&LP, which balances social and economic benefits of lake water quality with social and economic use of the wider catchment [3]. Without significant N and P reduction from land use, lake nutrient targets will not be achieved, hence this is an Action Plan priority.

Land management efforts to reduce nutrient and sediment loads to the lake have been in place since the 1970s. The focus has been on riparian protection, erosion control schemes, land retirement and conversion to forestry blocks. These actions have had some success in limiting total P loads, but have been less successful in limiting nitrate and soluble P. Soluble P is at elevated concentrations in some spring fed streams due to extraction from the underlying geology and the length of time it takes for water to flow through the aquifers.

The land management working group settled on nutrient reduction targets from catchment land use of 170 t yr^{-1} and 6 t yr^{-1} N and P respectively. To limit the further increase of nutrients from diffuse agricultural sources regulation has been set in the W&LP primarily aimed at properties over 0.4 ha exporting N loads greater than 10 kg ha^{-1} yr^{-1}. Nutrient loss from properties has been capped at 2001 – 2004 levels. Land has been benchmarked for this period in consultation with the land owner using modelling tools such as Overseer.

A Sustainable Land Use Implementation Board has been established with the objective of helping achieve the agricultural component of nutrient reduction. Nutrient trading schemes, best nutrient management practice, industry initiatives and directives, new technologies, subdivision, economic incentives, and property purchases are all potential mechanisms used to meet this objective.

Modelling has been undertaken to estimate N exports from the Rotorua sub-catchments. The Rotorua and Taupo Nitrogen model (ROTAN) balances surface and ground water using groundwater mean residence time coupled with land use nutrient export predictions. ROTAN can then be used to run land use change scenarios to predict changes in N inputs to the lake.

4 Conclusion

EnvBOP's lake monitoring regime uses the nutrient status of the lakes, hypolimnetic oxygen depletion, together with their macrophyte and algal conditions in determining lake health. Other factors are also taken into consideration such as, land development intensity, invasive aquatic weeds, lake use intensity and the concerns of local residents.

The major achievement that has enabled effective lake management has been the setting of quantitative limits (TLIs) for the trophic state of the lakes in a statutory document. A rule to limit nutrient loss is triggered if the TLI is exceeded and an action plan is required for lake remediation. This policy structure has been enabled by the Resource Management Act (1991).

Two important factors in achieving success in producing and executing proposed and accepted Action Plans are firstly extensive consultation with the people of the Rotorua district and secondly helping these people gain knowledge of basic limnology, particularly the agricultural community. Reduction of inputs from the agricultural sector is essential to meet lake water quality objectives. Many helpful actions from farmers are not mandatory but are carried out voluntarily as their understanding of eutrophication issues grow.

References

[1] Burger, D.F., Hamilton, D.P., Pilditch, C.A. & Gibbs, M.M., *Benthic nutrient fluxes in a eutrophic, polymictic lake*, Hyrobiologia, 584:13-25, 2007.

[2] Burns, N.M., Bryers, G. & Bowman, E., *Protocols for monitoring trophic levels of New Zealand lakes and reservoirs*. Published by NZ Ministry for the Environment, Wellington, 2000.

[3] Environment Bay of Plenty, *Proposed Lakes Rotorua & Rotoiti Action Plan*. Environment Bay of Plenty, Environmental Publication 2007/11, 2007.

[4] Gibbs, M., *Temperature and BOD differentials inshore and offshore in Lake Rotorua near Ohau Channel*, Taupo Research Laboratory, DSIR 14: Client Report No, TP64 to Department of Conservation, 1991.

[5] Gibbs, M., *Lake Okaro re-treatment with Z2G1 in August 2009*, NIWA Client Report: HAM2009-177, 2010.

[6] Gibbs, M., Hawes I. & Stephens, S., *Lake Rotoiti – Ohau Channel: assessment of effects of engineering options on water quality*. NIWA Client Report HAM2003-142, 2003.

[7] Gibbs, M., Bremner, D., van Kooten, M. & Özkundakci, D., *Comparison of four P-inactivation agents on Lake Rotorua sediments*, NIWA Client Report: HAM2008-105, 2008.

[8] Hamilton, D., McBride, C. & Uraoka, T., *Lake Rotoiti fieldwork and modelling to support considerations of Ohau Channel Diversion from Lake Rotoiti*, Centre for Biodiversity and Ecology Research, University of Waikato, 2005.

[9] Hoare, A., *The sensitivity to phosphorous and nitrogen of Lake Rotorua, New Zealand*, Progress in Water Technology 12: 897-904, 1980.

[10] Morgenstern, U., Reeves, R. Daughney, C. & Cameron, S., *Groundwater age, time trends in water chemistry, and future nutrient load in the lakes Rotorua and Okareka area*. Geological & Nuclear Survey Report 2004/17, 2004.

[11] Nurnberg, G.K., *The prediction of internal phosphorous loads in lakes with anoxic hypolimnia,* Limnology Oceanography 29:111-124, 1984.

[12] Rutherford, K., *Lake Rotorua Nutrient Target Loads*, NIWA Client Report HAM2003-155, 2003.

[13] Scholes P., *Rotorua Lakes Water Quality Report 2009*. Environment Bay of Plenty, Environmental Publication 2009/12, 2009.
[14] Stephens, S., *Modelling the diversion of the Ohau Channel outflow in Lake Rotoiti to the Kaituna River*, NIWA Client Report HAM2004-164, 2004.
[15] Vollenweider, R.A., *Advances in defining critical loading levels for phosphorous in lake eutrophication*, Mem. 1si. Ital. Idrobiol. 33: 53-83, 1976.
[16] Vincent, W.F., Gibbs, M.M., and Dryden S.J. *Accelerated eutrophication in a New Zealand lake: Lake Rotoiti, Central North Island.* New Zealand Journal of Marine and Freshwater Research 75:431-440.
[17] White P.A., Cameron S.G., Kilgour G., Mroczek E., Bingall G., Daughney C., & Reeves R.R., *Review of Groundwater in the Lake Rotorua Catchment*, Institute of Geological & Nuclear Sciences Client Report 2004/130, 2004.

Assessment of the ecological status of the shallow Lake Neor (Iran) using a macroinvertebrate community structure

R. M. Nadushan[1], H. Emadi[1], M. R. Fatemi[2] & M. Samanpajuh[1]
[1]*Department of Fisheries, Faculty of Marine Science and Technology, North Branch of Tehran, Islamic Azad University, Iran*
[2]*Department of Marine Biology, Faculty of Marine Science and Technology, Science and Research Branch, Islamic Azad University, Iran*

Abstract

Macroinvertebrates play a key role in freshwater lentic and lotic ecosystems. Lake Neor, located in the northwest of Iran (37° 55′ to 38° 1′ 45), 2700m above sea level, is one of the lakes with little information available regarding the limnological data. Therefore, the present study was planned and conducted during a period of one year focusing on macroinvertebrate fauna and assemblage. Sampling was done monthly at four sites from August 2008 to July 2009. Altogether, a total of 11 species belonging to three phyla, seven classes, 10 orders and 10 families were identified, indicating a low diversity. From these species, eight are recorded for the first time from this lake and from the country, including *Pisidium supinum, Fossaria humilis, Stagnicola catascopium, Quistadrilus multisetosus, Stylodrilus herngianus, Mooreobdella fervida, Podura aquatica, Dasycorixa hybrida.* From these species, *Gammarus fasciatus*, *Pisidium supinum* and *Quistadrilus multisetosus* were dominant, constituting 49%, 43% and 5% of the total benthic fauna respectively. Two predominant species, *Gammarus fasciatus* and *Pisidium supinum*, are indicator species that are sensitive to eutrophic conditions, reflecting the oligo-mesotrophic character of the lake. Dissolved oxygen, temperature, TOM and habitat homogeneity seemed to be the main environmental factors affecting the community structure of macroinvertebrate fauna in this lake. At present, in this shallow water body

WIT Transactions on Ecology and the Environment, Vol 135, © 2010 WIT Press
www.witpress.com, ISSN 1743-3541 (on-line)
doi:10.2495/WP100151

with an average depth of 3 and a maximum of 5 meters, the annual water level fluctuation is around 1 to 2 meters during the period of using water for agricultural purposes. These annually high water level fluctuations would result in an increase in nutrient recycling and nutrient release to the water column, which in turn might affect the trophic status of the lake. Hence, the present situation of the lake, mainly its trophic regime, clearly shows the need for developing a management and monitoring plan.

Keywords: shallow lake, macroinvertebrates, assessment, trophic status, Lake Neor.

1 Introduction

Macroinvertebrates play a key role in freshwater ecosystems in linking primary production, allochthonous input and top predators in freshwater ecosystems. They are usually diverse, are represented by many phyla and have different requirements for feeding, growth and reproduction. This often results in an extremely heterogeneous distribution of the macroinvertebrate community [1]. Composition and abundance of zoobenthos depend on multiple environmental factors (physical or/and chemical) and substratum type [2, 3]. Detritivores prevail in habitats rich in organic matter, e.g. in the vicinity of plants [4]. Bare sand is inhabited by psammophiles [3, 4], while habitats experiencing oxygen deficiencies are dominated by organisms resistant to hypoxia [5]. Plants are extremely important for composition and abundance of bottom fauna, as well as for epiphytic and mining organisms, providing a food source and protection against predators and excessive water movements. Use of macroinvertebrates as biotic indicators has been applied mostly to deep lake ecosystems and may provide a reliable indicator of eutrophication. According to Wetzel [1], their use in shallow lakes is not clear.

Lake Neor is an ancient shallow lake ecosystem situated in a topographically diverse landscape on the western flank of the Baghrudagh mountain range, NW Iran. Lake Neor formed tectonically during the late Eocene epoch. The basin has a cold mountainous climate with an annual precipitation 400 mm. Due to high latitude (2500-3200 meter above sea level), high precipitation, physico-chemical processes, slopes lower than 35%, erosion and sediment transfer has been high (430 $ton.ha^{-1}.year^{-1}$) and during recent decades sediment deposit and irrigation has impacted the lake's trophic state.

Neor Lake is a growing place for rainbow trout introduced to the lake many years ago for sport and commercial fishing; therefore, every year a few thousand fingerlings are released into the lake to support this fishery. These fishes are fed on the bottom dwelling *Gammarus fasciatus* and have a very rapid growth and very peculiar characteristics with pink flesh. The natural surroundings of the lake are also the habitat of birds, such as ruddy shelduck and migrant swallows and are a famous resort for many people who come to the lake during the summer, but the lake lacks an ecological and tourisms management plan to ensure its sustainability.

2 Materials and methods

Lake Neor, NW Iran (48′E, 38′N) is a shallow water body with a mean depth of 3 and a maximum of 5.5 m. It occupies an area of 420 acre and is 2700 m above sea level. There are 13 permanent and seasonal springs originating from the surrounding heights and a few underground springs within the lake. Four

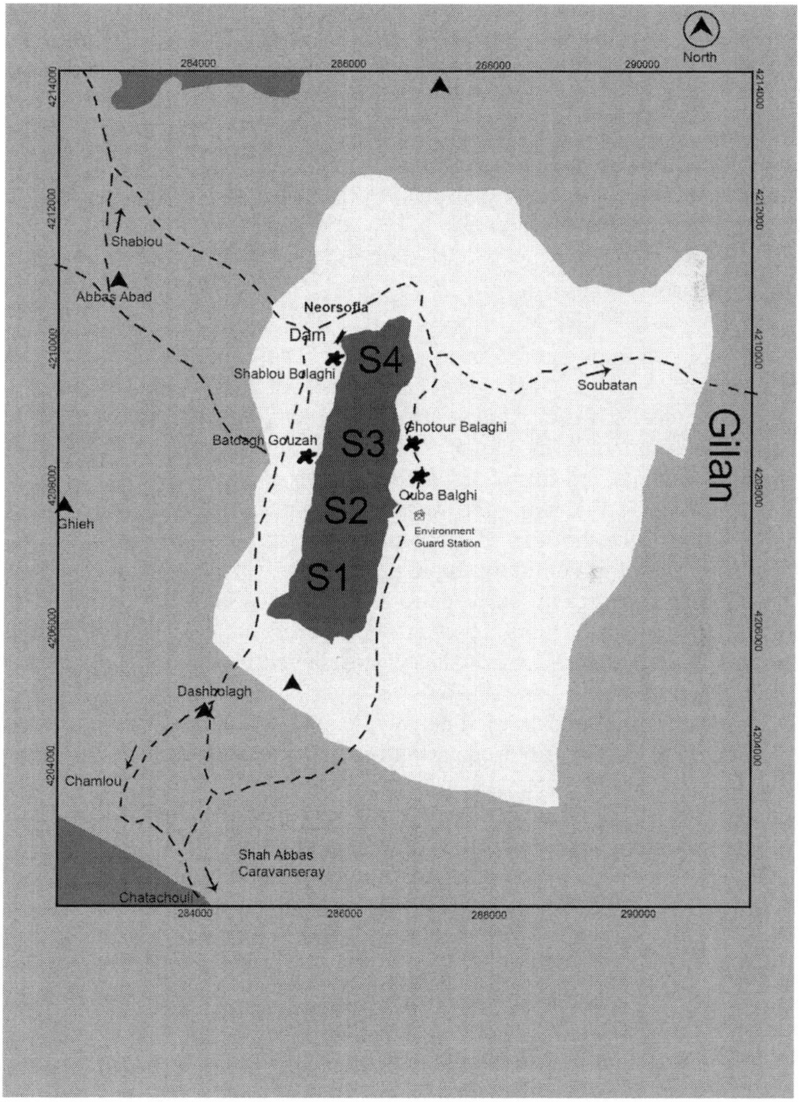

Figure 1: Schematic map of Lake Neor and location of sampling sites (not to scale).

sampling sites were surveyed in Lake Neor. All sites represented the sublittoral (S1 and S4) and the profundal zone (S3, S4). The littoral zone of Lake Neor is subjected to seasonal fluctuations and the water level falls to approximately 1-2 m below the mean level between the winter and summer periods because of irrigation. Regarding this impact, this research focused on the sublittoral and profundal benthos.

Monthly samples (in triplicate) were taken from August 2008 to July 2009 using an Ekman grab sampler (225 cm^2). All samples were sieved (mesh size 0.5 mm) and preserved in 10% formalin. Macroinvertebrates were sorted to the lowest practicable taxonomic level. Water samples were collected using a HydroBios water sampler (2 litre capacity) from the same stations at the surface and bottom zone (0.5m above bottom). Temperature, pH and DO were measured using a WTW meter. Water transparency was estimated with a Secchi disk. Organic matter content in sediments and grain size was determined by heating 1g of dried (24 h at 100 ºC) sediments at 550 ºC for 2 h.

3 Results

Variations, of environmental variables among sites and dates indicate some distinct temporal and spatial trends. The surface water temperatures varied from 0°C in January to 17.7°C in August and bottom temperatures followed the same variation (i.e. from 4°C in January to 16.8°C in August). A weak thermal stratification was noticed from December; the ice period started in January and continued through April, while the water mixing began early in May (sampling in March and April was done after breaking the ice cover) (Figure 2(a)). The lake's mixing regime is continuous cold polymictic.

The dissolved oxygen in surface waters fluctuates between 8.6 mg l-1 and 11.23 mg l-1, and in the bottom waters between 7.9 mg l-1 and 8.75 mg l-1; appearing DO values were always at saturation level throughout the period of no ice cover (Figure 2(b)).

Secchi-disk transparency varied between 0.25 and 0.48 m. The highest values may be related to the low phytoplankton biomass, while there was low seasonal variability throughout the year (Figure 2(c)).

Sediments in all sampling stations were fine with a diameter of less than 0.06 mm. Organic content of sediments is shown in Figure 3.

Altogether 11 species belonging to 3 phyla, 7 classes, 10 orders and 10 families were identified from quantitative samples. Adult *Dasycorixa hybrida* and *Podura aquatic*a were observed in springs inlets and sampling stations under ice respectively.

According to abundance *Gammarus fasciatus, Pisidium supinum* and *Quistadrilus multisetosus* were predominant and they contributed 49%, 43% and 5% of total density (their relative abundance were 49%, 43% and 5%). Relative abundance of *Stylodrilus heringianus* was 1% while the overall abundances of other species were low. Densities of Oligochaete species were maximal during the late spring and of other species were in late summer and early autumn, all with one peak in seasonal pattern.

(a)

(b)

(c)

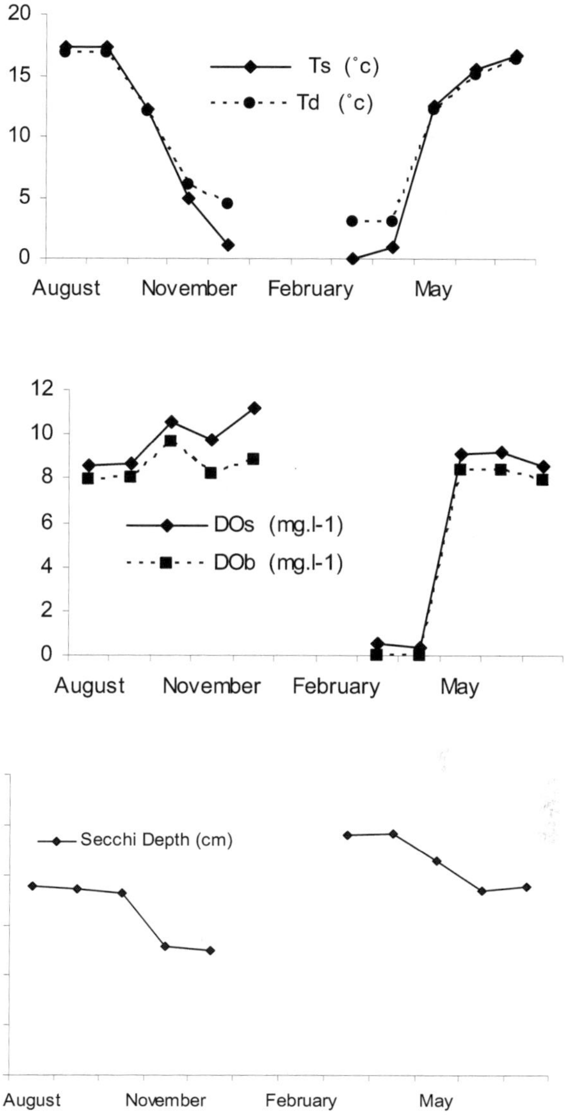

Figure 2: Monthly variation of temperature, dissolved oxygen (s: surface and d: depth near bottom) and Secchi depth of Lake Neor - August 2008 to July 2009.

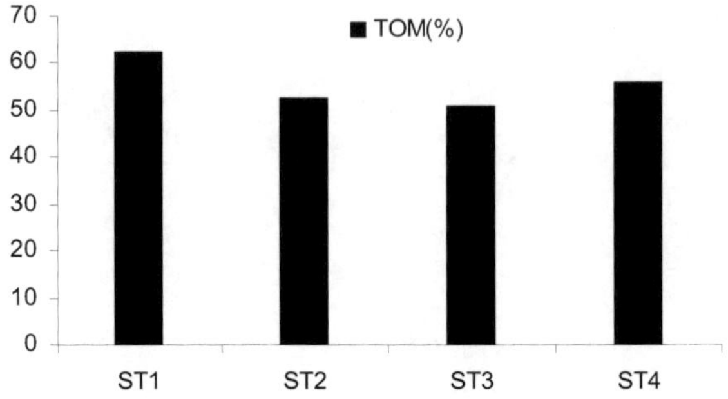

Figure 3: Sediment organic content of Lake Neor at the four sampling stations.

Table 1: Taxonomic composition and classification of the macroinvertebrates of Lake Neor - August 2008 to July 2009.

Phylum	Class	Order	Family	Genus	Species
Arthropoda	Malacostraca	Amphipoda	Gammaridae	*Gammarus*	*G. fasciatus*
	Insecta	Diptera	Chironomidae	*Chironomus*	*Chironomus sp.*
		Hemiptera	Corixidae	*Dasycorixa*	*D. hybrida*
	Entognatha	Collembola	Poduridae	*Podura*	*P. aquatic*
Annelida	Oligochaeta	Tubificida	Tubificidae	*Quistadrilus*	*Q. multisetosus*
		Lumbriculida	Lumbriculidae	*Stylodrilus*	*S. heringianus*
	Hirudinea	Arhynchobdellida	Erpobdellidae	*Mooreobdella*	*M.fervida*
Mollusca	Gastropoda	Pulmonata	Planorbidae	*Planorbis*	*P. carinatus*
		Basommatophora	Lymnaeidae	*Stagnicola*	*S. catascopium*
				Fossaria	*F.humilis*
	Bivalvia	Veneroida	Sphaeriidae	*Pisidium*	*P.supinum*

Stagnicola catascopium and *Fossaria humilis* were not found in samples from station 1 during the study period. Other taxa were observed at all months and all stations. Moreover, no differences in abundance and taxonomic composition (number of taxa) were found between months and the stations in Lake Neor (ANOVA, $P>0.05$).

4 Discussion

Lake Neor, located in northwest of Iran, 2700m above sea level is ice-covered 5 months and during ice free months, circulation is continuous, so this lake is a Continuous Cold Polymictic. In this study it was shown that benthic macrofauna of Lake Neor was dominated by *Gammarus fasciatus* and *Pisidium supinum*. *Quistadrilus multisetosus* and *Stylodrilus heringianus* had 5% and 1% of total density respectively and other organisms were less common.

Gammarus fasciatus is known to be particular abundant in shallow well oxygenated areas, survives well at water temperatures around 10–15°C, generally a dominant species in water bodies with thick macrophyte beds and organic rich sediments [6] and common in Oligo-mesotrophic conditions.

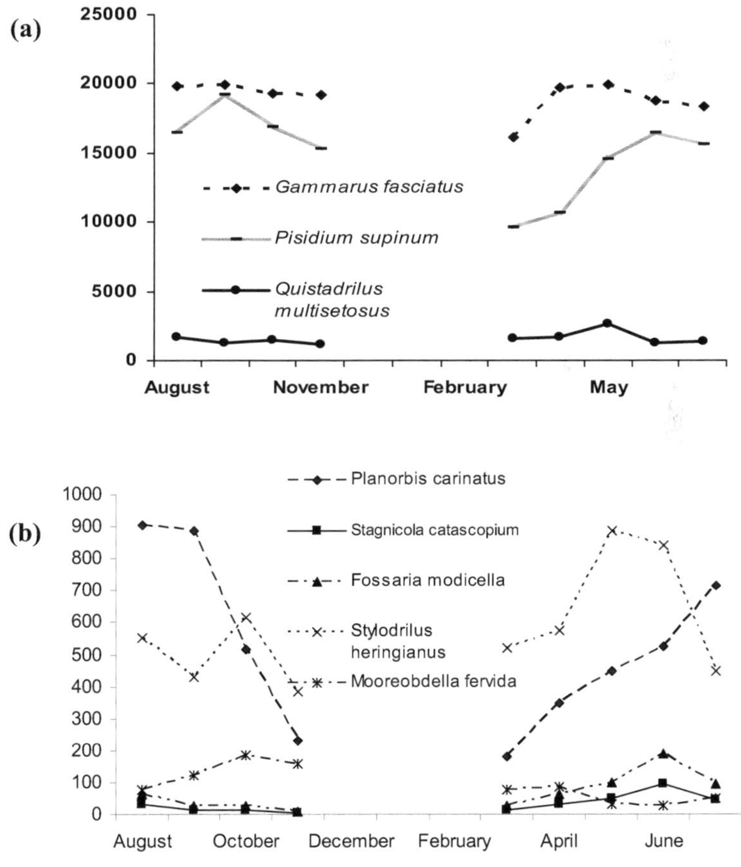

Figure 4: Monthly population densities of the most abundant benthic species in Lake Neor - August 2008 to July 2009.

P. supinum is second dominant species reflecting Oligo-mesotrophic state of Lake Neor since it is known to be very sensitive to eutrophication [7]. (It is a good indicator of water quality since it is very sensitive to eutrophication.). It filters suspended material from the water column but can directly ingest deposits from the sediments as well, so it can reach to high densities in organic rich sediments [8, 9]. The dominant species among Oligochaetes *was Quistadrilus multisetosus* in Lake Neor. It exists in settings with underground springs and macrophyte beds [10]. Underground springs are scattered all over in Lake Neor. *Stylodrilus heringianus* is another characteristic species indicating oligotrophy, usually with low abundance [11]. *Planorbis carinatus* was the dominant species among Gastropods in Lake Neor, especially in shoreline. This snail is known as common inhabitant of calm, clear water with rich vegetation, feed on dying or dead water plants and algae and an indicator of oligotrophic condition.

From the present work it is clear that taxonomic composition of the zoobenthos of the Lake Neor is typical for an oligo-mesotrophic shallow water body. It is dominated by detritivorous organisms living in shallow well oxygenated waters, such as *Gammarus fasciatus* and *Pisidium supinum* and *Planorbis carinatus*. Two other dominant species *Quistadrilus multisetosus* and *Stylodrilus heringianus* are pelophilous inhabitant of profundal zone of eutrophic and oligotrophic lakes respectively. On the other hand the total zoobenthos density in our study, especially for two dominant species is high and comparable with eutrophic lakes (Figure 4(a)). It was demonstrated that macrophytes increase habitat heterogeneity [3, 12], improve food conditions, stabilize bottom sediments and provide protection against wave action [3], and their presence stimulates development of diverse and abundant bottom fauna [2, 13, 14]. It seems organic-rich sediments in Lake Neor have very high quantity and quality of food especially in vicinity of littoral plants for benthic fauna (Figure 3). However water level fluctuations have reduced the diversity, or altered the composition of littoral habitats in this lake. Therefore low richness in Lake Neor may be a result of the very variable environmental conditions such as annual temporary exposure of its phytolittoral parts to air.

5 Conclusions

In this study, we could show the trophic status of the lake. Suitable physical characteristics, water turnover, dissolved oxygen, low depth have favoured in a high density of *Gammarus fasciatus* living in an oligo-mesotrophic condition. At present biological integrity of this lake in terms of species richness (and productivity) is relatively intact, just annual water level fluctuations (for irrigation) would result in phytolittoral habitats become inaccessible for benthic invertebrates especially with receding lake water level and an increase in organic matter in vicinity of these plants. This continuous stressful condition may result in an increase in TOM, nutrient recycling and nutrient releasing to the water column which in turn might affect the trophic status of the lake. Hence the present situation of the lake mainly its trophic regime, clearly shows need for developing a management and monitoring plan.

References

[1] Wetzel, R. G.; Limnology, Lake and river ecosystems. San Diego, USA: Academic Press. pp. 665-710, 2001.

[2] Heino, J, Lentic macroinvertebrate assemblage structure along gradients in spatial heterogeneity, habitat size and water chemistry, Hydrobiologia. (418), pp. 229–242, 2000.

[3] Tolonen, K.T., Hämäläinen, H., Holopainen, I.J. and Karjalainen, J., Influences of habitat type and environmental variables on littoral macroinvertebrate communities in a large lake system. Arch Hydrobiol, (152), pp. 39–67, 2001.

[4] Timm, T., Seire, A., Pall, P., Half a century of oligochaete research in Estonian running waters. Hydrobiologia (463), pp. 223–234, 2001.

[5] Brodersen, K.P., Pedersen, O., Lindegaard, C., Hamburger, K., Chironomids (Diptera) and oxy-regulatory capacity: an experimental approach to paleolimnological interpretation. Limnol. Oceanogr. (49), pp. 1549–1559, 2004.

[6] Risnoveanu, G., & Vadineanu, A Long-term functional changes within the Oligochaeta communities within the Danube River Delta, Romania. Hydrobiologia, (506/ 509), 399–405, 2003.

[7] Reckendorfer, W., Baranyi, C. Funk, A. & Schiemer, F., Floodplain restoration by reinforcing hydrological connectivity: expected effects on aquatic mollusc communities. Journal of Applied Ecology 43(3), pp. 474-484, 2006.

[8] Lopez G. R. & Holopainen, I. J. Interstitial suspension-feeding by Pisidium spp. (Pisidiidae: Bivalvia): A new guild in the lentic benthos? American Malacological Bulletin (5), pp. 21–30, 1987.

[9] Smit, H., Van Der Velden, J. A. &. Klink. A., Macrozoobenthic assemblages in littoral sediments in the enclosed Rhine-Meuse Delta. Netherlands Journal of Aquatic Ecology 28(2), pp. 199-212, 1994.

[10] Yıldız, S. & Balık S., The Oligochaeta (Annelida) Fauna of the Inland Waters in the Lake District (Turkey), E.U. Journal of Fisheries and Aquatic Science, (22), pp. 165-172, 2005.

[11] Milbrink, G. Timm T. & Lundberg S., Indicative profundal oligochaete assemblages in selected small Swedish lakes, Hydrobiologia, (468), pp. 53-61, 2002.

[12] Weatherhead, M.A., James, M.R., Distribution of macroinvertebrates in relation to physical and biological variables in the littoral zone of nine New Zealand lakes. Hydrobiologia (462), pp. 115–129, 2001.

[13] Bänziger, R., A comparative study of the zoobenthos of eight land–water interfaces (Lake of Geneva), Hydrobiologia (300/301), pp. 133–140, 1995.

[14] Cogerino, L., Cellot, B., Bournaud, M., Microhabitat diversity and associated macroinvertebrates in aquatics banks of large European river. Hydrobiologia (304), pp.103–115, 1995.

Section 5
Agricultural contamination

Experimental field research on nitrate balance in agricultural soils

A. M. Marinov[1], M. Pele[2], E. M. Draghici[3], G. Vasile[4] & M. Artimon[2]
[1]*Department of Hydraulics, Hydraulic Machinery and Environmental Engineering, Faculty of Power Engineering, University Politehnica of Bucharest, Romania*
[2]*Chemistry Department, University of Agronomic Sciences and Veterinary Medicine Bucharest (UASVMB), Romania*
[3]*Vegetable Crops Department, (UASVMB), Romania*
[4]*Agrochemistry Department, (UASVMB), Romania*

Abstract

High nitrate levels in groundwater endanger public water supplies. Although the European Community sets 50 mg NO_3^-/L as the maximum tolerated concentration in drinking water, this level is exceeded in many aquifers. The input concentrations of nitrates into the groundwater have to be reduced to acceptable levels. One of the measures to facilitate this reduction is the application of fertilizers according to the actual demand of plants and the existing quantity of nitrogen in the root zone prior to the application. Our work focuses on reducing soil contamination due to agricultural practices by analysis of irrigation and drainage with respect to transport of nitrogenous fertilizer through the soil. This paper describes our experimental field research developed in tunnels during the plants' (*Lactuca Sativa* L) life time. The concentration of nitrogen compounds in soil and the nitrogen uptake by plants were measured and compared for different fertilization schedules. We compare existing methods of fertilizer application, water management, and agricultural practices to improve nitrate pollution control in soils and we propose corresponding best management practices.

Keywords: soil pollution with nitrate, fertilization, irrigation, agricultural practices, experimental research.

WIT Transactions on Ecology and the Environment, Vol 135, © 2010 WIT Press
www.witpress.com, ISSN 1743-3541 (on-line)
doi:10.2495/WP100161

1 Introduction

A fundamental goal of subsurface water management is the protection of soil, water and groundwater resources against degradation and contamination. The natural capacity of aquifers to purify subsurface water must not be destroyed by contamination from industrial, agricultural, and domestic sources. Subsurface processes are extraordinarily slow, and toxic effects may only become evident several years or decades after a contaminant release. Rehabilitation of contaminated soils or aquifers requires many years, or even generations. Crop responses to nitrogen (N) fertilizer are impressive and rewarding to farmers, leading them to exaggerate application rates. The over fertilization and over watering that are associated with irrigated row-cropping increase the probability of developing groundwater pollution problems (Pereira and Quelhas dos Santos [4]).

Although nitrogen fertilizers take several chemical forms, processes occurring in the soil lead to their transformation into nitrate-nitrogen (N- NO_3^-). Nitrates are highly soluble in soil water and are not retained by the chemical complex in the soil. Soil microorganisms retain N- NO_3^- only temporarily, and mineralization occurs later with the return of N to nitrate form. Mobile nutrients such as nitrogen in nitrate form move easily with water flow to deeper soil layers and to the groundwater.

Our work focuses on reducing soil contamination due to agricultural practices by analysis of (a) plant growth and (b) irrigation and drainage with respect to the transport of nitrogenous fertilizer through the soil and toward the plants' roots. For a tunnel row-crop of lettuce (*Lactuca Sativa* L) we developed an experimental design aimed at analyzing the fertilization influence on the soil pollution with nitrogenous compounds and on the plant growth. We measured the physical soil properties, nitrogen concentration in the soil profile and the plant nitrate content throughout the lettuce lifetime. The acquired data will be used in a future study to calibrate and improve a mathematical model (Marinov and Diminescu [3]), for subsurface pollution forecasting.

2 Monitoring nitrogen migration in a field experiment

The transport, storage, exchange and transformation processes of heat and solutions in the soil and groundwater zone are called migration processes. These processes are physically, chemically and biologically influenced by the soil, water, plants and pollutants properties.

Our research analyses the nitrogen migration process in six plant growth scenarios developed in an agricultural tunnel. Fig. 1 shows the lettuce field after three weeks of life. The goal of our research is to determine the influence of the fertilizer application rate and the influence of covering the field with plastic materials (mulch) on the nitrogen transfer in soil and plants during the lettuce's lifetime.

Usually the mulch method conserves up to 80% of the soil's moisture allowing irrigation water to be more efficiently used; significantly moderates soil temperature fluctuations; improves soil health, and plant productivity; avoids soil

Figure 1: Lettuce field in the third week of plant life.

erosion problems; reduces or eliminates herbicide use; permits more efficient fertilizer use, with less leaching, and protects groundwater quality (Draghici [2]).

Initial soil measurements included grain-size distribution curve, field capacity, bulk density, soil specific gravity, and the volumetric water content at saturation.

Air and soil temperature [°C] and light intensity [klux] in the tunnel were measured daily (at 9 am and 2 pm), and the average daytime values are shown in fig. 2.

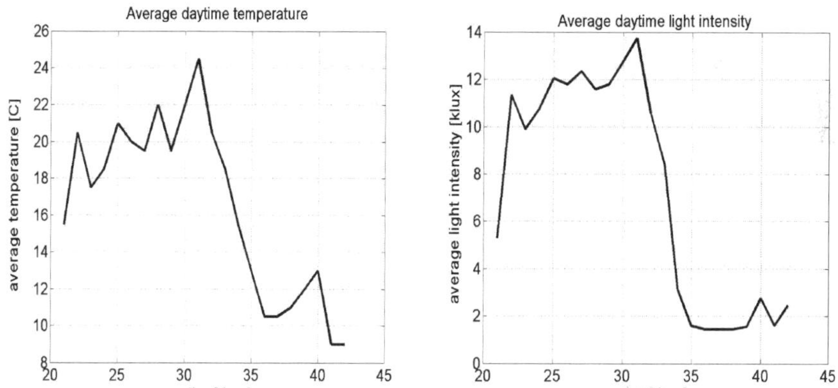

Figure 2: Average daytime values for air temperature and light intensity.

The following parameters were measured weekly during plant lifetime in six (F1-F6) separate subfields shown in fig. 3:

- for the soil profile, at different depths: volumetric water content, water pressure (suction), nitrogen concentration;
- for the plants: nitrogen concentration in leaves, nitrogen concentration in roots, plant growth parameters (leaf count, size and weight; plant diameter, height and volume, length of roots).

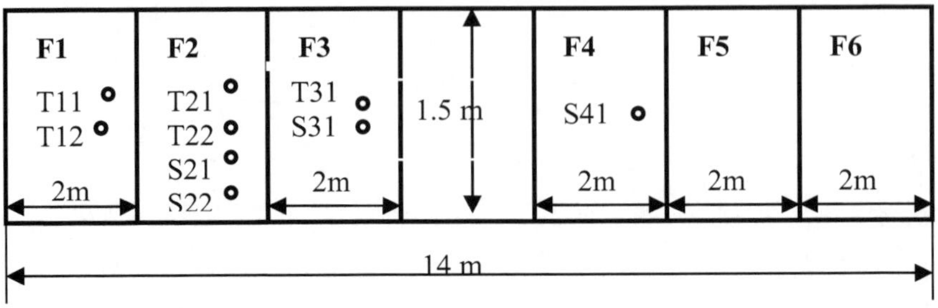

Figure 3: Field description and sampling instruments position. T11, T12, T21, T22, T31 are tensiometric tubes, S21, S22, S31, S41 are extraction tubes SPS200 (soil water sampling).

Each field has a 3 m^2 surface and 5*9 lettuces planted on 29.10.2009 (day 1). The fertilizer (Ammonium nitrate: 17% N- NO_3^- and 17% N-NH_4+) dissolved in water was applied with the irrigation water over each field in the fourth week of plant lifetime (24.11.2009-day 26). A second fertilization identical with the first one was applied ten days later on 04.12.2009 (day 36). Fertilizer concentrations and soil covering procedures are detailed in table 1.

Table 1: Experimental design.

Field nr.	Field symbol	Irrigation water	Fertilizer conc./m^2	N-NO_3	N-NH_4	Covering procedure
		L/m^2	g/L m^2	g/Lm^2	g/Lm^2	Plastic cover
1	F1	1.67	0	0	0	NO
2	F2	1.67	5.00	0.850	0.850	NO
3	F3	1.67	2.33	0.397	0.397	NO
4	F4	1.67	5.00	0.850	0.850	YES
5	F5	1.67	2.33	0.397	0.397	YES
6	F6	1.67	0	0	0	YES

3 Experimental research

Experiments were performed to investigate transient water and N transport in the soil profile, during and after the fertilization and to compare differences in nitrogen N- NO_3^- and N-NH_4^+ concentrations due to different agricultural practices.

Distributed soil samples were taken in each field from 0 to 44 cm depth. Six scenarios (table 1) were comparatively analysed. Lettuce growth, N- NO_3^- and total nitrogen (Ntotal) accumulation in leaves were studied in each case.

3.1 Soil properties measurement

Soils absorb and retain water, which may be withdrawn by plants during periods between rainfall and irrigations. Water holding capacity refers to the amount of water held between field capacity (at the moist end) and wilting point (at the dry end). Plant available water is that portion of the water holding capacity that can be absorbed by plant and is generally considered to be 50 percent of the water holding capacity. This plant-available water depends on soil properties.

We measured the soil samples total volume – V_t, pores water volume – V_w, soil particles volume – V_s, dry weight of the sample – m_s, and m_w – the mass of water. We calculated: bulk density – ρ_b at different depth, grain density – ρ_s, volumetric water content at saturation, – $\theta_s(\%)$, gravimetric water content at saturation – ω_s (%), volumetric water content in different fields (spatial and temporal variation) – $\theta(z,t)(\%)$, gravimetric water content in different fields (spatial and temporal variation) – $\omega(z,t)(\%)$.The water pressure in the pores of the unsaturated soil was measured using tensiometric tubes and an electronic pressure transducer SMS2500S (fig. 4). The electronic device provides directly the negative relative air pressure (p_{air}) inside the top of tensiometric tube. If d is the height of the water column above the soil surface and ρ_w the water density, the pressure (suction) in N/m^2, at a depth z below soil surface, is given by:

$$p = p_{air} + \rho_w g(z+d)\,, \tag{1}$$

while the same relative pressure p expressed in cm H_2O is:

$$h = (p_{air}/\rho_w g) + (z+d) < 0\,. \tag{2}$$

Suction is usually expressed as:

$$pF = \log(\text{abs}(h)) \tag{3}$$

Figure 4: Soil properties measurement.

The soil volumetric water content was measured using a moisture sensor HMS 9000, coupled to a reading device MICROTERM 4800 (fig. 4). The sensor measures the soil electrical capacity. Between the dielectric soil permittivity (ε) and its volumetric water content (θ) a linear relationship:

$$\varepsilon = A\theta + B \quad (4)$$

can be expected. The constants A and B have to be calculated for each soil type as part of the system calibration. Volumetric water content – θ, can be read on the display, after a correct calibration of the device.

3.2 Nitrogen concentration in soil and plant

The mineral soluble compounds represented by nitrates and ammonia salts are the most mobile form of nitrogen in soil and their quantitative determination is used to characterize the amount of accessible nitrogen in soil.

3.2.1 Sample preparation

A mixture formed from 10 grams of soil and 50 mL of distilled water was stirred for one hour and then filtered off. The resulting solution (extract) was used for quantitative determination of nitrate and ammonium ions. Each sample was analyzed in duplicate and the result was expressed as mg/kg fresh weight.

The lettuce leaves were washed with distilled water for dust removal. The samples were sliced and dried on a sheet of paper in order to eliminate the excess of moisture. After these procedures, precisely weighted samples were grounded and treated with acetic acid 2% according to the ratio 1:20. Colored samples were cleared with animal charcoal. The filtrate was used for nitrate determination (Benton et al. [1]). Each sample was analyzed in duplicate and the result was expressed in mg/kg fresh weight.

3.2.2 Standards and calibration

All chemicals used were of analytical reagent grade. A 100 mL standard stock solution containing 100 ppm N- NO_3^- was prepared by dissolution in distilled water 0.7219g of KNO_3,. A 100 mL standard stock solution that contains 100 ppm N-NH_4^+ was prepared using 0.4715 g of $(NH_4)_2SO_4$ dissolved in distilled water. Working standards were prepared by dilution of the stock solution. The calibration curves for nitrate and ammonium nitrogen were linear for the studied concentration ranges.

3.2.3 Analytical methods

The ammonium nitrogen from water and soil was determined with the spectrophotometric method employing the Nessler reagent (Pansu and Gautheyrou [5]). This method is based on the formation of a yellow-orange colored compound (oxidimercurammonium iodide) as a result of reaction between Nessler reagent and ammonium ions from sample. The color intensity of the newly formed compound is proportional with the ammonium nitrogen concentration in the soil.

$$NH_4^+ + 2[HgI_4]^{2-} + 4OH^- \rightarrow HgO{\cdot}Hg(NH_2)I + 7I^- + 3H_2O \quad (5)$$

The nitrate nitrogen dosage from soil, water and plant were measured with the spectrophotometric method using phenol-2.4-disulphonic acid in basic medium. All the measurements were made at 410 nm using the VIS spectrophotometer Metertek SP830 Plus. The resulting yellow colored compounds are called nitroderivates. These have absorption maxima at 410 nm and their color intensity is proportional with the sample nitrates concentrations.

3.2.4 Total nitrogen – the Kjeldahl method

The Kjeldahl procedure can be divided into three main steps, as follows: digestion, distillation and titration. In the first step, organic nitrogen is transformed into ammonium sulphate by heating the sample with H_2SO_4. For this step, potassium sulphate and copper sulphate are used as catalysts to increase the boiling point of the solution. The chemical decomposition is complete when the solution becomes clear and colorless. The digested sample is alkalinized with NaOH and the nitrogen is distilled off as ammonia trapped in a boric acid solution. The amount of ammonia nitrogen is then quantified by titration with a standard H_2SO_4 solution. A reagent blank is carried through the analysis and the volume consumed for this blank is subtracted from each determination.

4 Experimental results

4.1 Physical characteristics and soil pollution

Using the measured values: V_t, V_w, V_s, m_s, and m_w (defined in section 3.1), we calculated in table 2: ρ_b (g/cm^3) at different depth, ρ_s (g/cm^3), θ_s (%), and ω_s (%).

Table 2: Soil properties in the experimental field.

Properties	bulk density	grain density	volumetric water content at saturation	gravimetric water content at saturation
Definition	$\rho_b = \frac{m_s}{V_t}$	$\rho_s = \frac{m_s}{V_s}$	$\theta_s(\%) = \frac{V_w}{V_t}100$	$\omega_s = \frac{m_w}{m_s}100$
Values	1.25-1.5 g/cm^3	2.47 g/cm^3	50.8	39

The soil pressure and moisture content variation in the soil profile (fig. 5) show the influence of irrigation on the water content and water velocity in the soil. The irrigations performed on 24.11.2009 and on 04.12.2010 produce an increase of humidity in the soil profile. Salads roots develop in the top 15 cm, this is where the plants take their water from. Because of plant water uptake and gravitational flow, humidity increases deeper in the soil, with smallest values above 15 cm, and larger values (close to saturation) between 15 and 40cm.

Air humidity is very large in the tunnel and the large day-night variations produce water vapor condensation and insure humidity close to the field capacity even between irrigations.

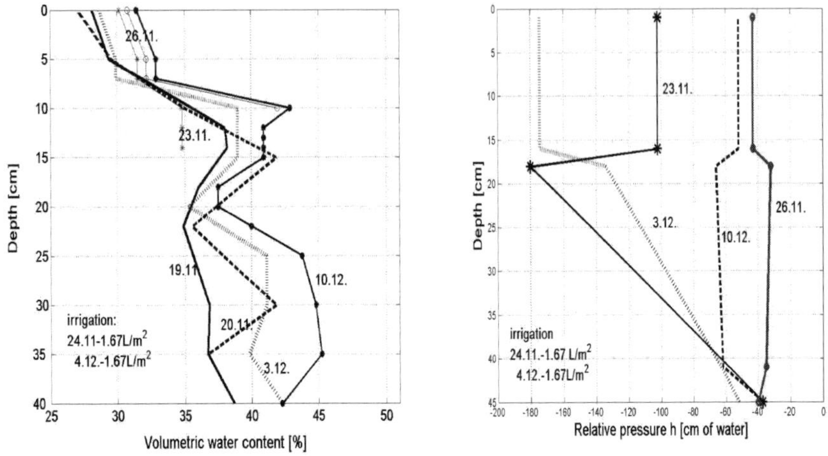

Figure 5: Volumetric water content and suction in the soil profile during the experiment.

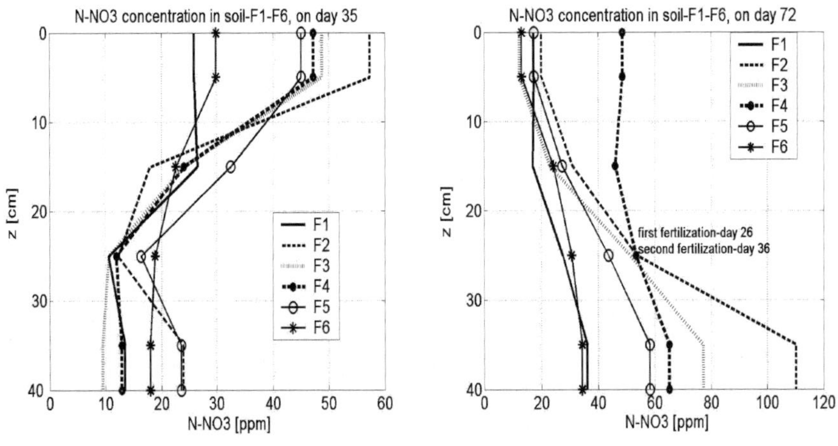

Figure 6: The $N\text{-}NO_3$ concentration in soil profile, in fields F1, F2,..F6, on day 35 and on day 72.

Fig. 6 shows the nitrate distribution in soils on day 35 (35 days after planting the salad and right before the second fertilization) and on day 72 (36 days after the second fertilization and 20 days after the plants were collected). On day 72 nitrate reaches deeper in the soil (compared to day 35), achieving maximum values between 30-40 cm depth. Maximum nitrogen fertilization and no plastic cover results in maximum nitrate at depth for field F2, while no nitrogen fertilization and plastic cover results in a minimum concentration for field F6.

4.2 Plant growth description

During the plant life we measured the number and dimensions of plant leaves, plant diameter, height and volume, and root length and we tried to correlate the plant growth with the nitrogen concentration in leaves, roots, and soil.

(a) (b)

Figure 7: Lettuce growth level. (a) Typical plant on the first day. (b) Fifth week in field F6 with mulch

To analyze the acquired data we plotted N- NO_3^- (fig. 8) and total nitrogen concentration in plant (fig. 9) during its lifetime, N- NO_3^- and total nitrogen concentration in the soil profile, for the six fields. To compare the behavior of the plants in the six agronomical conditions and the fertilization method's impact on soil pollution we define a set of parameters which describe the nitrogen accumulation in plant and in the soil profile, and the fertilization efficiency.

We calculate the N- NO_3^- and total nitrogen accumulation rate in plant's leaves, in a Δt time interval between measurements:

$$v_N = \frac{C(t+\Delta t) - C(t)}{\Delta t}, \tag{4}$$

where $C(t)$ and $C(t+\Delta t)$ are the concentrations at the beginning and at the end of each time interval.

We compared $v_{N-NO3}(t)$ from fig. 8 and $v_{N\,total}(t)$, from fig. 9, with the plant growth rate (the rate of increase of plant mass) in a time interval Δt, (fig. 9), defined as:

$$v_g = \frac{m(t+\Delta t) - m(t)}{\Delta t}, \tag{5}$$

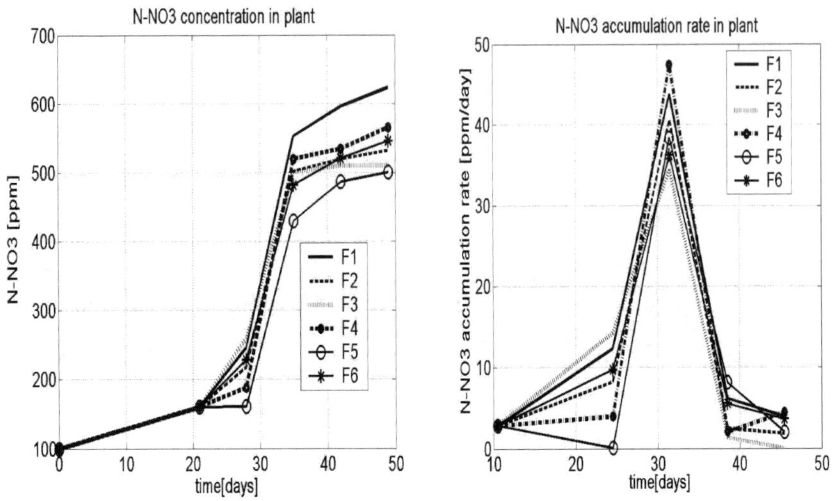

Figure 8: N-NO3 concentration in plant and N-NO_3 accumulation rate in plant.

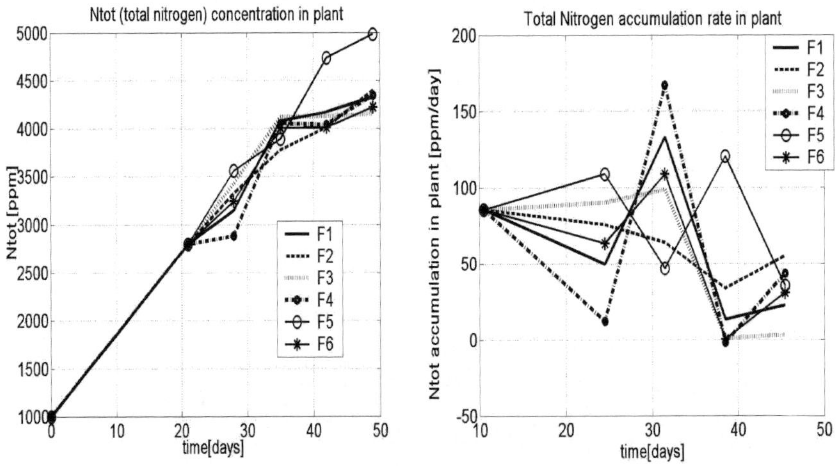

Figure 9: Total nitrogen concentration in plant and total nitrogen accumulation rate in plant.

where $m(t)$ is plant mass at time t and $m(t+\Delta t)$ is plant mass at time $t+\Delta t$. We calculated an efficiency of plant growth in a Δt interval of time, like a ratio between the plant growth rate v_g and nitrogen accumulation rate v_N :

$$Efficiency\ of\ growth = \frac{v_g}{v_N} = \frac{m(t+\Delta t) - m(t)}{C_{N-NO3}(t+\Delta t) - C_{N-NO3}(t)}. \qquad (6)$$

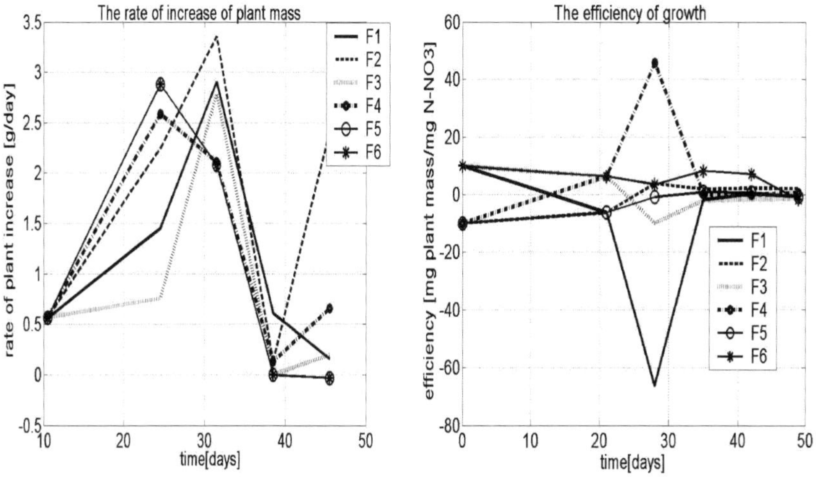

Figure 10: The rate of increase of plant mass and the efficiency of growth.

We studied the time dependence of the efficiency of plant growth in each field (fig. 10).

At every time step (t) we calculate the average plant mass for the 6 fields F1..F6

$$\overline{m}(t) = \frac{\sum_{j=1}^{6} m(t, j)}{6}, \tag{7}$$

where $m(t, j)$ is the leaf mass for a representative lettuce in field j on day t.

The average mass of N-NO_3 for the six fields is:

$$\overline{m}NO_3(t) = \frac{\sum_{j=1}^{6} mNO_3(t, j)}{6}, \tag{8}$$

where $mNO_3(t, j)$ is the mass of N-NO_3 contained in a typical lettuce in field j.

We defined total plant growth sensitivities to changes in nitrate or changes in total nitrogen, in the field j, $TPGS(j)$ [g of plant/mg N] as:

$$TPGS_{N-NO3}(j) = \frac{m(50, j) - m(1, j)}{mNO3(50, j) - mNO3(1, j)}, \tag{9}$$

$$TPGS_{N\,tot}(j) = \frac{m(50, j) - m(1, j)}{mN\,tot(50, j) - mN\,tot(1, j)} \tag{10}$$

where $m(50, j)$ and $m(1, j)$ are the plant mass at the end (t=50 days) and at the beginning of the experiment (t=1 day), respectively, $mNO3$ (50, j) and $mN\ tot$ (50, j) are the mass of N-NO_3 and total nitrogen contained in lettuce leaves at the end of experiment. Ideally we want these plant growth sensitivities to be large, i.e., we want an agricultural practice in which the plant growth is most sensitive to the addition of fertilizer.

To quantify the best fertilization method among the six fields depicted in table 1, we generate a *Score matrix,* considering four important criteria for a good fertilization method:

1. $TPGS_{N-NO3}(j)$ needs to be large as explained above;

2. $TPGS_{N\,tot}(j)$ needs to be large;

3. The concentration of N- NO_3^- [ppm] in the first 15 cm under the soil surface (the roots zone) needs to be large at the end of experiment. This ensures that enough nitrate is left in the soil to be useful for future crops.

4. The concentration of N- NO_3^- [ppm] between 15 cm and 40 cm depth needs to be small at the end of experiment to avoid groundwater pollution.

Each of the six fertilization methods is given a score which reflects how well they accomplish the above criteria (1 for the best method and 6 for the worst). The winning field and the best fertilization solution will be the one with the minimum total score.

According to our *score matrix*, the winning field was field F4, suggesting that the best management practice is to use mulch and two fertilizations.

Table 3: Score matrix.

Criterium	U.M	Cond.	F1	F2	F3	F4	F5	F6
$TPGS_{N-NO3}(j)$	g/mg N	Max	1.52	2.19	1.21	1.60	1.49	1.48
Score	-	-	3	1	6	2	4	5
$TPGS_{N\,tot}(j)$	g/mg N	Max	0.22	0.32	0.18	0.23	0.21	0.21
Score	-		3	1	6	2	4	5
N-NO_3-0-15cm	ppm	Max	36.2	110.3	77.2	65.3	58.3	34.5
Score	-		2	6	5	4	3	1
N-NO_3-15-40cm	ppm	Min	17.4	20.1	12.5	48.6	17.5	13.3
Score	-	-	4	2	6	1	3	5
Total Score	-	-	11	10	23	9	14	16

5 Conclusions

Our study quantifies the impact of nitrate fertilization on plant growth and on the transport of nitrate in soil. In particular, we studied the development of plants and the accumulation of N- NO_3^- and total nitrogen (Ntotal) in plant leaves for fifty days, in six different fields in which we applied different nutrient fertilization methods. The following ideas emerge from our experimental results:

- the growth rate of the plants which are not covered by plastic (fields F1, F2, F3) depend directly on daytime light intensity and temperature (fig. 2). For fields F1, F2 and F3 the growth rate of plants (fig. 10) reaches its maximum on day 32

(6 days after the first fertilization), corresponding to the peak in light intensity and temperature. By contrast, covering the field with mulch decorrelates the growth rate from these physical parameters. Maximum growth rate in fields F4, F5 and F6 therefore does not coincide with the maximum of light;

- the second fertilization ensures higher growth rate in field F2 compared to F3 and F4 (fig 10). The lack of a plastic cover and thus more light results in higher growth rates for field F2 compared to F4. By contrast, v_g decreases after day 32 and remains low in fields F1 and F6 in which we did not add fertilizer;
- the highest efficiency of plant growth (eqn. (6)) is observed in field F4 on day 27, after the first fertilization. After 35 days the maximum efficiency is in field F6, showing clearly the advantage of covering the field with mulch. After day 49 the efficiency became greater in F2 but the nitrate concentration at 30-40 cm depth in this field is the biggest compared with the other field as shown in fig. 6;
- initial nitrate in soil is actively used by plants until day 26, the day of the first fertilization. As expected from previous research, Draghici [2], the plastic cover ensures a maximal efficiency of initial nitrate utilization for fields F4, F5, F6 (fig. 10).

As a result of score matrix analysis, the mulch method with maximum nitrate fertilization, corresponding to field F4, emerges as the best method for nitrogen fertilization of a salad crop. This method insures a minimal nitrate concentration in the plant, the lowest nitrate concentration in the soil below 15 cm, and a large nitrate concentration left in the top 15 cm of soil, to be used by potential future crops.

Acknowledgements

This research was developed in the frame of a Romanian National Research Program IDEI, grant 189/2007, sponsored by CNCSIS, Romania.

References

[1] Benton, J., Jones, J., Wolf, B., Mills, H., Plant Analysis Handbook-A Practical Sampling, Preparation, Analysis, and Interpretation Guide, Micro-Macro Publishing, 1991.

[2] Draghici, E.M. The relation between the pot size and some quality index at lettuce nurseling, *Lucrari stiintifice USAMVB Horticultura*, seria B LI, pp. 68-71, 2008.

[3] Marinov, A.M., Diminescu, M.A., Experimental research and mathematical modelling of soil and groundwater contamination with nitrogen compounds, *Proc. of the IX-th. Int. Conf. on Water pollution,* eds. C.A. Brebbia, WIT Transactions, UK, pp.115-126, 2008.

[4] Pereira, L.S. & Quelhas dos Santos, J., Fertilizer and water application, and control of nitrate pollution. Management issues, *NATO Advanced Research Workshop on Nitrate Contamination: Exposure, Consequences, and Control,* Lincoln, Nebraska, USA, pp.141-162, 1990.

[5] Pansu, M., Gautheyrou, J., Handbook of Soil Analysis (Mineralogical, Organic and Inorganic Methods), Springer - Verlag, Berlin, Heidelberg, 2006.

Generation and size distribution of sediment eroded in a small-scale catchment of the Western Cape (South Africa)

N. Z. Jovanovic[1], R. D. H. Bugan[1], C. Petersen[1], W. P. De Clercq[2], A. Rozanov[2] & H. Botha[3]

[1]CSIR, Natural Resources and Environment, South Africa
[2]Department of Soil Science, University of Stellenbosch, South Africa
[3]Department of Process Engineering, University of Stellenbosch, South Africa

Abstract

Rainfall, runoff (overland flow), total sediments mobilized and their size distribution were measured for a number of events over three years (from 2006 to 2008) in a small-scale catchment of the Berg River (Western Cape). The aim was to develop local-scale rainfall/runoff, sediment generation and particle size distribution relationships that could be incorporated into existing hydrological catchment models for predicting erosion, transport and deposition of sediments with sorbed contaminants. The results indicated that rainfall intensity was weakly correlated to rainfall amounts ($R^2 = 0.54$), whilst runoff (between 12.4% and 18.2% of rainfall) was weakly correlated to the product of rainfall amount and peak intensity at two sites with different soil properties and slopes ($R^2 = 0.29$ and 0.44), depending on antecedent moisture condition, land use and management practices. Sediment concentration was inversely correlated to runoff volume. However, more runoff produced larger total sediment loads. The bulk of sediment particles mobilized via runoff water had diameters in the range between 4.76 and 15.95 μm (measured with a Saturn DigiSizer 5200 particle size analyzer). Sediment size distribution was relatively uniform for all events recorded, did not depend on sediment concentration, runoff, rainfall amount and intensity, and it was similar to the texture of the soil A-horizons (source of sediments). Spatial information on soil A-horizon textural properties could therefore be used to infer the nature of sediments mobilized in the catchment.

WIT Transactions on Ecology and the Environment, Vol 135, © 2010 WIT Press
www.witpress.com, ISSN 1743-3541 (on-line)
doi:10.2495/WP100171

Sediment size distribution in the catchment can be used, in conjunction with sorption data, to estimate the transport of sorbed contaminants, e.g. nutrients and pesticides. In addition, sediment size distribution can be used, in conjunction with chemical and isotopic properties of mineral particles, to trace the origin and history of sediment transport in catchments.

Keywords: erosion, overland flow, particle size distribution, sediment concentration, Western Cape.

1 Introduction

The most widely used erosion models are applicable to mapping spatial distribution of erosion risk or to estimating erosion rate [1]. In recent years, focus has shifted towards the development of models for predicting sediment generation, transport and deposition in catchments, in order to assess the consequences of erosion in terms of water pollution and sedimentation downstream [2]. It was recognized that control of water pollution requires an understanding of the dynamic processes occurring during individual rainfall events, the size and timing of peak discharges of water and sediment from hillslopes to rivers, as well as the nature of the sediments detached and that of the associated contaminants. Recent advances indicated that the size of mobilized soil particles depends on the two mechanisms of detachment, namely splash and wash, where particles detached through the kinetic action of raindrops are generally coarser compared to those detached through the mechanical action of overland flow [3]. The size of mobilized particles and their sorptive characteristics have implications to the transport of contaminants attached to sediments, for example nutrients and pesticides [4]. Total mobilized sediments and their particle size distribution depend on the source of sediments, i.e. the textural and aggregate properties of the soil surface. The texture of sediment particles was generally found to be finer compared to the texture of the original soil [3], or similar [5].

In this study, undertaken in the mid-reaches of the Berg river catchment (Western Cape, South Africa), the hypothesis was that the texture of sediment particles in overland flow depends mainly on the texture of parent material from where these particles were mobilized, regardless of rainfall and runoff intensities and volumes. If verified, this relation could be incorporated into existing hydrological catchment models and facilitate predictions of sediment particle sizes originating from heterogeneous areas. In order to test the hypothesis, rainfall, overland flow (referred to further as runoff), total sediments mobilized and their particle size distribution were measured for a number of events over three years (from May 2006 to October 2008) at runoff plots established in the field in a small-scale catchment. The aim was to develop rainfall/runoff, sediment generation and particle size distribution relationships that could be incorporated into existing hydrological catchment models for predicting erosion, transport and deposition of sediments with sorbed contaminants.

2 Material and methods

The study site (Goedertrou farm in the Berg river catchment) is located in the Western Cape, South Africa, about 5 km North-East of the town of Riebeek West (fig. 1). The area is semi-arid and receives about 350 mm a^{-1} of rainfall. Rainfall occurs mainly in winter (from April until October).

Fig. 1 shows the location of the experimental site in the Western Cape and the experimental scheme in the Goedertrou catchment (~20 ha), on a Google Earth map. Two standard Wischmeier runoff plots (22.3 m x 2 m) parallel to one

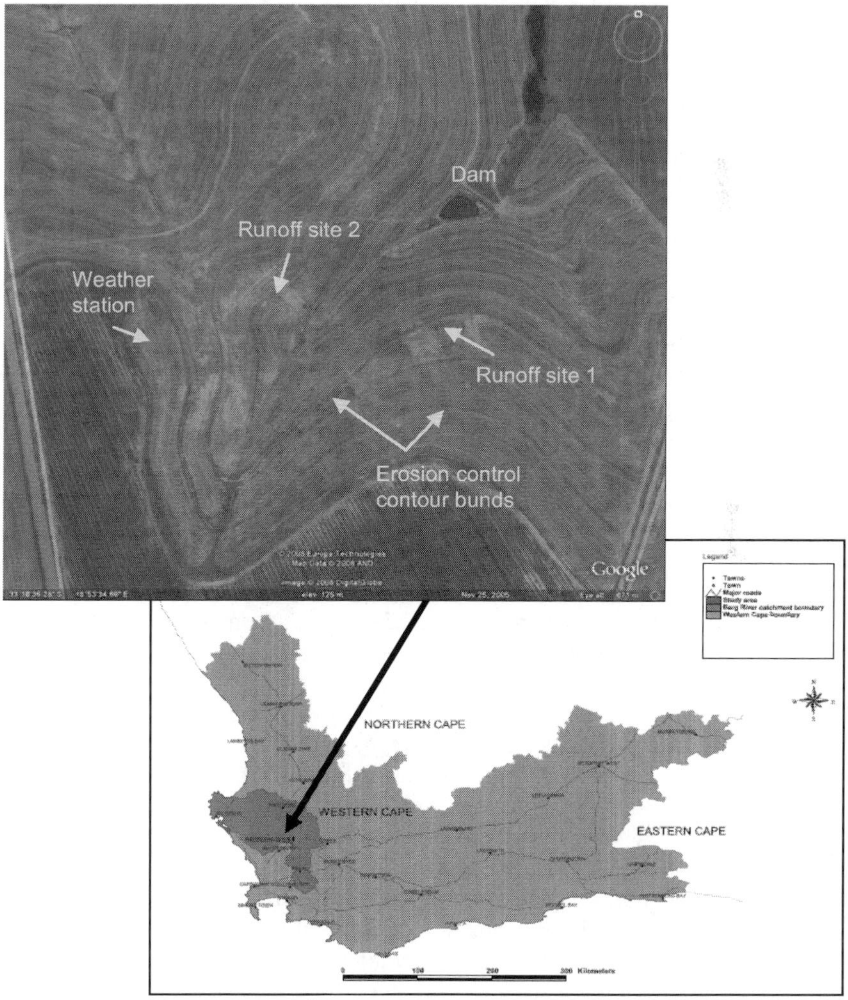

Figure 1: Location of the experimental site in the Western Cape and Google Earth map of the Goedertrou experimental catchment.

another, were established at each runoff site 1 and 2. The runoff sites were chosen to represent two different, but typical hydrological units, each located between two erosion control contour bunds. Site 1 was North-oriented, whilst site 2 was South-East-oriented. The soils at the two sites are also different: 0.4 m deep sandy clay loam Cambisol [6] overlying Malmesbury shale at site 1; deep clay loam Luvisol [6] at site 2. The slope of site 1 is 9.1%, whilst the slope at site 2 is 12.4%. During 2006 and 2007, the Goedertrou farm was left uncultivated to allow re-growth of wheat and medic grass (crops planted during previous seasons) to support grazing. During 2008, the field was cultivated to wheat. An automatic weather station was installed in the catchment (fig. 1) to measure and log rainfall, solar radiation, temperature, relative humidity, leaf wetness, as well as wind speed and direction.

Measurements of rainfall, runoff, total sediments mobilized and their particle size distribution were carried out at each runoff plot (fig. 1, runoff sites 1 and 2). A flow splitter was installed in order to divide water and sediment flow from the Wischmeier runoff plots into two halves. The first portion from the flow splitter lead into a tipping bucket fitted with a magnetic switch, in order to record overland flow electronically. The tipping bucket was calibrated so that every tip corresponded to 1 L of water. Water volume data were collected every 10 min and stored with an MCS data logger (Mike Cotton Systems, Cape Town). The first flow portion was wasted thereafter. The second flow portion from the splitter lead into two sediment traps. The flow splitter was calibrated in the field to determine the exact fraction of water diverted onto the tipping bucket and sediment traps. Half of the flow was diverted onto the tipping bucket, and the other half into the sediment traps. The splitter was levelled during each field visit to ensure the calibration was correct at all times. The sediment tanks (traps) were emptied and washed after each field visit. Water volumes were also measured manually during field visits, by measuring the litres of water collected in the tanks.

The first sediment tank in the sequence was used to trap coarser sediments. Water and suspended solids then overflowed into the second sediment tank in the sequence, where finer sediments were collected. The sediment tanks were used to collect water and sediment for sampling and laboratory analysis. Water and sediment samples were taken approximately bi-weekly during the 2006, 2007 and 2008 winter seasons. Sediment and water samples were collected in 100 mL plastic bottles after scraping the bottom of the tanks with a clean spade.

Total sediment mass was measured in the laboratory by evaporating water from sub-samples of known volume. The remainder of the samples was then used to measure sediment particle size and distribution with a Saturn DigiSizer 5200 particle size analyzer (Micromeritics Instrument Corporation). The Saturn DigiSizer makes use of a laser beam and measures the intensity of light scattered at various angles by organic and/or inorganic particles suspended in a liquid sample. Scattered light intensities depend on particle size, shape, refractive index and wavelengths of incident light, and they are measured by a charge-coupled device that contains over one million detector elements. The particle size distribution in a water sample is calculated from the scattered light intensity based on the Mie theory [7], and by applying the mathematical procedure of deconvolution. Particle diameters are calculated from volumes assuming particles are spheres. The instrument measures in the particle diameter range from 0.1 to 1000 μm, where each order of magnitude is split into 40 pre-defined or user-defined particle size classes. Three tests were run for each sample in order to check the repeatability of results. Ultrasonic probes were not used so as to preserve aggregates and allow measurement of particles mobilized under natural conditions. For this reason, caution was exercised in handling the samples (no shaking and agitation) in order to prevent dispersion of aggregates.

Data were analyzed with a Saturn 5200 V1.09 software (Micromeritics Instrument Corporation). The software is used to enter measurement settings (analysis conditions and material properties), perform a large number of data analyses as well as to report results. It can generate a large number of reports in graphical and numerical format with different levels of detail. The output results considered in this study for each water sample were the following:

- Mode of the frequency distribution of particle diameters (calculated from volumes, assuming particles are spheres) in each sample. This output variable represented the highest frequency of occurrence of a particle size class, or the peak of frequency distribution comprising at least 5% of the distribution.
- Particle diameters representing 90th, 50th and 10th percentiles of the cumulative volume distribution of all particles in each sample.

Rainfall at runoff sites 1 and 2 was recorded every 10 min with tipping bucket rain gauges and data were stored with MCS data loggers. The data loggers were enclosed in a box and powered via battery and solar panel. Volumetric soil water contents at sites 1 and 2 were measured electronically with Echo sensors installed at 10 cm and 40 cm soil depth and connected to Echo loggers (Decagon Devices Inc.). The purpose of the electronic measurement of soil water content was to describe soil moisture conditions antecedent to runoff events.

3 Experimental results

Three years of rainfall, runoff, sediment concentration and sediment size distribution data were analyzed in order to develop local-scale, empirical relationships for sediment mobilization and particle size distribution. Data included all measurements taken from 2006 to 2008, and from both sediment collection tanks. Data were processed separately for site 1 and 2, as these sites represent different soil types and slopes. Although field visits for sampling and downloading data were planned after major rainfall events, in some instances the data represented more than one rainfall event occurring in the period antecedent to sampling (period since the previous field visit and sampling). Data for some rainfall and runoff events were not available due to malfunction of data loggers, damage to equipment caused by storms and grazing cattle. In the events when rainfall data were not available from loggers installed at the runoff plot sites, records obtained from the weather station were used.

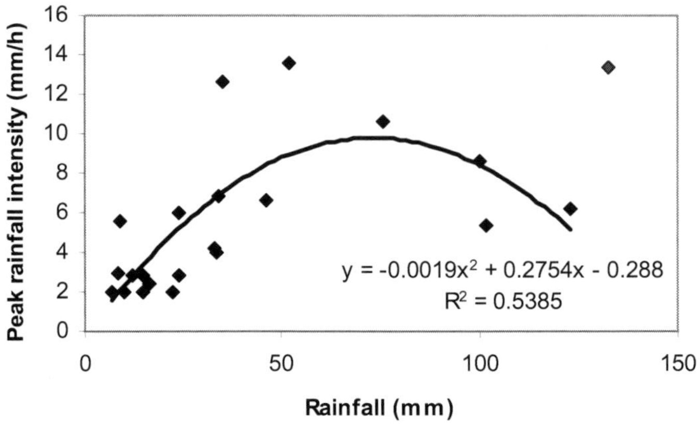

Figure 2: Relationship between rainfall and rainfall intensity (data refer to events that occurred from 2006 to 2008; number of observations = 23).

A second order polynomial gave the best fit for the relationship between rainfall and peak rainfall intensity (R^2 = 0.54) (fig. 2). Fig. 2 gives a good representation of the nature of rainfall in the area, generally consisting of prolonged low-intensity drizzle. Highest peak rainfall intensities generally occurred for total rainfall of about 70 mm. Frontal systems, which delivered most rainfall, generally had peak intensities < 10 mm h^{-1}. Maximum rainfall intensity in the data set was 13.4 mm h^{-1}. This was an exceptional event (total rainfall = 132.6 mm that occurred between 3 to 12 July 2008).

Runoff at site 1 was 12.4% of rainfall for all events recorded (ranging from 0 to 60.6% for individual events), whilst runoff at site 2 was 18.2% of rainfall (ranging from 0 to 69.1%). The linear relationships between runoff and the product of rainfall amount and peak intensity are presented in fig. 3 for sites 1 and 2. The relatively low R^2 coefficients in fig. 3 are an indication that other factors may affect the relationship between runoff and rainfall, such as antecedent moisture condition, land use and management practices.

Total sediment concentrations measured in both tanks were plotted as a function of runoff volumes. The data are shown in fig. 4 for sites 1 and 2. The inverse relationship indicates that high concentrations of sediments were generally measured when small runoff events occurred (low rainfall amounts and/or intensity). With high rainfall and runoff volumes, sediment concentrations were generally low due to dilution effects. This indicates that most sediment mobilization may occur at the beginning of a runoff event. Concentrations of sediments in runoff water were between 0 and 22.0 g L^{-1} (2.6 g L^{-1} on average).

In terms of total sediment load, more eroded material was mobilized from the runoff plots during high rainfall and runoff events at both sites (fig. 5). It should be noted, however, that sediment load data in fig. 5 originated from

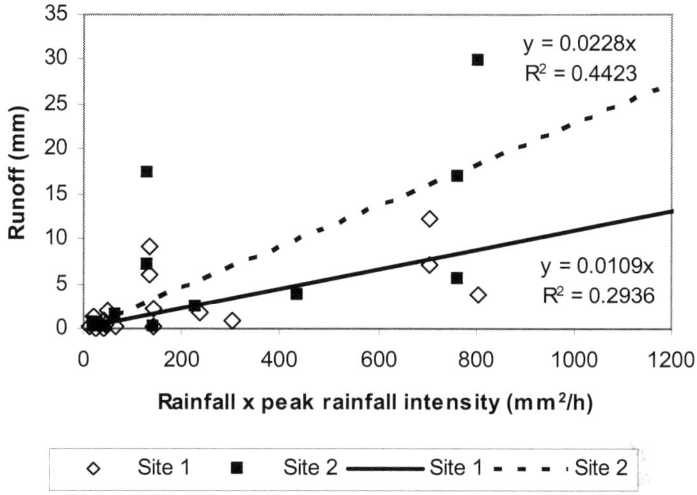

Figure 3: Relationship between runoff and the product of rainfall amount and peak intensity for site 1 (number of observations = 20) and site 2 (number of observations = 15) (data refer to events occurred from 2006 to 2008).

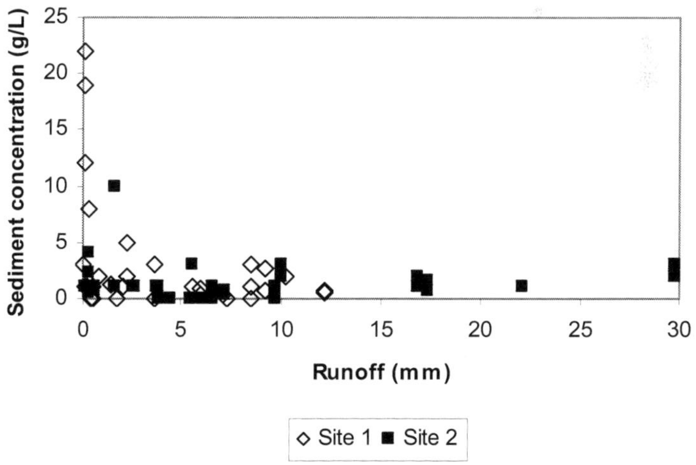

Figure 4: Relationship between sediment concentration and runoff for site 1 (number of observations = 36) and site 2 (number of observations = 39) (data refer to events occurred from 2006 to 2008).

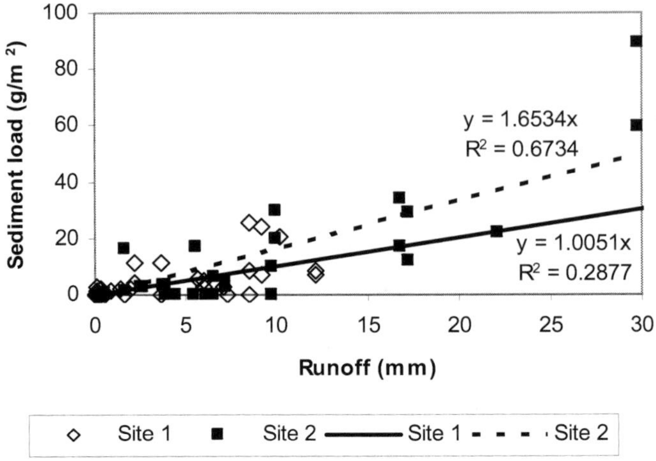

Figure 5: Relationship between sediment load and runoff for site 1 (number of observations = 36) and site 2 (number of observations = 39) (data refer to events occurred from 2006 to 2008).

measurements of sediments trapped in the collection tanks, representing the cumulative result of one or more events, whilst the dynamics of sediment concentrations and loads during individual runoff events was not investigated at the runoff plots. Site 2 generally produced lower sediment concentrations in runoff water (fig. 4), but more runoff and higher sediment loads compared to site 1 (fig. 5). Total annual sediment mobilization ranged between 0.02 and 0.85 Mg ha^{-1} a^{-1} (0.26 Mg ha^{-1} a^{-1} on average).

The particle size analyzer outputs numerical and graphical representations of particle volume frequency percentage and cumulative finer volume percentage. The particle size analyzer was set to detect and record any peaks including at least 5% of particle size (diameter) distribution in a sample. In most cases, the peaks included >70% of the particle size distribution. In samples where more than one peak was identified, the peak that accounted for the highest percentage of particles was considered. The peak frequency of particle diameters within a certain size class is the mode of frequency distribution of all particle sizes, and it was used as an indicator of the most frequent size of sediments mobilized for each sample.

The modes of particle diameter frequency distribution are shown in fig. 6 as a function of sediment concentrations for all samples analyzed at sites 1 and 2. Modes of particle diameters were always in the silt range between 4.76 and 15.95 μm, regardless of sediment concentration and therefore rainfall amount, intensity and runoff. On average, slightly smaller modes of sediment particles were measured at site 2 (9.35 μm) compared to site 1 (9.49 μm) (fig. 6).

Another indicator of the size of sediments mobilized was the cumulative finer particle volume percentage over the full range of particle diameters. The shape of

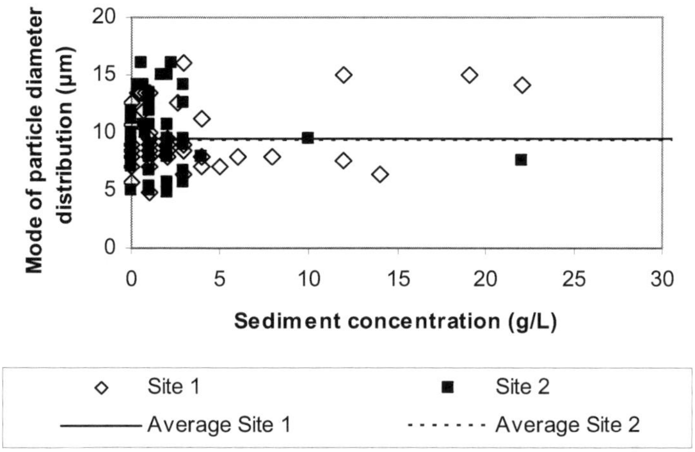

Figure 6: Relationship between mode of particle diameter frequency distribution obtained with the particle size analyzer and sediment concentration for site 1 (number of observations = 43) and site 2 (number of observations = 44) (data refer to events occurred from 2006 to 2008).

this curve is determined by the percentages of particles finer than a certain diameter occurring in the sample. For the purpose of this study, we considered the 90th, 50th (median) and 10th percentiles of the cumulative particle size distribution in each sample. The results are shown in fig. 7 for sites 1 and 2. Again, the data were spread within a narrow range regardless of sediment concentration and therefore rainfall amount, intensity and runoff. Sediment diameters mobilized from site 1 covered a slightly wider range compared to site 2 (fig. 7).

From the analyses of particle size distribution, it appeared that the size of mobilized sediment particles does not depend on the magnitude and intensity of rainfall and runoff events. The main factors were the source of sediments (texture of the top soil) and the slope of the site determining the gravity force for transport of particles of a certain size. This was confirmed through particle size analyses of soil A-horizons in the vicinity of the runoff plots. The modes of particle diameter frequency distribution of soil A-horizons were 12.67 µm for site 1 and 7.55 µm for site 2. The 90th, 50th (median) and 10th percentiles of the particle size distribution in the soil A-horizons were 75.63, 13.69 and 1.49 µm for site 1, and 77.22, 6.30 and 0.91 µm for site 2, respectively. The values of modes and percentiles for soil A-horizons were therefore in the range of those measured for mobilized sediments (fig. 6 and 7). Similar results were also obtained from controlled rainfall simulation tests done on 13 different soil types in the catchment [8]. This indicated that the nature of the soil, as the source of sediments, plays a dominant role in the nature of particles mobilized.

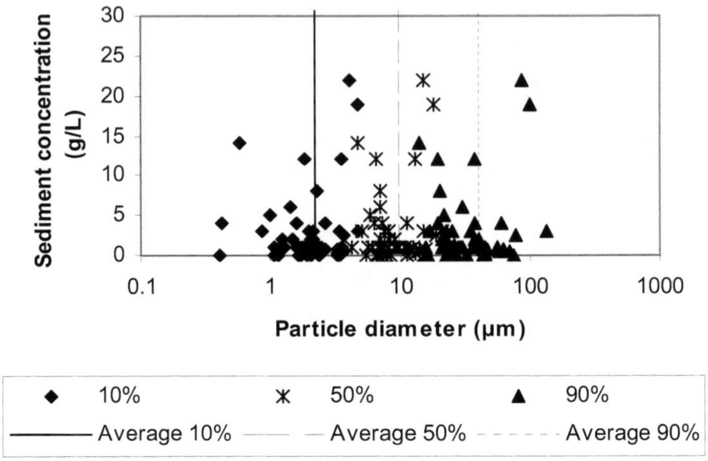

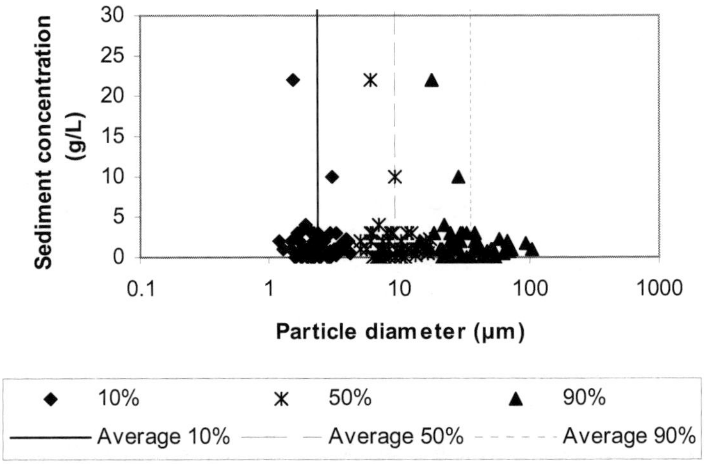

Figure 7: Relationship between 90th, 50th (median) and 10th percentiles of the particle size distribution in runoff samples, measured with the particle size analyzer, and sediment concentration for site 1 (top graph; number of observations = 43) and site 2 (bottom graph; number of observations = 44) (data refer to events occurred from 2006 to 2008).

4 Conclusions

The following main conclusions were drawn from this study:

- Rainfall generally consists of prolonged low-intensity drizzle in the study area. Rainfall intensity was weakly correlated to rainfall amounts ($R^2 = 0.54$).

- Runoff ranged between 12.4% and 18.2% of rainfall for all events recorded. Runoff was weakly correlated to the product of rainfall amount and peak intensity (R^2 = 0.29 for site 1 and R^2 = 0.44 for site 2).
- An inverse relationship was observed between sediment concentration and runoff volume. However, more runoff volume produced larger total sediment loads. Total annual sediment mobilization ranged between 0.02 and 0.85 Mg ha^{-1} a^{-1} (0.26 Mg ha^{-1} a^{-1} on average).
- The bulk of sediment particles mobilized via runoff water had diameters in the range between 4.76 and 15.95 μm. Particle size distributions resembled the texture of the soil A-horizons, regardless of sediment concentration, runoff, rainfall amount and intensity.

The findings of this study open up opportunities for a wide range of applications. For example, spatial information on soil A-horizon textural properties could be used to infer the nature of sediments mobilized in large-scale catchments. This information, along with the empirical relationships between rainfall, runoff and sediment load found in this study, can be incorporated into existing hydrological catchment models, and facilitate the estimation of total sediment load in the Berg river catchment. Similarly, the texture of sediments deposited downstream of their source could be used, in conjunction with chemical and isotopic properties of mineral particles, to identify areas in the catchment that are prone to erosion. Knowledge of sediment particle size distribution can also be used to estimate transport of sorbed pollutants (e.g. nutrients or pesticides) via sediment particles, provided sorption coefficients are known.

Acknowledgements

The authors wish to acknowledge the funding of the Water Research Commission (Pretoria, South Africa) and the contributions of M. Fey (University of Stellenbosch, Department of Soil Science), T. Scheepers, J. Twahirwa, I. Harun, A. Ali, G. Frantz, P. Meyer and B. Lawrence (University of the Western Cape, Department of Earth Sciences).

References

[1] Wischmeier, W.H. & Smith, D.D., *Predicting Rainfall Erosion Losses*, USDA Agricultural Handbook No. 537, 1978.

[2] Shirazi, M.A., Boersma, L., Johnson, C.B. & Haggerty, P.K., Predicting physical and chemical water properties from relationships with watershed soil characteristics. *Journal of Environmental Quality,* **30**, pp. 112-120, 2001.

[3] Issa, O.M., Bissonais, Y.L., Planchon, O., Favis-Mortlock, D., Silvera, N. & Wainwright, J., Soil detachment and transport on field- and laboratory-scale interrill areas: Erosion processes and the size-selectivity of eroded sediment. *Earth Surface Processes and Landforms,* **31(8)**, pp. 929-939, 2006.

[4] Wu, Q., Riise, G., Lundekvan, H., Mulder, J. & Haugen, L.E., Influences of suspended particles on the runoff of pesticides from an agricultural field at Askim, SE-Norway. *Environmental Geochemistry and Health,* **26(2)**, pp. 95-302, 2004.

[5] Beuselinck, L., Steegen, A., Govers, G., Nachtergaele, J., Takken, I. & Poesen, J., Characteristics of sediment deposits formed by intense rainfall events in small catchments in the Belgian Loam Belt, *Geomorphology*, **32(1-2)**, pp. 69-82, 2000.

[6] FAO, *World Reference Base for Soil Resources*, Food and Agricultural Organization of the United Nations, Rome, Italy, 1998.

[7] Mie, G., Beiträge zur Optik trüber Medien, speziell kolloidaler Metallösungen, *Annalen der Physik (Leipzig),* **330**, pp. 377-445, 1908.

[8] De Clercq, W.P., Fey, M.V. & Jovanovic, N., *Land Use Impacts on Salinity in Western Cape Waters*. Water Research Commission Report No. K5/1503: Pretoria, South Africa, 2010.

Determination of LC50 of Diazinon toxin and linear anionic detergents on *Rutilus frisii kutum*

A. Tehranifard[1], M. Fazeli[2] & J. Yarabbi[3]
[1]*Department of Marine Biology, Islamic Azad University of Lahijan, Iran*
[2]*Tarbiatmodares University, Iran*
[3]*Adamson University, Philippines*

Abstract

Diazinon is a pesticide that is commonly used in the rice fields of northern Iran The harmful effect of Diazinon and two anionic detergents were determined on *Rutilus frisii kutum*, individually and together on a laboratory scale. Fish were adapted to laboratory circumstances, distributed species wise in 500 litre tanks. Experiments were performed according to the TCR standard procedures. The LC50- 96 h was 0.34 mg/l for Diazinon. The LC50- 96 h in same species fingerlings were 4.69 and 12.24 mg/l under liquid and powder detergents, respectively. The LC50- 96 h were 7.27 and 0.9 mg/l for a mixture of Diazinon and liquid detergent and a mixture of Diazinon and powder detergent, respectively. The result showed that the toxicity of the mixture of Diazinon and powder detergent was greater than other mixtures. Generally, the detergent concentration toxicity rate (LC50- 96 h) in the ecosystem is less than the laboratory scale, but detergent can mix with some chemicals in the ecosystem and become stronger. The mortality rate will increase under this condition.
Keywords: LC50- 96 h, acute toxicity, anionic detergent, Diazinon, Rutilus frisii kutum.

1 Introduction

Today, pesticides are used for many different purposes, such as agriculture, human and animal health protection, pest control in forest and aquatic environments, and the protection of buildings and other structures. The uses of pesticides are increasing because the target pests become stronger, and therefore, large amounts of detergents contaminate soil and water annually (Young [5]). Pesticides sources in the environment include those resulting from direct pesticide application for a

WIT Transactions on Ecology and the Environment, Vol 135, © 2010 WIT Press
www.witpress.com, ISSN 1743-3541 (on-line)
doi:10.2495/WP100181

specific purpose, such as those used for weed and insect control in aquatic environments, and those entering indirectly from spray drift, atmospheric precipitation runoff and erosion from agricultural lands, effluent discharges from sewers and factories, accidental spills and volatilization. The quantity of a pesticide moving with runoff and sediment to an aquatic environment depends on many factors. These include topography, intensity and duration of runoff or irrigation, soil erodibility, land management and cropping practices [2].

Pollution magnitude varies with aquatic environment properties, such as surface area and depth, and hydraulic characteristics. However, the highest pesticide residues are in rivers; lower residues occur in estuaries and the lowest are in the oceans. Pollution magnitude in lakes and reservoirs depends on their proximity to an agricultural or industrial area (Tamim and Weigmann [2]). Biological communities are sensitive to their chemical environment, although the degree of sensitivity varies among species and communities. The use of pesticides in Iranian agriculture is common, but from an environmental protection of point of view it is not controlled as effectively as it ought to be. If the fields are too near to the aquatic ecosystem the pesticides may be washed out from the treated soil and pollute the aquatic ecosystem by runoff. This is particularly dangerous in rice fields. In this case the water is directly tested and in connection with rivers or standing waters [3].

There are approximately 600,000 hectares of rice fields, and houses near the rivers use detergents daily in the north of Iran. There is no ecotoxicological data about the harmful effect of pesticides and detergents individually or in mixed concentrations on aquatic organisms in Iran.

For this reason we consider that the investigation of some special Anionic detergents and Diazinon are important. This is the first ecotoxicological effects investigation on fish in Iran.

This research should correctly imply an at least approximate equivalence of fish species in their susceptibility to chemical compounds under the conditions described in the official guidelines (OECD [1]).

2 Material and methods

2.1 Diazinon

Is commonly used in rice fields of northern Iran were bought from pesticides shop in Bandar Anzali. Active ingredient: EC 60% (Diazinon)

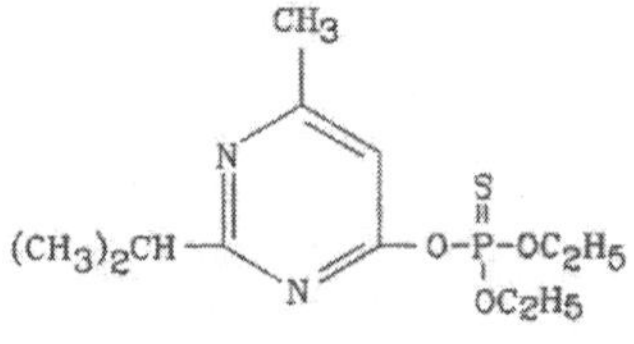

Figure 1: Chemical formula. Water solubility: in water at room temperature 40 mg/l. Toxicity: tech (rat): Oral LD50 300-400 mg/kg. 3.5mg/L (4hr). Rabbit. inhalation: LC50.

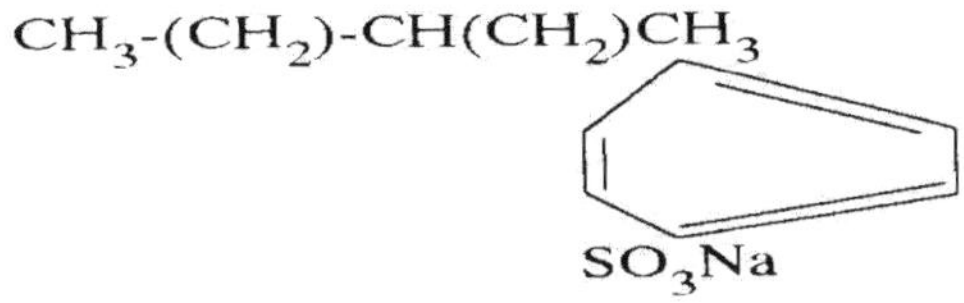

Figure 2: L.A.S formula.

2.2 Test organism

Rutilus frisii kutum (White fish), was bought from Shahid Ansari fish culture and propagation station.

At the Gilan Fisheries Research Center, which was given the task of studying the effect of pesticide and detergents on fish were performed in similar equipments. The temperature of the room was regulated to 25±2°C. There were two light sources of 70×100cm (20w and 60cm long). Duration of illumination (14h/day) was automatically regulated [4].

Fish were adapted to laboratory circumstances distributed species wise in 500 litre tanks. Experiments for determining Diazinon, 24-hr, 48-hr, 72-hr, 96-hr LC10, LC50 and LC90 were performed according to the TCR standard procedures (TCR [4]).

Different test concentrations plus a control were used. 1-2 g fingerling fish (*Rutilus frisii kutum)* was used for each concentration series. A total of these replicates were carried out for each one. Ten fingerling fish (1-2g weight) were placed in a 30 litre aquarium containing 20 litres tap water for each test concentration and control. Pesticides stock solution was prepared with distilled water and four concentrations of them plus a control were chosen. Each treatment consisted of three replicates. During of the test (4 day) the organisms were not fed and the medium was not renewed during the bioassay. Observations were made at each 24-h and the mortality results recorded. Control survival was always 100%. LC10, LC50, LC90, (the concentration of the toxicant that reduced the survival rate to 10%, 50% and 90%) were calculated using linear regression analysis [5].

3 Results

3.1 Influences of the investigated pesticide Diazinon on *Rutilus frisii kutum*

According to three different experiments with various concentrations of Diazinon on *Rutilus frisii kutum* (0.3 to 5 mg/l) in comparison with the control, mortality was increased especially with the high concentration. LC50-96hr was 0.34 mg/l.

The results are summarized in Table 1 and Figure 3.

Table 1: Effect of Diazinon on *Rutilus frisii kutum.*

Diazinon	Lethal concentration (p.p.m)	24hr	48hr	72hr	96hr
Rutilus frisii kutum	LC10	0.24	0.14	0.18	0.14
	LC50	1.7	0.08	0.49	0.34
	LC90	12.057	4.954	1.34	0.8

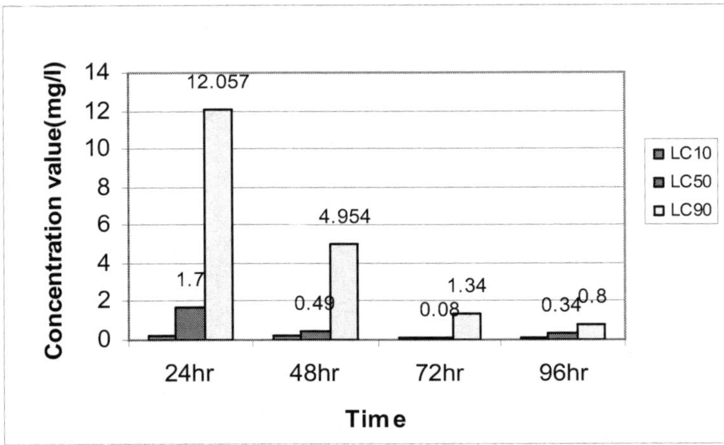

Figure 3: Influence of Diazinon on *Rutilus frisii kutum.*

Table 2: Effect of powder detergent on *Rutilus frisii kutum.*

Powder detergent	Lethal concentration (p.p.m)	24hr	48hr	72hr	96hr
Rutilus	LC10	11.54	9.88	9.54	9.07
frisii	LC50	14.67	12.95	12.65	12.24
kutum	LC90	18.63	16.98	16.77	16.50

3.2 Influence of anionic detergents on *Rutilus frisii kutum*

3.2.1 Influence of powder detergent

Concentrations of powder detergent in three experiments were from 10mg/l to maximum 24mg/l in comparison with the control. LC50-96hr was 12.24mg/l [1]. The effective concentrations are summarized in Table 2 and Figure 4.

3.2.2 Influence of liquid detergent

Concentrations of liquid detergent in three experiments were from 5mg/l to maximum 25 mg/l in comparison with the control. LC50-96hr was 8.93mg/l [1].

The effective concentrations are summarized in Table 3 and Figure 5.

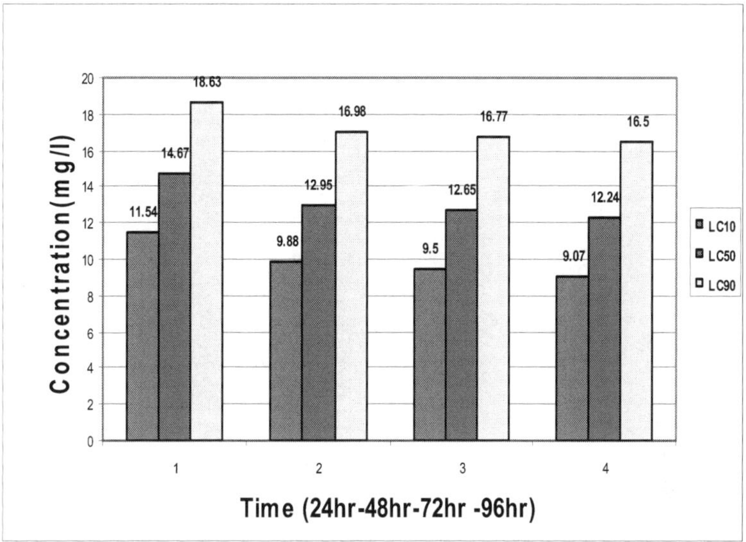

Figure 4: Influence of powder detergent on *Rutilus frisii kutum.*

Table 3: Effect of liquid detergent on *Rutilus frisii kutum.*

Liquid detergent	Lethal concentration (p.p.m)	24hr	48hr	72hr	96hr
Rutilus frisii kutum	LC10	7.80	6.05	7.24	6.59
	LC50	11.68	11.06	10.08	8.93
	LC90	17.49	13.2	16.89	15.40

3.3 Influence of mixed pesticide and anionic detergents on *Rutilus frisii kutum*

3.3.1 Mixed Diazinon and powder detergent

According to three different experiments with various concentrations of mixed Diazinon and Powder Anionic Detergent (1-4mg/l) the mortality of fish increased especially in high concentration.LC50-96hr was 0.9 mg/l [1].

The results summarized in Table 4 and Figure 6.

3.3.2 Mixed liquid detergent and Diazinon

According to three different experiments with various concentration of mixed Diazinon and Liquid Anionic Detergent (4-20 mg/l) the mortality of fish increased especially in high concentration.LC50-96hr was 7.28 mg/l. Mixed Diazinon and Powder Detergent with comparison of mixed Diazinon and Liquid Detergent was highly toxic[4].

The results summarized in Table 5 and Figure 7.

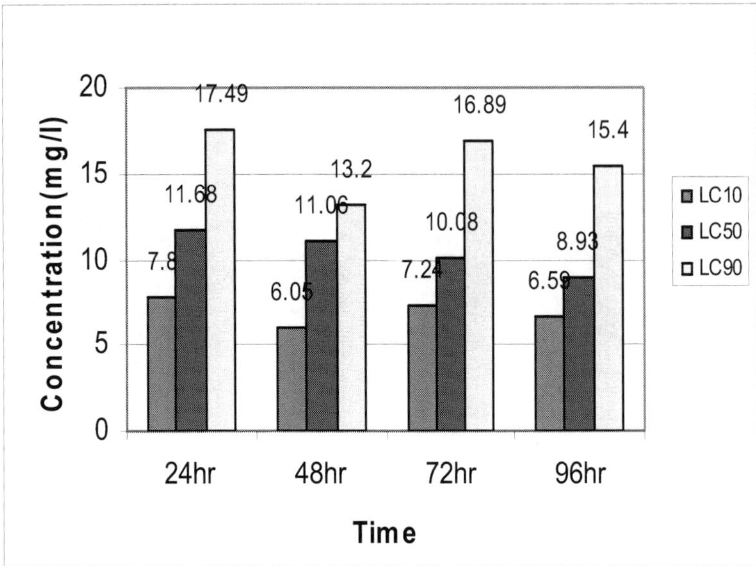

Figure 5: Influence of liquid detergent on *Rutilus frisii kutum.*

Table 4: Effect of mixed Diazinon and powder detergent on *Rutilus frisii kutum.*

Mixed Diazinon & powder detergent	Lethal concentration (p.p.m)	24hr	48hr	72hr	96hr
Rutilus frisii kutum	LC10	1.1553	1.0055	0.9446	0.5152
	LC50	1.7313	1.5677	1.4762	0.8970
	LC90	2.5945	2.4444	2.3067	1.5615

4 Discussion

This study indicates that effect of the pesticide and two detergents individually and together were different. The results showed that the LC50-96h was 0.34 mg/l for Diazinon. The LC50- 96 h for the fingerling of the same species under the effect of liquid and powder detergents, were 8.93and 12.24 mg/l respectively. The LC50- 96 h under the effect of a Diazinon and liquid detergent and the mixture of Diazinon and powder detergent were 7.27 and 0.9 mg/l respectively was 0.34 mg/l. According to the results, Diazinon was highly toxic, and it is LC50 – 96hr. It is clear that if Diazinon enters lagoons, water bloom will occur. In spring and summer, if the farmers are going to use pesticides, we have a serious problem with water bloom in fish ponds, especially in Anzali lagoon.

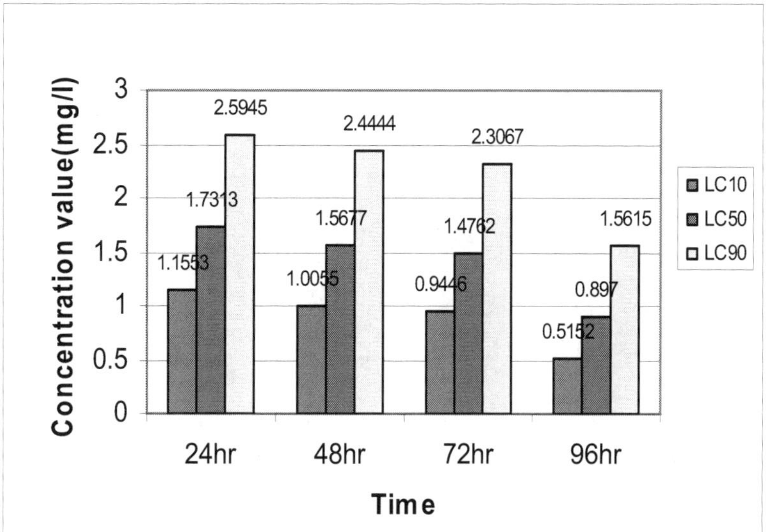

Figure 6: Influence of mixed Diazinon and powder detergent on *Rutilus frisii kutum*.

Table 5: Influence of mixed Diazinon and liquid detergent on *Rutilus frisii kutum*.

Mixed Diazinon & liquid detergent	Lethal concentration (p.p.m)	24hr	48hr	72hr	96hr
Rutilus frisii kutum	LC10 LC50 LC90	8.1995 12.2831 18.4061	7.4043 10.9760 16.2708	3.8637 7.632 16.009	3.6386 7.2759 13.703

Thousands of fish may die annually and it is very dangerous for our aquatic ecosystems. Results showed that the most important reason for this problem is the mixture of detergents and pesticides. These mixtures are very toxic in general; it can be said that sometimes the concentration rate of detergents in a given ecosystem is less than the calculated LC50-96hr in a laboratory condition, but when the same concentration rate of detergent in ecosystem is mixed with another chemical material, it is very much possible that the percentage rate of mortality in fishes rises even higher. During the first 15 minutes of exposure, there was a rapid increase in the rhythm of opercula movements. Fish would also swim to the surface, gasping for air. A side from increased opercula rates, exposed fish also exhibited dark colorations on its dorsal side, along the entire

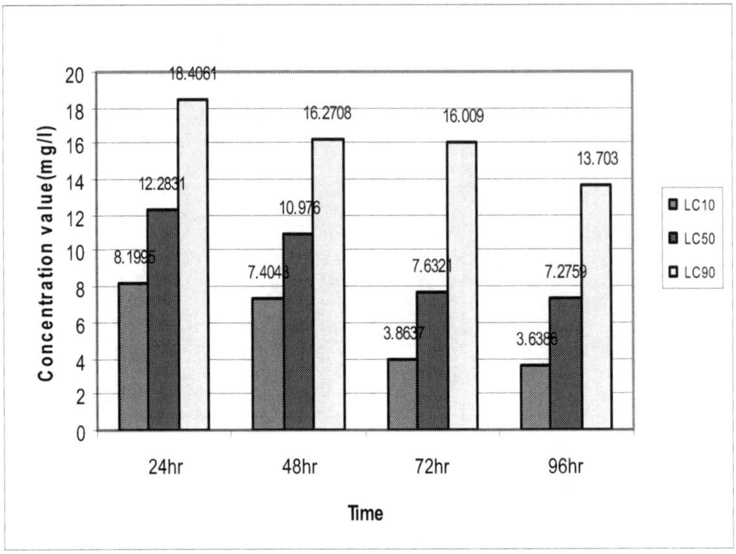

Figure 7: Influence of mixed Diazinon and liquid detergent on *Rutilus frisii kutum.*

dorsal length. Then the fish started to lose equilibrium, swimming in a vertical head down position, and setting on the bottom vertical side up. It would lay quietly, opercula movement becoming irregular with occasional bursts of frenzied swimming. Fish were also less sensitive to noise or movement. Death occurred within 3-9hr after the body darkening reaction started. This pigment modification of the dorsal surface of both investigated fish typically resembled fish under stress (Tamse and Gacutan [3]). Eyeballs become expanded and abnormal behaviour of fish increased with the increase of concentration of pesticide and detergents. In the all experiments dissolved oxygen in the aquatic medium of fish was always near to 80-100% of air saturation. The pH values have significantly decreased from 7.57 at concentration in aquatic medium to minimum 6.1.Hardness of water in the experiments was increased from 326 to 658 mg/l. When test were finished (96hr), the control aquaria showed a very clean aquatic medium, but aquaria with pesticide and detergents individually and in mixed showed a very turbid medium. This turbidity was probably caused by the fish residues or the suspended compounds or its degradation products on water.

The objective of ecotoxicological laboratory tests and risk evaluation after all is the protection of the natural ecosystems and their inhabitants.

References

[1] OECD, Guidlines for testing of chemicals. Part C. Test Guidline.2031984a.

[2] Tamim, M.Y., and D.L. Weigmann, Pesticides: a continuing dilemma. Journal water pollution control federation. Washington D.C. pp: 1199-1205, 1988.
[3] Tamse, C.T., and R.Q. Gacutan Acute toxicity of Nifurpirnol, a fish chemotherpeutant to milk fish (*Chanos chanos*) fingerlings pp: 346-350, 1994.
[4] T.R.C., OECD Guidelines for testing of chemicals section 2. Effects on biotic systems pp: 1-39, 1984.
[5] Young, A.L., Minimising the risk associated with Pesticide use. An overview in pesticides – minimizing the risks. N.N. Ragsdale and R.J. Kuhr (Eds). ACS Symp. Ser. 336. Amer. Chem. Soc. Washington. D.C. 1987.

Section 6
Biosystems

Microbial parameters as indicators of mariculture impact on the marine environment

G. Caruso & G. Zappalà
Istituto per l'Ambiente Marino Costiero (IAMC), National Research Council, Italy

Abstract

Microbial abundance and extracellular activity rates (leucine aminopeptidase, LAP, beta-glucosidase, β-GLU, and alkaline phosphatase, AP) were studied in water and sediment samples for two years in three Sicilian mariculture sites. Microbial activities were measured with specific fluorogenic substrates; heterotrophic culturable bacteria (CFU) were counted on Marine agar plates. The amount of particulate organic matter (POM) and its composition in terms of bio-polymeric fractions (proteins, carbohydrates, lipids) were also monitored as indicators of trophic conditions. Cage sites were sampled before and after the set up of each off-shore mariculture plant, and compared to "control" sites. The study aimed at monitoring the impact caused by fish farming activities on the marine environment and at assessing the suitability of these microbial parameters as indirect indicators of organic pollution.

Significant variations occurred both in microbial activity rates and in abundances at cage stations. Compared with the trophic parameters, the microbial parameters showed a more direct response to environmental perturbation due to fish farming. This suggested their use as indicators in assessing fish farming impact on both the pelagic and benthic compartments.

Keywords: mariculture, environmental impact, microbial parameters, extracellular enzyme activity, heterotrophic culturable bacteria, water, sediment.

1 Introduction

The increase in mariculture recorded in recent years has led to a consistent number of studies about the potential impact of rearing activities on the marine environment. The massive inputs of organic wastes (i.e. excretes, faeces, feed

www.witpress.com, ISSN 1743-3541 (on-line)
doi:10.2495/WP100191

wastes and medicinal substances) are the main cause of pollution occurring both in the pelagic and benthic compartments [1, 2]. Only a few studies have focused on the consequences of aquaculture on the functioning of the ecosystems [3-6] and on the need of new descriptors of environmental impact [7, 8]. Microbial decomposition of organic polymers through extracellular enzymes plays a major role in the turnover of organic matter in aquatic ecosystems [9]. It is well known that microorganisms are able to respond to organic matter enrichment and to the variation of the trophic conditions by modulating their activity rates [10, 11]. Previous researches [5, 6, 12, 13] have pointed out that both total prokaryotic abundance and enzyme activity rates undergo significant changes in association with mariculture or aquaculture practices.

A multidisciplinary study was performed to investigate the impact of off-shore mariculture in three Mediterranean sites (Castellammare Gulf, Capo d'Orlando and Portopalo), where experimental plants using off-shore cages were set up. A two-year investigation was carried out to quantify the impact of fish farming on the marine environment, both at pelagic and benthic levels. Particularly, some extracellular enzyme activities (leucine aminopeptidase, LAP, beta-glucosidase, β-GLU, alkaline phosphatase, AP), involved in the decomposition of proteins, polysaccharides and organic phosphates respectively, were estimated, together with the heterotrophic bacterial density and the content of particulate organic matter (POM) and of its biopolymeric fraction (proteins, carbohydrates and lipids). The research aimed to detect possible variations in the distribution of the examined parameters due to productive activities, and to evaluate the suitability of the microbial parameters (abundance and activity) as indicators of fish farming impact, in comparison with the parameters conventionally related to the trophic status, such as POM and organic biopolymers.

2 Materials and methods

2.1 Study areas and experimental design

The study sites were three marine areas used for mariculture. In two of them, Castellammare Gulf (site A) and Capo d'Orlando (site B), located along the North-Western and North-Eastern Tyrrhenian coast respectively, two experimental mariculture plants were set up, while in the third site (Portopalo, site C), located along the South-Western coast of Sicily, off-shore fish farming was performed using a previously-established productive plant (Sicily Fish Farm). The microbiological controls were performed on both seawater (0.5 m depth) and sediment (surface layer, 0-1 cm) samples. Four stations ("cage stations"), located near the rearing structures, were sampled per each of the sites A and B; two other stations, located 1 Km away from the cages, were used as "control stations". Four samplings were performed, two covering the period of cage set up, in 2004 (site A: May and June; site B: May and July), and the remaining two one year later, when the mariculture plant was already in an active productive phase. At site C only two samplings were carried out, both in July 2004 and 2005 (pre-impact and post-impact periods, respectively).

2.2 Studied parameters

2.2.1 Microbial parameters

Extracellular enzyme activities were determined using specific fluorogenic substrates. Water sub-samples (10 ml) were added with increasing amounts of the substrates L-leucine-7amido-4-methyl-coumarin hydrochloride (Leu-MCA), 4-methylumbelliferyl (MUF)-β-glucoside and MUF-phosphate for LAP, β-GLU and AP, respectively [14]. Concentrations ranging from 0.1 to 200 μM were used for LAP and AP, while for β-GLU slightly lower concentrations (0.1-100 μM) were applied; saturation was generally reached when a final concentration of 200 μM was used. Measurements were performed with a Turner TD700 fluorimeter, equipped with specific filters (excitation/emission wavelengths: 380/440 nm for LAP and 365/455 nm for AP and β-GLU) after 3 hours of incubation at "*in situ*" (± 1°C) temperature. Known concentrations (range: 20-80 μM) of methylcoumarine (MCA) and methylumbelliferone (MUF), were used as the standards for Leu-MCA and MUF-β-glucoside and MUF-phosphate respectively. The maximum velocity (V_{max}) of hydrolysis of the substrate was calculated through the Lineweaver-Burke transformation of each substrate concentration plotted *versus* the corresponding velocity of hydrolysis and reported as nM h^{-1}.

Sediment samples were re-suspended and diluted in sterile seawater, in a 1:10 (weight: volume) ratio. V_{max} was reported as μM h^{-1} and normalised to the dry weight of sediment.

Heterotrophic culturable bacteria (CFU) were counted by culture method on Marine agar 2216 (Difco) plates incubated at 22°C for 7 days. They were reported as the number of Colony Forming Units (CFU) per ml or g.

2.2.2 Biochemical parameters (POM and its biopolymeric fractions)

POM concentration was estimated after filtration of water samples on a Whatman GF/F membrane, and filters were processed at 980°C in a Perkin-Elmer CHN-Autoanalyzer 2400, using acetanilide as the standard [15]. The biopolymeric fraction [proteins (PRT), carbohydrates (CHO), and lipids (LIP)] was also determined [16-18].

2.3 Statistical analysis

Analysis of variance (ANOVA) was performed to evaluate the statistical significance of "*time*" (pre-impact *versus* post-impact periods) or "*space*" factors (cage *versus* control stations). Prior to analysis, logarithmic data transformation was carried out to attain their normal distribution. Pearson correlation coefficient *r* between microbial and biochemical parameters was also calculated.

3 Results

The mean values of enzyme activities and CFU obtained at each examined site are reported in Table 1, separately for water and sediment compartments.

3.1 Water

Mean temperature values ranged from 17.80 to 26.16 °C, 26.03 to 27.02 °C, 18.70 to 20.99 °C, in Castellammare Gulf, Capo d'Orlando and Sciacca, respectively; at the same sites, mean salinity values ranged from 36.24 to 37.72, 37.86 to 38.00, 36.35 to 37.80, respectively. Fig. 1 shows the mean enzyme activity and CFU values obtained in the waters of the sites A, B and C, at the cage and control stations. Fig. 2 shows the mean concentrations of both POM and its biopolymeric fractions measured at the three sites A, B, C.

The patterns of enzyme activities and POM evidenced a large variability depending on the studied site. At site A both the enzyme activity rates and CFU showed, on average, higher values at cage stations compared to control ones, although significant differences between the stations were found for LAP ($F = 5.98$, $P<0.05$) and CFU ($F = 11.37$, $P<0.01$) only. Comparing pre- with post-impact periods, significant changes were recorded in the mean values of LAP and AP, which were both particularly active at this site, and increased at cage stations about 38 and 23 times compared to their pre-impact values ($F = 7.12$, 19.86, $P<0.01$, respectively).

Mean β-GLU and CFU values also increased significantly over time at cage stations ($F = 8.34$, 7.87, $P<0.01$, respectively), confirming the rapid responsiveness to organic inputs of both microbial metabolism and abundance.

Mean POM values showed the abundance of organic matter at site A, which was higher, although not significantly, at control stations than at cage ones. A similar distribution was observed for the biopolymeric fraction. Peaks in the values of biochemical parameters occurred during spring; a significant inter-annual reduction was observed from pre-impact to post-impact periods ($F = 10.87$, 6.23, 6.36, $P<0.01$, for POM, PRT and CHO). This result could be related to the increase of both LAP ($r = -0.82$, $P< 0.01$, $n=8$ with POM) and β-GLU ($r = -0.63$, $P<0.05$, $n=8$ with POM), suggesting that the microbial community was actively metabolising the available organic matter.

At site B (Fig. 1), mean enzyme activities and CFU were lower than those found at site A throughout the experimental period. A different behaviour was observed for β-GLU, whose activity levels were high at this site. Enzyme activity values were generally higher at cage stations than at control ones, with significant differences detected both for LAP and CFU ($F = 6.31$ and 5.64, $P<0.05$, respectively). ANOVA confirmed a significant increase over years for LAP, AP, CFU ($F= 4.83$, 5.04, 4.50, $P<0.05$, respectively), and mostly for β-GLU ($F= 24.83$, $P<0.01$), suggestive of the cage impact on the environment.

Concerning the biochemical parameters, lower mean POM concentrations were recorded at site B compared to site A, both at cage and control stations. This site showed the minimum PRT and CHO concentrations at the control stations.

At cage stations, mean values of POM increased over time, particularly PRT, as demonstrated by ANOVA ($F= 6.83$, $P<0.01$ between pre- and post-impact periods), suggesting the water enrichment due to the release of organic wastes;

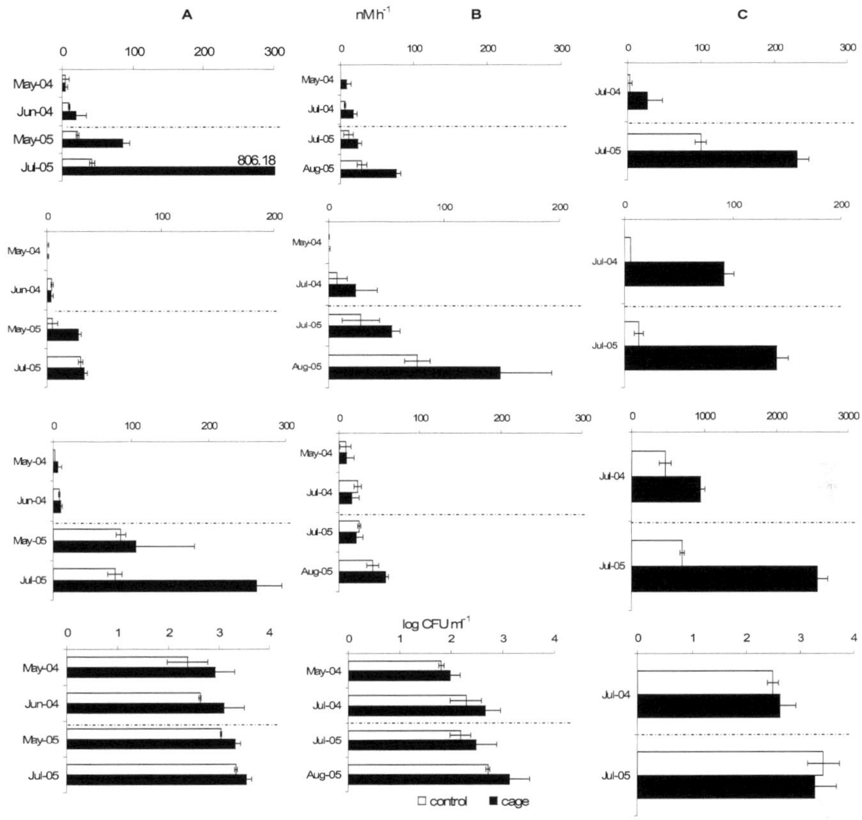

Figure 1: Water. Mean values ± s.d. of enzyme activity rates and heterotrophic bacteria obtained over time at cage and control stations of sites A, (Castellammare Gulf), B (Capo d'Orlando) and C (Sciacca). First line: LAP; second line: β-GLU; third line: AP; fourth line: CFU. Dotted lines separate the pre-impact period from the post-impact one.

conversely, mean CHO and LIP values remained almost unchanged over time (Fig. 2).

Site C was characterised by peaks of both AP and β-GLU and by LAP values which were intermediate between those detected at sites A and B. Similarly to other sites, both enzyme activities and CFU values were higher at cage stations, than at control ones. Only AP values displayed statistically significant spatial differences ($F = 8.24$, $P < 0.05$), while an inter-annual increase was observed for LAP, AP and CFU values ($F = 15.23$, $P < 0.01$; 6.34, $P < 0.05$; 16.49, $P < 0.01$ respectively).

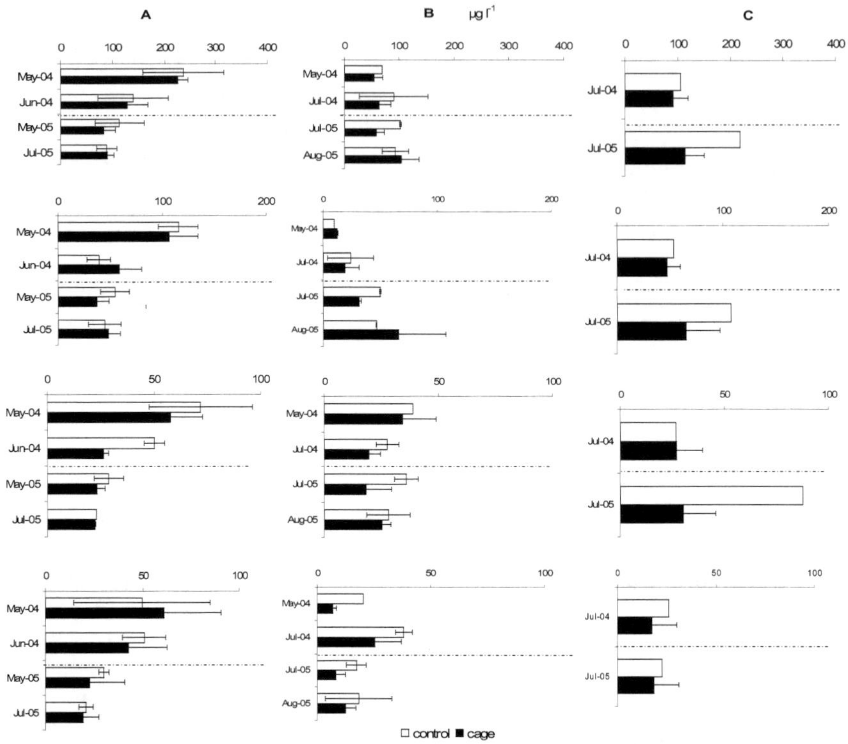

Figure 2: Water. Mean values ± s.d. of POM and its biopolymeric fractions. First line: POM; second line: PRT; third line: CHO; fourth line: LIP. Dotted lines separate the pre-impact period from the post-impact one.

Regarding the biochemical parameters (Fig. 2), POM concentrations in site C were intermediate between those recorded at sites A and B. At cage stations, both PRT and CHO reached their maximum values compared to those found in the other sites. Spatial and temporal variations were not significant.

3.2 Sediment

Fig. 3 shows the mean enzyme activity values and CFU obtained in the sediments of the sites A, B and C, at cage and control stations, while the mean concentrations of POM and biopolymeric fractions at the same sites is illustrated in Fig. 4. In sediments, mean enzyme activity values were, on average, three orders of magnitude higher than those found in waters. Particularly, site A was characterised by high AP, which showed activity values higher at cage stations compared to control ones, although no significant spatial differences were observed (Table 1). A similar result was found for LAP ($F= 26.77$, $P < 0.01$, cage *vs.* control) and CFU (Table 1). A significant increase of both AP and CFU values between pre-impact and post-impact periods was recorded (F= 9.31, P<0.01, 3.31, P<0.05, respectively).

POM concentrations measured at site A were higher at cage stations, perhaps due to the low hydrodynamism and low LAP and β-GLU activity rates of this site, which led to the sinking of organic matter in the sediment underneath the cages. At these latter, POM and its biopolymeric fractions (PRT, CHO, LIP) reached values of 3025.03 μg g^{-1}, 2366.99 μg g^{-1}, 482.13 μg g^{-1}, and 204.74 μg g^{-1} , respectively.

All the biochemical parameters showed at cage stations values significantly higher than those found at control ones (F = 12.23, P<0.01, 6.26, P <0.05, 11.29, 10.71, P<0.01, for POM, PRT, CHO and LIP, respectively). No significant inter-annual variations were observed by ANOVA.

At site B LAP and AP reached activity values which were about one order of magnitude higher than those measured at the site A. At the cage stations, both microbial activities and CFU were always higher than those measured at control

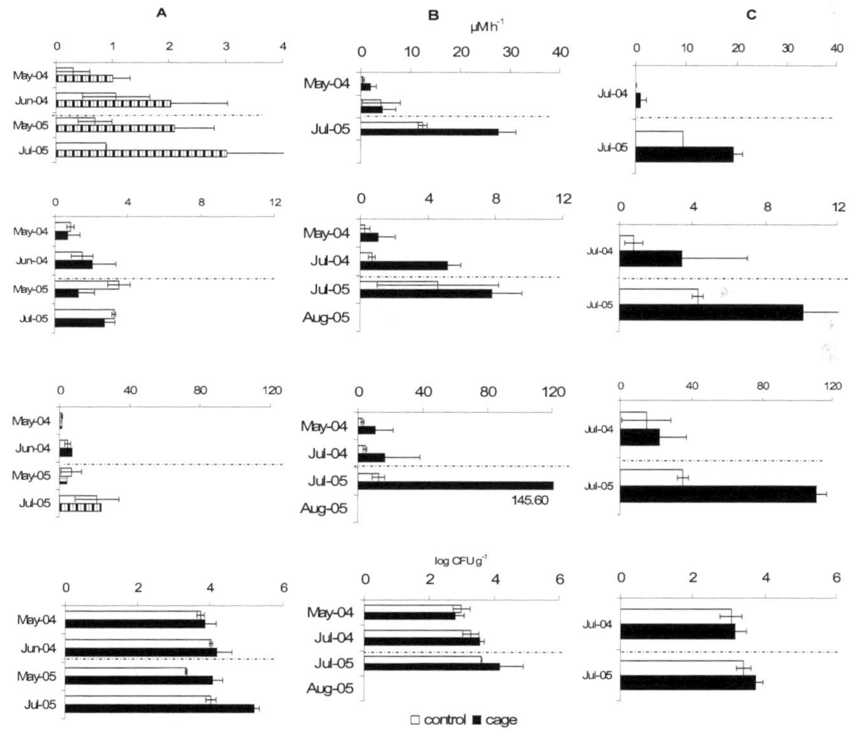

Figure 3: Sediment. Mean values ± s.d. of enzyme activity rates and heterotrophic bacteria obtained over time at cage and control stations of sites A, B and C. First line: LAP; second line: β-GLU; third line: AP; fourth line: CFU. Dotted lines separate the pre-impact period from the post-impact one.

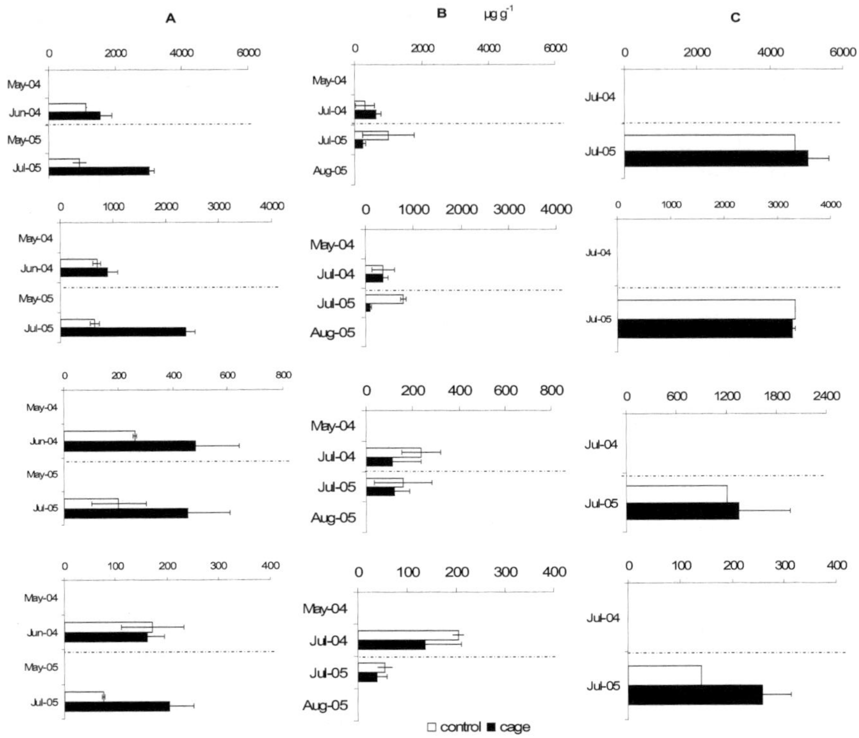

Figure 4: Sediment. Mean values ± s.d. of POM and its biopolymeric fractions. First line: POM; second line: PRT; third line: CHO; fourth line: LIP. Dotted lines separate the pre-impact period from the post-impact one.

stations; spatial significant differences were, however, recorded for β-GLU only ($F= 9.40$, $P < 0.01$). Conversely, significant variations over time were observed for LAP, β-GLU and CFU ($F= 10.79, 7.82, 8.08$, $P<0.01$, respectively).

Both POM and its biopolymeric fractions displayed averaged concentrations 2 or 3 times (for CHO or PRT, respectively) lower than those found at the site A; they sometimes were higher at the control stations. This distribution pattern could be due to the high hydrodynamism of this site, which was responsible of the spreading of organic matter also in areas not directly affected by the cage presence. Moreover, the reduction of PRT concentrations observed at the cage stations during the post- impact period could be related to higher decomposition mediated by LAP ($r = -0.95$, $P<0.01$, $n= 4$).

At site C, all the microbial parameters, and particularly AP, showed higher values at the cage stations with respect to those measured at the control ones. Site C was characterised by the lowest CFU values. Significant increases over time were found for LAP, AP and CFU at cage stations ($F= 28.36$, $P < 0.01$, 8.07

Table 1: Range of variation (before-after farming) of the mean values measured for each parameter in the waters and sediments of the studied sites. Water values expressed in nM h^{-1} (LAP, β-GLU, AP), ml^{-1} (log CFU), μg ml^{-1} (other parameters); sediment values expressed in μM h^{-1} (LAP, β-GLU, AP), g^{-1} (log CFU), μg g^{-1} (other parameters).

Parameter	Water		Sediment	
	Control	Cage	Control	Cage
LAP	4.38 - 41.86	4.69 - 806.18	0.31 - 1.07	1.02 - 3.03
β-GLU	0.76 - 29.19	0.77 - 32.24	0.86 - 3.53	0.68 - 2.75
AP	2.20 - 86.78	5.72 - 261.71	1.26 - 21.71	1.41 - 24.14
Log CFU	2.38 - 3.34	2.92 - 3.56	3.35 - 4.03	3.87 - 5.21
POM	88.51 - 237.38	83.17 - 225.63	911.05 - 1119.77	1529.63 – 3025.03
PRT	38.98 - 115.53	37.50 - 106.84	634.94 - 685.86	885.41 - 2366.99
CHO	22.92 - 71.94	22.34 - 57.57	200.40 - 262.05	453.31 - 482.13
LIP	20.92 - 50.83	19.10 - 61.23	72.72 - 171.87	162.10 - 204.74
LAP	0.15 - 28.97	8.52 - 76.33	0.48 - 12.40	2.01 - 27.64
β-GLU	0.42 - 77.46	0.51 - 148.85	0.27 - 4.61	1.07 - 7.82
AP	8.06 - 42.03	9.54 - 58.30	2.94 - 12.65	10.50 - 145.60
Log CFU	2.19 - 2.75	1.98 - 3.13	2.99 - 3.60	2.79 - 4.17
POM	69.22 - 103.48	53.65 - 103.29	310.13 - 1001.44	243.61 - 617.45
PRT	10.01 - 50.16	12.32 - 66.29	369.78 - 787.43	84.85 - 369.69
CHO	27.83 - 38.97	18.22 - 34.45	159.88 - 236.45	111.50 - 121.11
LIP	17.24 - 38.39	6.89 - 25.54	54.13 - 203.90	37.66 - 136.27
LAP	2.95 - 99.61	26.28 - 231.67	0.11 -9.30	0.95 - 19.32
β-GLU	5.64 - 13.22	91.99 - 140.32	0.77 - 4.30	3.40 - 10.11
AP	451.12 - 688.37	950.84 - 2572.20	14.15 - 35.26	21.96 - 111.12
Log CFU	2.49 - 3.44	2.62 - 3.29	3.07 - 3.41	3.27 - 3.74
POM	105.96 - 218.13	91.93 - 113.90	4680.38	5052.95
PRT	53.22 - 107.93	47.51 - 65.14	3335.68	3283.19
CHO	26.72 - 87.59	27.04 - 30.19	1204.45	1344.81
LIP	22.61 - 26.02	17.38 - 18.57	140.18	257.22

and 9.94, $P < 0.05$, respectively); this indicated a significant impact of mariculture activities on the benthic compartment. At this site, the highest concentrations of POM and of its biopolymeric fractions were reached.

4 Discussion

With the onset of mariculture activity, significant variations of the enzymatic activity rates were detected at all the three sites, particularly at the cage stations. The load of both organic substrates and nutrients released by the productive practices (faecal pellets, food residuals) induced a significant enhancement of the decomposition processes of the microbial community. LAP and AP were the enzyme activities most significantly influenced by mariculture activity; this result probably depended on changes occurring in the metabolic profiles of the microflora caused by the change in environmental conditions. It is well known that, in aquatic environments, microbial metabolism is significantly affected by the availability and quality of organic polymers [9]. Microbial enzyme activities measured in this study fall in a range of values comparable to those reported in eutrophic areas [19, 20], showing a significant response to the presence of allochthonous inputs released by the cages. The most significant changes in the examined parameters were detected at site A, both at pelagic and benthic compartments, and at site C, at the benthic level.

Significant differences in the activity rates from the pre-impact to the post-impact periods were sometimes observed also at the control stations, particularly at site B, suggesting that also these stations were affected by mariculture effects, due to the high hydrodynamism. Conversely to enzyme activities, CFU remained almost homogeneous at the different sites and did not show significant variations.

The concentrations of POM found in this study were comprised in a range typical of other Mediterranean areas [21], showing in the Castellammare Gulf a seasonal variability ($r = 0.786$, $P<0.05$, $n= 6$ with temperature). Compared to trophic parameters, microbial variables exhibited at all the examined sites a response more directly correlated to the presence of the disturbance factor, while organic matter did not show evident accumulation. This result could be due to the lability of organic inputs, which were quickly degraded and therefore did not cause irreversible alteration of the environmental equilibrium. Moreover, the inverse relationships found between the activity rates and POM content at the cage stations led us to hypothesise that organic inputs were quickly decomposed by enzyme activities or that the microbial community responded more to dissolved, rather than particulate, organic polymers.

The results pointed out that microbial variables can find a practical application in monitoring studies being sensitive indicators, already at an earlier phase, of changes occurring in the environment. In conclusion, the microbial variables could be proposed as a new useful tool to describe the environmental status and to monitor the interactions of productive activities on the environment, paying also attention to the pelagic compartment and not only to the benthic one.

Water parameters here presented were measured on samples taken the traditional way, with boats and sampling bottles; automatic water samplers integrated in buoy-based automatic measuring systems [22-25] or in towed ondulating vehicles [26] can give not only punctual, but time-space synoptic data to be used in complex early warning and environment monitoring networks, to better assess the mariculture impact and to help protect the capital investment from the arrival of pollutants that could damage the farmed fish or make them harmful to human health.

Acknowledgement

The authors thank Dr. A. Modica (ENI Syndial, Priolo) for the trophic data.

References

[1] Rosenthal, H., Weston, D., Gowen, R, Black, E., Report of the *ad hoc* study group on "Environmental impact of Mariculture". *ICES Statutory Meeting*, CM, 1897/F:2, 1987.

[2] Holmer, M., Black, K., Duarte, C.M., Marbà, N., Karakassis, I. (eds). *Aquaculture in the ecosystem*, Springer-Verlag: New York, 2008.

[3] Danovaro, R., Corinaldesi, C., La Rosa, T., Luna, G.M., Mazzola, A., Mirto, S., Vezzulli, L., Fabiano, M., Aquaculture impact on benthic microbes and organic matter cycling in coastal Mediterranean sediments: a synthesis. *Chemistry and Ecology,* **19**, pp. 59–65, 2003.

[4] La Rosa, T., Mirto, S., Favaloro, E., Savona, B., Sara', G., Danovaro, R., Mazzola, A., Impact on the water column biogeochemistry of a Mediterranean mussel and fish farm. *Water Research*, **36**, pp. 251–259, 2002.

[5] Vezzulli, L., Chelossi, E., Ricciardi, G., Fabiano, M., Bacterial community structure and activity in fish farm sediments of the Ligurian Sea (Western Mediterranean). *Aquaculture International,* **10**, pp. 123–141, 2002.

[6] Bedwell, M.S. & Goulder, R., Bacterial outputs from ponds and tanks used in intensive fish farming. *Letters Applied Microbiology,* **23**, pp. 412–416, 1996.

[7] Zenetos, A., Streftaris, N., Larsen, L.H., *An indicator-based approach to assessing the environmental performance of European marine fisheries and aquaculture*. European Environmental Agency: Copenhagen, Technical Report, **87**, 2002.

[8] Danovaro, R., Gambi, C., Luna, G.M., Mirto, S., Sustainable impact of mussel farming in the Adriatic Sea (Mediterranean Sea): evidence from biochemical, microbial and meiofaunal indicators. *Marine Pollution Bulletin,* **49**, pp. 325–333, 2004.

[9] Chrost, R. J., Microbial ectoenzymes in aquatic environments. *Aquatic microbial ecology: biochemical and molecular approaches*, eds. J. Overbeck & R.J Chrost, Springer-Verlag: New York, pp. 47–78, 1990.

[10] Thompson, A.J. & Sinsabaugh, R.L., Matric and particulate phosphatase and aminopeptidase activity in limnetic biofilms. *Aquatic Microbial Ecology*, 21, pp. 151– 159, 2000.

[11] Rath, J., Schiller, C., Herndl, G.J., Ectoenzymatic activity and bacterial dynamics along a trophic gradient in the Caribbean Sea. *Marine Ecology Progress Series*, **102**, pp. 89–106, 1993.

[12] Caruso, G., Genovese, L., Mancuso, M., Modica, A., Effects of fish farming on microbial enzyme activities and densities: comparison between three Mediterranean sites. *Letters Applied Microbiology*, **37**, pp. 324-328, 2003.

[13] La Rosa, T., Mirto, S., Mazzola, A., Danovaro, R., Differential response of benthic microbes and meiofauna to fish-farm disturbance in coastal sediments. *Environmental Pollution*, **112**, pp. 427–434, 2001.

[14] Hoppe, H.G., Use of fluorogenic model substrates for extracellular enzyme activity (EEA) measurement of bacteria. *Handbook of Methods in Aquatic Microbial Ecology,* ed. P.F. Kemp, B.F. Sherr, E.B. Sherr, & J.J. Cole, Boca Raton: Lewis Publishers, pp. 423–431, 1993.

[15] Iseki, K., MacDonald, R.W., Carmack, E., Distribution of particulate matter in the south-eastern Beaufort Sea in late summer. *Proc. of the NIPR Symposium for Polar Biology,* **1**, pp. 35–46, 1987.

[16] Hartree, E. F., Determination of proteins: a modification of the Lowry method that gives a linear photometric response. *Analytical Biochemistry*, **48**, pp. 422–427, 1972.

[17] Dubois, M., Gilles, K., Hamilton, J.K., Rebers, P.A., Smith, F., Colorimetric method for determination of sugar and related substances. *Analytical Chemistry*, **28**, pp. 350–356, 1956.

[18] Bligh, E.G. & Dyer, W., A rapid method for total lipid extraction and purification. *Canadian Journal of Biochemistry and Physiology*, **37**, pp. 911–917, 1959.

[19] Caruso, G. & Zaccone, R., Estimates of leucine aminopeptidase activity in different marine and brackish environments. *Journal Applied Microbiology,* **89**, pp. 951–959, 2000.

[20] Zaccone, R. & Caruso, G., Microbial hydrolysis of polysaccharides and organic phosphates in the northern Adriatic Sea. *Chemistry and Ecology*, **18**, pp. 85–94, 2002.

[21] Finenko, Z.Z. & Zaika, V.E., Particulate organic matter and its role in the productivity of the sea. *Marine Food Chains*, ed. J.H. Steele, Oliver & Boyd: Edinburgh, pp. 32–44, 1970.

[22] Zappalà, G., Caruso, G., Crisafi, E., The SAM integrated system for coastal monitoring. *Proc. of the Coastal Environment IV*, ed. C.A. Brebbia, WIT Press, Southampton (UK), pp. 341-350, 2002.

[23] Zappalà, G., Caruso, G., Azzaro, F., Crisafi, E., Integrated Environment Monitoring from Coastal Platforms. *Proc. of MEDCOAST 03*, ed. E. Ozhan, Middle East Technical University: Ankara, Turkey, **3**, pp. 2007-2018, 2003.

[24] Zappalà, G., Caruso G., Crisafi, E., Coastal pollution monitoring by an automatic multisampler coupled with a fluorescent antibody assay. *Proc. of Coastal Environment V*, eds. C.A. Brebbia, J.M., Saval Perez, L. Garcia Andion & Y. Villacampa. WIT Press, Southampton (UK), pp. 125-133, 2004.

[25] Zappalà, G., Azzaro, F., A new generation of coastal monitoring platforms. *Chemistry and Ecology*, **20**, pp. 387-398, 2004.

[26] Zappalà, G., Marcelli, M., Piermattei, V., Development of a sliding device for extended measurements in coastal waters. *WIT Transactions on Ecology and the Environment*, eds. D. Prats Rico, C.A. Brebbia, Y. Villacampa Esteve. WIT Press, Southampton (UK), **111**, pp. 187-196, 2008.

Toxic effect of polycyclic aromatic hydrocarbon metabolites on fish bone metabolism

K. Hayakawa[1], N. Suzuki[2], K. Kitamura[3], K. Bekki[1], J. Nakano[1], M. Yoshita[1], A. Toriba[1], T. Kameda[1] & N. Tang[1]
[1]*Institute of Medical, Pharmaceutical and Health Sciences, Kanazawa University, Japan*
[2]*Noto Marine Laboratory, Kanazawa University, Japan*
[3]*Division of Health Sciences, Graduate School of Medical Science, Kanazawa University, Japan*

Abstract

Marine water is contaminated with polycyclic aromatic hydrocarbons (PAHs) from rivers and the atmosphere. Oil spilled from tankers or petroleum factories also causes marine pollution with PAHs. It is well known that PAHs, such as benzo[*a*]pyrene, show carcinogenicity and/or mutagenicity. In developing teleost fish, furthermore, it has been reported that spinal deformity was induced in Pacific herring and pink salmon by PAHs, although the detail mechanism of toxicity in teleosts is not elucidated yet. As a toxic pathway of PAHs in animals, the metabolic activation of PAHs in the presence of P450 is considered. In teleosts, as well as in mammals, it has been reported that PAHs are converted into monohydroxylated polycyclic aromatic hydrocarbons (OHPAHs) by an enzyme: cytochrome P4501A1. Thus, OHPAH might have a toxic effect in teleosts. In the present study, we examined the estrogenic activity of OHPAHs using the yeast-two hybrid assay system with human estrogen receptor (ER) α. As a result, we detected estrogenic activity in 4-hydroxybenz[*a*]anthracene (4-OHBaA). Then, the direct effect of 4-OHBaA on osteoclasts and osteoblasts in teleosts was examined using the assay system with fish scales, which contain osteoclasts, osteoblasts, and the bone matrix of two layers (bony layer: a thin, well-calcified external layer; a fibrillary layer: a thick, partially calcified layer). When the scales of goldfish and wrasse were incubated with 4-OHBaA, which

WIT Transactions on Ecology and the Environment, Vol 135, © 2010 WIT Press
www.witpress.com, ISSN 1743-3541 (on-line)
doi:10.2495/WP100201

showed agonistic activity in the yeast-two hybrid assay system described above, the suppressing effect was observed for both osteoclasts and osteoblasts. The osteoclastic and osteoblastic activities did not change with 1-hydroxypyrene, which has no binding activity to human ERα. Thus, we conclude that PAH metabolites have a toxic effect on fish bone metabolism and that careful attention should be given to aquatic PAH contamination.

Keyword: polycyclic aromatic hydrocarbons, oil spill, yeast-two hybrid assay, osteoclasts, osteoblasts, teleost scale.

1 Introduction

Polycyclic aromatic hydrocarbons (PAHs) are ubiquitous environmental pollutants. They are formed through incomplete combustion of fossil fuel, wood and other organic materials including automobile exhaust, domestic heating and industrial processes.

Storm water runoff and atmospheric deposition of PAHs are now the largest sources of aquatic PAH contamination [1, 2]. Furthermore, an oil spill from an oil tanker, such as the Exxon Valdez and Nakhodka, directly induces contamination by PAHs in a marine environment [3-5]. In the Nakhodka C-heavy oil, 210 μg/g of benz[*a*]anthracene having four aromatic rings was detected [5]. In developing teleosts, it has been reported that spinal deformity was induced in Pacific herring and pink salmon by PAHs [6, 7] although the detail mechanism of toxicity in teleosts has not elucidated yet. As a toxic pathway of PAHs in animals, the metabolic activation of PAHs in the presence of P450 is considered. In teleost fish, as well as in mammals, it is known that PAHs are converted into monohydroxylated polycyclic aromatic hydrocarbons (OHPAHs) by the enzyme cytochrome P4501A1 [7-10]. Thus, PAH metabolites such as OHPAH might have a toxic in teleosts.

In the present study, we examined estrogenic activity of OHPAHs using the yeast-two hybrid assay system with human estrogen receptor (ER) α. Then, the direct effect of 4-hydroxybenz[*a*]anthracene (4-OHBaA) on osteoclasts and osteoblasts in teleosts was examined using the assay system with fish scale, which contains osteoclasts, osteoblasts, and the bone matrix of two layers (bony layer: a thin, well-calcified external layer; a fibrillary layer: a thick, partially calcified layer) [11, 12]. Using the scale assay, furthermore, we compared the results with those of 1-hydroxypyrene (1-OHPy), which had no binding activity to ERα as a negative control.

2 Methods

2.1 Chemicals

Chemical structures of OHPAHs, quinoid PAHs (PAHQs) PAH ketones (PAHKs) and tested in this report are shown in Figs. 1 and 2. These chemicals were purchased from Kanto Chemical (Tokyo, Japan), Aldrich (Milwaukee, WI,

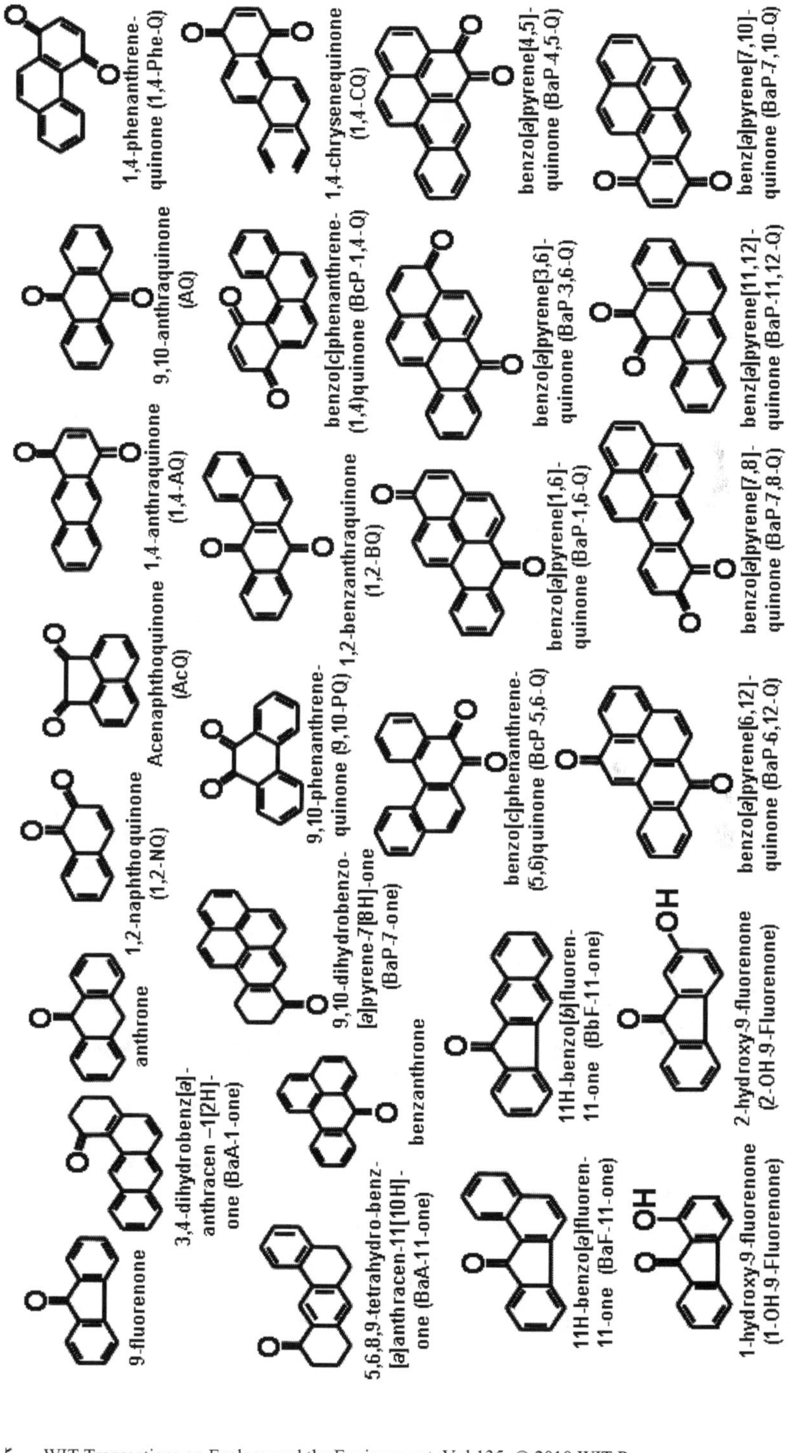

Figure 1: Chemical structures of OHPAHs.

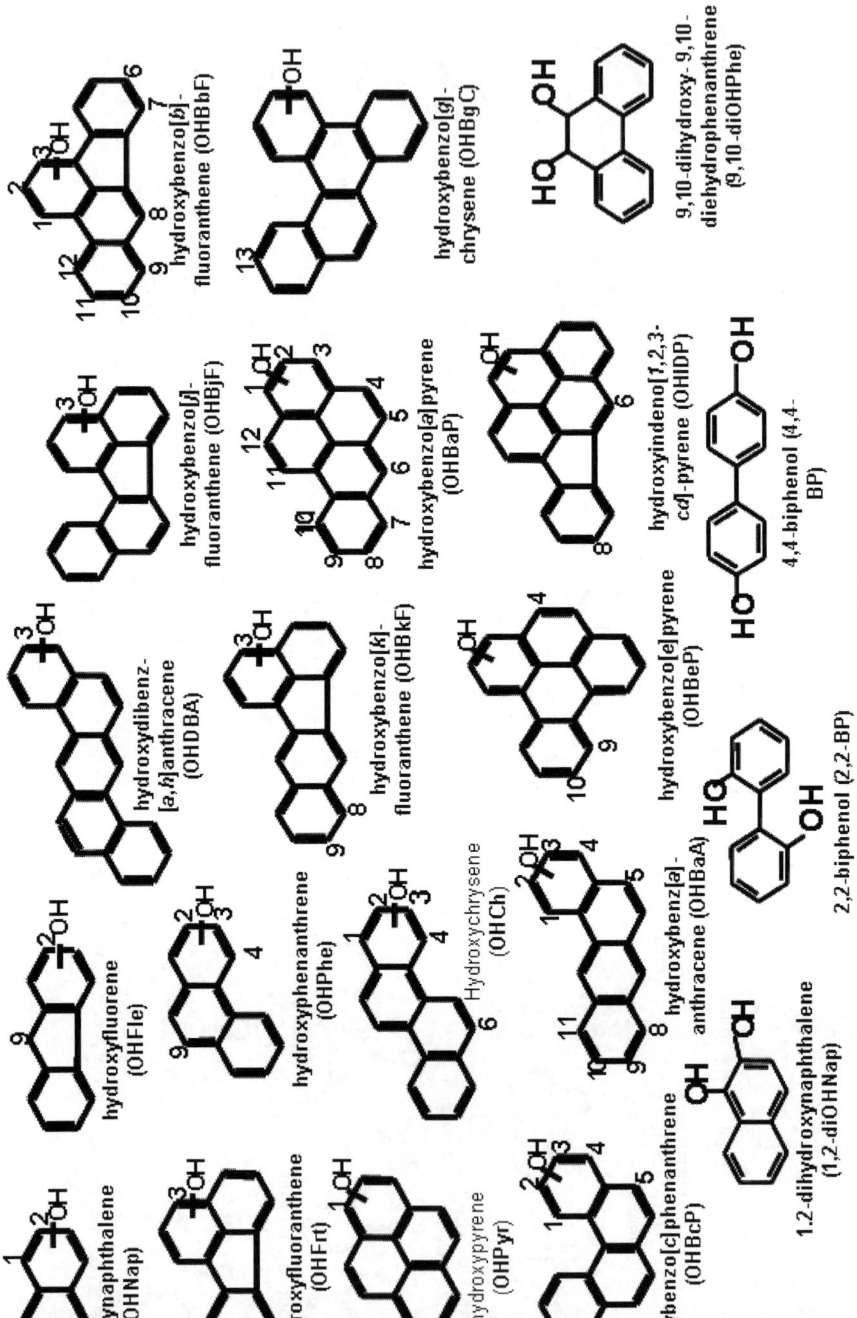

Figure 2: Chemical structures of PAHQs and PAHKs.

USA), Chiron AS (Trondheim, Norway), and NCI Chemical Carcinogen Repository (Kansas City, MO, USA). Test compounds were dissolved in ethanol and stored at -20°C until use. All other chemicals were of the highest quality available from commercial sources.

2.2 Assay of estrogenic activities

Yeast cells expressing human ERα were prepared according to the previous report [13]. Estrogenic activity of OHPAHs was evaluated by the yeast two-hybrid assay method using the above yeast cells as described [14]. To examine the agonistic activity of OHPAHs, the yeast cells were grown overnight at 27°C with shaking in synthetic defined medium free from tryptophan and leucine, and treated with each test compound in the concentration range from 1×10^{-9} M to 1×10^{-6} M at 27°C for 4 hrs. After the incubation, the treated cells were collected and enzymatically digested with 1 mg/ml Zymolyase 20T at 37°C for 20 min. 2-Nitrophenyl-β-D-galactoside (ONPG) was added to the lysate to a final concentration of 4 mg/ml. After incubation at 27°C for 20 min, the reaction was stopped by the addition of 1 M Na_2CO_3. The yeast debris was removed by centrifugation and β-galactosidase activity was assayed by measuring the absorbance of supernatant at 415 nm. Three independent experiments were run, and the mean values of the three were shown in this report. Relative effective potency of estrogenic activity (REP_E) was calculated as the value of the relative concentration of the test compound that gave the same activity of 17β-estradiol (E_2).

2.3 Animals

A previous study [15] indicated that sensitivity to calcemic hormones was higher in mature female than in mature male teleosts. Therefore, female goldfish (*Carassius auratus*) were purchased from a commercial source (Higashikawa Fish Farm, Yamato-koriyama, Japan), and their scales were used in an *in vitro* assay. As a marine teleost, female wrasse (*Pseudolabrus sieboldi*) caught in Tsukumo Bay of Noto Peninsula were used. These fish were kept under normal conditions until the start of experiments.

All experimental procedures were conducted in accordance with the Guide for the Care and Use of Laboratory Animals of Kanazawa University.

2.4 Effects of OHPAHs and E_2 on osteoclastic activity in cultured scales of goldfish and wrasse

The 4-OHBaA and 1-OHPy (NCI Chemical Carcinogen Repository), and E_2 (water-soluble type, Sigma-Aldrich, Inc., MO, USA) were purchased and used in the present study.

A 1% penicillin-streptomycin mixture (ICN Biomedicals, Inc., OH, USA) was added to Eagle's modified minimum essential medium (MEM; ICN Biomedicals, Inc.). HEPES (Research Organics, Inc., OH, USA) (20 mM) was added to MEM and adjusted to pH 7.0. After filtration, MEM was used in this experiment. Scales were collected from goldfish and wrasse under anesthesia with ethyl 3-aminobenzoate, methanesulfonic acid salt (Sigma-Aldrich, Inc.). The scales were cut into halves. One half of a piece was then placed into a well of a 24-well microplate in MEM (1 ml) supplemented with OHPAHs or E_2 (10^{-7}, 10^{-6}, and 10^{-5} M). The other half was also placed into a well of a 24-well microplate in an OHPAH/E_2-free medium as a control. Eight scales were used for each dose. These scales were incubated for 6 hrs at 15°C. After incubation, each incubated scale was washed with saline and transferred to its own well in a 96-well microplate. An aliquot of 100 μl of 20 mM tartrate in a 0.1 M sodium acetate buffer (pH 5.3) was added to each well. This microplate was frozen at -85°C immediately and kept at -20°C until analysis.

To analyze the tartrate-resistant acid phosphatase (TRAP) activity, an aliquot of 100 μl of 20 mM para-nitrophenyl-phosphate and 20 mM tartrate in a 0.1 M sodium acetate buffer (pH 5.3) was added to each well of a melted solution in the microplate. This plate was incubated at 20°C for 30 min while being shaken. After incubation, the reaction was stopped by adding 50 μl of a 3 N NaOH-20 mM EDTA solution. A colored solution of 150 μl was transferred to a new plate, and the absorbance was measured at 405 nm. The absorbance was converted into the amount of produced para-nitrophenol (pNP) using a standard curve for pNP. Detail methods were described in Suzuki *et al.* [11].

2.5 Effects of OHPAHs and E_2 on osteoblastic activity in cultured scales of goldfish and wrasse

Goldfish and wrasse were anesthetized with ethyl 3-aminobenzoate, methanesulfonic acid salt (Sigma-Aldrich, Inc.), and scales were collected from them. The scales were cut into halves. One half of a piece was then placed into a well of a 24-well microplate in MEM (1 ml) supplemented with OHPAHs or E_2 (10^{-7}, 10^{-6}, and 10^{-5} M). The other half was also placed into a well of a 24-well microplate as a control. Eight scales were used for each dose. These scales were incubated for 6 hrs at 15°C. After incubation, each incubated scale was washed with saline and transferred to its own well in a 96-well microplate. An aliquot of 100 μl of an alkaline buffer (100 mM Tris-HCl, pH 9.5; 1 mM $MgCl_2$; 0.1 mM $ZnCl_2$) was added to each well. This microplate was frozen at -85°C immediately and kept at -20°C until analysis.

The ALP activities were measured in the same manner as for the measurement of TRAP activity.

2.6 Statistical analysis

The statistical significance was assessed by the student's *t*-test. The selected significance level was $P < 0.05$.

3 Results

3.1 Estrogenic activity

The results are indicated in Figure 3. A significant increase of β-galactosidase activity ($REP_E > 1 \times 10^{-3}$) was observed for 3-, 4- and 10-OHBaAs and

2-hydroxychrysene (2-OHCh). These were all 4-ring OHPAHs. Among them, 4-OHBaA exhibited the strongest estrogenic activity ($REP_E = 7.5 \times 10^{-3}$). 2-hydroxyfluorene (2-OHFle), 2-hydroxuphenanthrene (2-OHPh) (3-ring), 3-hydroxyfluoranthene (3-OHFrt), 1-OHPy, 1-OHCh, 3-hydroxybenzo[*k*]fluoranthene, 12-hydroxybenzo[*b*]fluoranthene (4-ring) and 4-hydroxybenzo[*e*]pyrene (5-ring) were also active ($1 \times 10^{-3} > REP_E \geq 1 \times 10^{-4}$). The estrogenic activities of the other OHPAHs, PAHQs and PAHKs were much weaker or undetectable.

3.2 Effects of OHPAHs and E_2 on osteoclastic activity in cultured scales of goldfish and wrasse

The results are summarized in Table 1. In goldfish, the TRAP activity of the scales was significantly lowered by 4-OHBaA treatment at 6 hrs of incubation (10^{-7} and 10^{-6} M, $P < 0.05$, 10^{-5} M, $P < 0.01$), although the TRAP activity did not change from that of the control with 1-OHPy. E_2 significantly increased the TRAP activity at 6 hrs (10^{-6} and 10^{-5} M, $P < 0.05$).

In the scales of wrasse after 4-OHBaA (10^{-7} to 10^{-5} M) treatment, the TRAP activity also decreased at 6 hrs (10^{-6} M, $P < 0.05$, 10^{-5} M, $P < 0.01$). There was no change in the TRAP activity in 1-OHPy-treated scales. In wrasse as well as goldfish, a significant difference (10^{-5} M, $P < 0.01$) was obtained between E_2-treated scales and control scales.

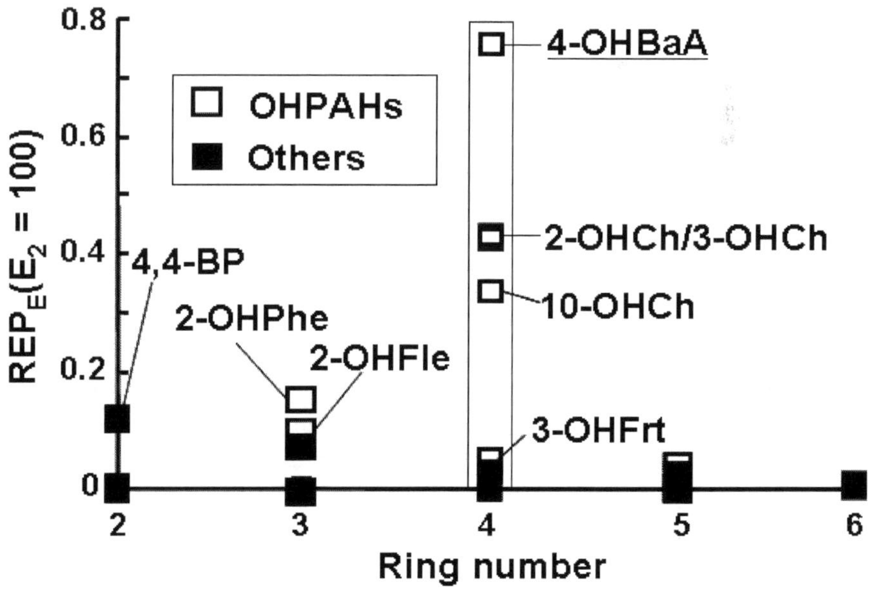

Figure 3: Relationship between the number of rings of OHPAHs and REP_E. REP_E indicates the relative effective potency of estrogenic activity.

Table 1: The toxicity of 4-OHBaA in osteoclasts and osteoblasts of the scales.

Goldfish (Fresh water teleost)	Osteoclasts	Osteoblasts
4-OHBaA	inhibition (10^{-7} M)	inhibition (10^{-5} M)
E_2	activation (10^{-6} M)	activation (10^{-6} M)
1-OHPy	no change	no change
Wrasse (Seawater teleost)		
4-OHBaA	inhibition (10^{-6} M)	inhibition (10^{-5} M)
E_2	activation (10^{-5} M)	activation (10^{-5} M)
1-OHPy	no change	no change

4-OHBaA: 4-hydroxybenz[*a*]anthracene
E_2: 17β-estradiol
1-OHPy: 1-hydroxypyrene

3.3 Effects of OHPAHs and E_2 on osteoblastic activity in cultured scales of goldfish and wrasse

The results are summarized in Table 1. In goldfish scales, at 6 hrs of incubation, the alkaline phosphatase (ALP) activity significantly decreased from the control values as a result of 4-OHBaA treatment (10^{-5} M, $P < 0.01$), while the ALP activity did not change with 1-OHPy treatment. E_2 significantly increased the ALP activity at 6 hrs of incubation (10^{-6} M, $P < 0.01$; 10^{-5} M, $P < 0.05$).

Similar changes were induced by 4-OHBaA in the scales of wrasse. The ALP activity significantly decreased in 4-OHBaA-treated wrasse scales at 6 hrs (4-OHBaA: 10^{-5} M, $P < 0.01$), although the ALP activity did not change from that of the control with 1-OHPy. In wrasse, a significant difference (at 6 hrs: 10^{-5} M, $P < 0.01$) was obtained between E_2-treated scales and control scales.

4 Discussion

The present study indicates that OHPAHs such as 4-OHBaA have an estrogenic activity using the yeast-two hybrid assay system with human ERα. In rat cytosol as well, 2-hydroxybenz[*a*]anthracene bound strongly to ER [16]. In the ERα reporter assay using a human breast cancer cell line (MCF-7), 3-OHBaA and 9-OHBaA had binding activity to ER [8]. Our data are supported by these reports.

In both freshwater and seawater teleosts, 4-OHBaA suppressed both osteoclastic and osteoblastic activities in the scales using our *in vitro* assay system while E_2 activated both osteoclastic and osteoblastic activities in the scales. E_2-specific binding [17] was detected in the scales of rainbow trout. However, a different action from E_2-treated scales was obtained in OHPAH-treated scales. In the case of bisphenol-A, which is bound to mammalian ERα [18], we previously reported that both osteoclastic and osteoblastic activities in the goldfish scales were significantly suppressed by bisphenol-A treatment [19]. In bony tissues including the scale of teleost, the β subtype of the ER is mainly expressed [20], while α subtype of ER is expressed in mammals [8]. This may be one reason for the anti-estrogenic-like action in OHPAH. To examine the effects of OHPAHs on bone tissues of teleosts in detail, plans are underway to conduct micro-array analyses using the scales of zebrafish.

The present study indicated that the strength of the inhibition activity in goldfish was stronger than that in wrasse. In goldfish, even at 10^{-7} M of 4-OHBaA, the osteoclastic activity was significantly inhibited. Exchange of calcium in the scales of freshwater teleosts may be faster than that in marine teleosts because they live in a low-calcium environment. In fact, the response of estrogen and calcitonin in the scales of freshwater teleosts was higher than that in those of marine teleosts [15, 21-23]. In the case of the scales, therefore, OHPAHs were more toxic in freshwater teleosts.

An oil spill from an oil tanker such as the Exxon Valdez and Nakhodka directly induces contamination by PAHs in a marine environment. In the C-heavy oil of the Nakhodka, 210 μg/g of BaA having four aromatic rings was detected [5]. This concentration is similar to that in the present experiment. Immediately after the oil spill, the high level of PAHs influenced marine animals, including fish. For a long time (more than 14 years), the toxicity of PAHs originating from an oil spill affected many marine animals (for a review, see Peterson *et al.* [24]). These facts, in conjunction with the findings of our study, indicate that careful attention should be given to aquatic PAH contamination.

5 Conclusion

We found that OHPAHs such as 4-OHBaA had an estrogenic activity using the yeast-two hybrid assay system with human ERα. This chemical suppressed both osteoclastic and osteoblastic activity in the scales of both freshwater and seawater teleosts. We believe that these phenomena are a cause of the disruption of the bone metabolism and the induction of spinal deformities in teleosts.

Acknowledgements

This study was supported in part by grants to K. Hayakawa (Grant-in-Aid for Scientific Research of ExTEND2005 and Global Environment Research Fund D-091 by the Ministry of Environment in Japan; Grant-in-Aid for Scientific Research (B) No. 21390034 and for Exploratory Research No.21651018 by Japan Society for the Promotion of Science), to N. Suzuki (Kurita Water and Environment Foundation; Grant-in-Aid for Scientific Research (C) No. 21500404 by JSPS), and to K. Kitamura (Grant-in-Aid for Scientific Research (C) No. 21500681 by JSPS).

References

[1] Lima, A. L. C., Eglinton, T. I. & Reddy, C. M., High-resolution record of pyrogenic polycyclic aromatic hydrocarbon deposition during the 20th century. *Environmental Science and Technology*, **37**, pp.53-61, 2003.
[2] Li, D. & Daler, D., Ocean pollution from land-based sources: East China Sea, China. *Ambio*, **33**, pp.107 113, 2004.
[3] Bue, B. G., Sharr, S. & Seeb, J. E., Evidence of damage to pink salmon populations inhabiting Prince William Sound, Alaska, two generations after the Exxon Valdez oil spill. *Transactions of the American Fisheries Society*, **127**, pp. 35-43, 1998.
[4] Heintz, R. A., Rice, S. D., Wertheimer, A. C., Bradshaw, R. F., Thrower, F. P., Joyce, J. E. & Short, J. W., Delayed effects on growth and marine survival of pink salmon *Oncorhynchus gorbuscha* after exposure to crude oil during embryonic development. *Marine Ecology Progress Series*, **208**, pp. 205-216, 2000.
[5] Hayakawa, K., Nomura, M., Nakagawa, T., Oguri, S., Kawanishi, T., Toriba, A., Kizu, R., Sakaguchi, T. & Tamiya, E., Damage to and recovery of coastlines polluted with C-heavy oil spilled from the Nakhodka. *Water Research*, **40**, pp. 981-989, 2006.
[6] Barron, M. G., Carls, M. G., Heintz, R. & Rice, S. D., Evaluation of fish early life-stage toxicity models of chronic embryonic exposures to complex polycyclic aromatic hydrocarbon mixtures. *Toxicological Science*, **78**, pp. 60-67, 2004.
[7] Billiard, S. M., Timme-Laragy, A. R., Wassenberg, D. M., Cockman, C. & Di, Giulio, R. T., The role of the aryl hydrocarbon receptor pathway in mediating synergistic developmental toxicity of polycyclic aromatic hydrocarbons to zebrafish. *Toxicological Sciences*, **92**, pp. 526-536, 2006.
[8] Charles, G. D., Bartels, M. J., Zacharewski, T. R., Gollapudi, B. B., Freshour, N. L. & Carney, E. W., Activity of benzo[*a*]pyrene and its hydroxylated metabolites in an estrogen receptor-α reporter gene assay. *Toxicological Sciences*, **55**, pp. 320-326, 2000.
[9] Jaruchotikamol, A., Jarukamjorn, K., Sirisangtrakul, W., Sakuma, T., Kawasaki, Y. & Nemoto, N., Strong synergistic induction of CYP1A1 expression by andrographolide plus typical CYP1A inducers in mouse hepatocytes. *Toxicology and Applied Pharmacology*, **224**, pp. 156-162, 2007.
[10] Mortensen, A. S. & Arukwe, A., Targeted salmon gene array (SalArray): A toxicogenomic tool for gene expression profiling of interactions between estrogen and aryl hydrocarbon receptor signaling pathways. *Chemical Research in Toxicology*, **20**, pp. 474-488, 2007.
[11] Suzuki, N., Kitamura, K., Nemoto, T., Shimizu, N., Wada, S., Kondo, T., Tabata, M. J., Sodeyama, F., Ijiri, K. & Hattori, A., Effect of vibration on osteoblastic and osteoclastic activities: Analysis of bone metabolism using goldfish scale as a model for bone. *Advance in Space Research*, **40**, pp. 1711-1721, 2007.

[12] Suzuki, N., Somei, M., Seki, A., Reiter, R. J. & Hattori, A., Novel bromomelatonin derivatives as potentially effective drugs to treat bone diseases. *Journal of Pineal Research*, **45**, pp. 229 234, 2008.
[13] Morito, K., Hirose, T., Kinjo, J., Hirakawa, T., Okawa, M., Nohara, T., Ogawa, S., Muramatsu, M. & Masamune, Y., Interaction of phytoestrogen with estrogen receptor α and β. *Biol. Pharm. Bull.*, **24**, pp. 351-356, 2001.
[14] Hirose, T., Morito, K., Kizu, R., Toriba, A., Hayakawa, K., Ogawa, S., Inoue, S., Muramatsu, M. & Masamune, Y., Estrogenic/antiestrogenic activities of benzo[*a*]pyrene monohydroxy derivatives. *Journal of Health Science*, **47**, pp. 552-558, 2001.
[15] Suzuki, N., Suzuki, T. & Kurokawa, T., Suppression of osteoclastic activities by calcitonin in the scales of goldfish (freshwater teleost) and nibbler fish (seawater teleost). *Peptides*, **21**, pp. 115-124, 2000.
[16] Ebright, R. H., Wong, J. R. & Chen, L. B., Binding of 2-hydroxybenzo[*a*]pyrene to estrogen receptors in rat cytosol. *Cancer Research*, **46**, pp. 2349-2351, 1986.
[17] Persson, P., Shrimpton, J. M., McCormick, S. D. & Björnsson, B. Th., The presence of high-affinity, low-capacity estradiol-17β binding in rainbow trout scale indicates a possible endocrine route for the regulation of scale resorption. *General and Comparative Endocrinology*, **120**, pp. 35-43, 2000.
[18] Singleton, D. W., Feng, Y., Yang, J., Puga, A., Lee, A. V. & Khan, S. A., Gene expression profiling reveals novel regulation by bisphenol-A in estrogen receptor-α-positive human cells. *Environmental research*, **100**, pp. 86-92, 2006.
[19] Suzuki, N. & Hattori, A., Bisphenol A suppresses osteoclastic and osteoblastic activities in the cultured scales of goldfish. *Life Sciences*, **73**, pp. 2237-2247, 2003.
[20] Yoshikubo, H., Suzuki, N., Takemura, K., Hoso, M., Yashima, S., Iwamuro, S., Takagi, Y., Tabata, M. J. & Hattori, A., Osteoblastic activity and estrogenic response in the regenerating scale of goldfish, a good model of osteogenesis. *Life Sciences*, **76**, pp. 2699-2709, 2005.
[21] Persson, P., Sundell, K. & Björnsson, B. Th., Estradiol-17β-induced calcium uptake and resorption in juvenile rainbow trout, *Oncorhynchus mykiss*. *Fish Physiology and Biochemistry*, **13**, pp. 379-386, 1994.
[22] Persson, P., Takagi, Y. & Björnsson, B. Th., Tartrate resistant acid phosphatase as a marker for scale resorption in rainbow trout, *Oncorhynchus mykiss*: Effects of estradiol-17β treatment and refeeding. *Fish Physiology and Biochemistry*, **14**, pp. 329-339, 1995.
[23] Guerreiro, P. M., Fuentes, J., Canario, A. V. M. & Power, D. M., Calcium balance in sea bream (*Sparus aurata*): The effect of oestradiol-17β. *Journal of Endocrinology*, **173**, pp. 377-385, 2002.
[24] Peterson, C. H., Rice, S. D., Short, J. W., Esler, D., Bodkin, J. L., Ballachey, B. E. & Irons, D. B., Long-term ecosystem response to the Exxon Valdez oil spill. *Science*, **302**, pp. 2082-2086, 2003.

Manganese absorption by water hyacinth and its study as a raw material in nanotechnology

M. C. Jiménez-Moleón[1], M. T. Mota-González[1]
& J. A. Ascencio-Gutiérrez[2]
[1]*CIRA, Universidad Autónoma del Estado de México, México*
[2]*ICF, Universidad Nacional Autónoma de México, México*

Abstract

Metal presence is a serious pollution problem in natural or treated water. The water hyacinth (*Eichhornia crassipes*) is an aquatic plant known for its high capacity to absorb metals. However, to use water hyacinths in a sustainable pollution removal process, the parameters that affect the process of metal absorption and the disposal of the contaminated organic matter must be studied; these being the aims of this work.

Water hyacinth assays were carried out in distilled water and the influence of Mn^{2+} initial concentrations (C_o: 10, 50, 100mg/L) and initial pH (pH_o: 3, 5, 7) were analyzed. Parameters such as pH, electric conductivity (EC), and residual Mn^{2+} were also measured. At the end of every assay, Mn was quantified in the different sections of the water hyacinth (root “R”, petiole “P” and leaf “L”) by atomic absorption spectrometry, and the presence of metal aggregates was detected by TEM microscopy.

E. crassipes did not neutralize its growing medium. The EC was not a good control parameter for absorption, increased with time. The water hyacinth was capable of efficiently removing Mn^{2+} from the solutions. It can be seen two phases of absorption (a rapid one followed by a slow absorption phase nearly linear). With increasing C_o, the absorbed Mn^{2+} increased, but the percentage of absorption during the rapid phase diminished. The final percentage of absorption increased with pH_o independently of C_o. The water hyacinth is capable of effectively translocating Mn. The bioconcentration factor (BCF) diminished with C_o, and increased with pH_o.

WIT Transactions on Ecology and the Environment, Vol 135, © 2010 WIT Press
www.witpress.com, ISSN 1743-3541 (on-line)
doi:10.2495/WP100211

Finally, aggregates of Mn^0 were detected in nanometric size (between 5 and 20nm) in all the sections, being suggested that *E. crassipes* can turn as raw material in nanotechnology.

Keywords: absorption, bioconcentration factor, E. crassipes, initial concentration, initial pH, Mn, nanotechnology, translocation, water hyacinth.

1 Introduction

Floating aquatic plants are an effective means of removing metals in water, as they have the ability to capture and accumulate them in their tissues. Although it has been declared as the worst aquatic weed worldwide [1], the macrophyte that seems to have the greatest potential for use in controlling polluted water, most cited and most dominant in tropical regions, is the water hyacinth, *Eichhornia crassipes* (Mart.) Solms. This has been the subject of several investigations to determine the adsorption capacity of organic and inorganic pollutants [2–4], although few cases have presented the influence of experimental conditions on the process. This article presents the effect on the absorption of Mn^{2+} that this macrophyte has at different initial metal concentrations and pH in neutral and acidic range, as the mobility of metal in solution is greatly reduced at basic pH values by the generation of hydroxides.

Despite the growing interest in soft technologies, because of its low operating and maintenance costs, they still represent an environmental problem, particularly when it comes to the removal of heavy metals in solution is the disposing should be given to the plant once it has succeeded in removing contaminant from the medium, at the expense of being transformed into a residual contaminated organic material. The present proposal is to examine the deposited metal within the water hyacinth, with the aim of analyzing whether they correspond to aggregates of nanometric size that can be used in one of the many nanotechnology fields, helping to consolidate the newly opened bridge between biotechnology and science of materials [5, 6], achieving a very favorable alternative solution to the problem of disposal of the water hyacinth biomass in which metal contaminants are concentrated.

2 Methodology

To realize the contact experiments, healthy plants with similar characteristics (size, wet weight, number of leaves, size and volume of roots) were collected in a water body of the valley of Toluca, State of Mexico, Mexico: Bordo de San Martín La Puerta.

In order to remove as much material and substances attached to the roots of the plants as possible 3 pretreatments were tested:

a) Rinse the plants with tap water, carefully trying to remove as much material attached to the root of the plants;

b) Wash roots gently with a sponge; and

c) Acid bath (pH 5).

After any of the above procedures, the water hyacinth was put in contact with a nutrient solution (900mL: 0.85 g/L $Ca(NO_3)_2 \cdot 4H_2O$, 2.00 g/L KH_2PO_4 and 3.69 g/L $MgSO_4 \cdot 7H_2O$ [7]) for 3 days. This nutrient solution was changed every 24 hours. Finally, to assess the most suitable pretreatment, three individual plants from each of the procedures specified above were exposed to a solution containing 10 mg Mn^{2+}/L and it was elected the pretreatment that allowed the least amount of residual metal after 10 days: pretreatment a). Thus, plants were washed thoroughly with tap water and kept in a nutrient-rich solution for 3 days, changing it every 24 hours.

Bioassays were conducted individually (three repetitions), at room temperature in 1L-vessels with 900 mL of distilled water. The initial concentrations of Mn^{2+} studied were 10, 50 and 100 mg/L and, with each concentration, the initial pH was adjusted to 3, 5 and 7, with ultrapure NaOH or HCl. The monitoring parameters were residual Mn^{2+} in water, pH and Electric Conductivity (EC), these two measures were polarographic while the Mn^{2+} concentrations were analyzed by atomic absorption, with equipment Varian, model Spectra 600, conditioned by graphite furnace or flame, depending on the concentration range to determine. Water samples were filtered with Whatman No. 41 filter paper (pore size 20-25 μm) and 1 drop of ultrapure $acid_3$ was added for preservation, which took place at 4°C until analysis.

At the end of each assay, the water hyacinths were sectioned in roots (R), petioles (P) and leaves (L), processing the main part of each segment to determine the amount of metal absorbed, while a small part was prepared for analyzing the possible presence of metallic aggregates generated inside the plant. Since the variation obtained in the residual Mn^{2+} in the different replicates was very consistent between them and, as discussed above, the plants were selected with similar characteristics, for determining the concentration of Mn within the sections of the plants, all the different parts of the water hyacinth corresponding to the repetitions of a determined assay were put together. Each section was dried in a porcelain dish in an oven at 100°C until constant weight, approximately 15 h, then the biomass was milled and the samples digested. It is important to note that several digestion techniques were used [8–11] including two techniques described in the Mars 5 microwave manual, it was necessary to make a change to one of these last techniques (table 1), to obtain transparent and colorless samples.

Table 1: Experimental conditions for microwave digestion of samples.

Stage	Power (%)	Ramp (min)	Temperature (°C)	Retention (min)
1	100	10:00	200	5:00
To wait the temperature diminished to 50 °C. Approximated time: 30 min.				
2	100	10:00	200	5:00

Finally, the presence of clusters of Mn^0 was studied through electron energy loss spectroscopy (EELS), optical microscopy and transmission electronic microscopy (TEM) using a Jeol JEM20 for TEM observations at low magnifications, and a Tecnai 200 TEM high-resolution images.

3 Results and discussion

With regard to the monitoring parameters behavior, during the different tests, regardless of the initial value, pH tended to stabilize slightly above 5, perhaps related to the equilibrium of CO_2 in water, so the water hyacinth did not neutralize its growth medium, against what has been reported in the literature [2, 12].

The minimum value of the EC responded to the initial measure and the maximum value was recorded in the last days of the testing, that is, the EC increased with time, demonstrating that it is not a good parameter for monitoring metal absorption in distilled water. In fact, the increase in EC was particularly strong towards the end of the trials, probably due to leaching of substantial quantities of inorganic and organic nutrients at the beginning of the decay phase [11, 13].

Fig. 1 shows the absorption of Mn^{2+} over time for the different initial conditions studied. In general, as it was reported in the literature for other metals [14–16], that when increasing C_0, the absorption rate decreased, so the lower concentrations required less contact time to achieve the same absorption percentage, probably because of the saturation of the absorption sites in the roots at higher concentrations.

On the other hand, the amount of metal absorbed by the end of the bioassays was clearly influenced by the initial pH value, as shown in fig. 2, increasing the pH_0, the amount absorbed also increased with an almost linear dependence. The absorption was less at lower pH values, might be due to the root potential damage and to the reduction in the number of sites in anionic form.

In fig.1, the existence of two absorption phases was also observed: an initial fast absorption phase followed by a slow almost linear one. The first absorption phase lasted 7 hours approximately, in which it was removed, on average, 51.1%, 29.4% and 18.8% for the initial concentrations of 10 mg/L, 50 mg/L and 100 mg/L, respectively, showing that the percentage of metal absorbed during the fast phase decreased with increasing C_0, the same trend as published in Cd absorption assays [14].

It is notable that, with the middle initial concentration, 50 mg Mn^{2+}/L, after the slow phase could be observed that the absorption rate was again increased, suggesting that the absorption process may have a cyclical behavior. Thus, as it is generally accepted, if the fast phase is where the metal absorption from the bulk takes place, while during the slow phase, what prevails (rather than the removal of metal from the liquid phase) is the release of absorption sites in roots due to the translocation of the metal to the aerial parts of the plant, it would be logical to think that once these new absorption sites appeared into the roots, the process rate would increase again, by a new phase of net absorption from the

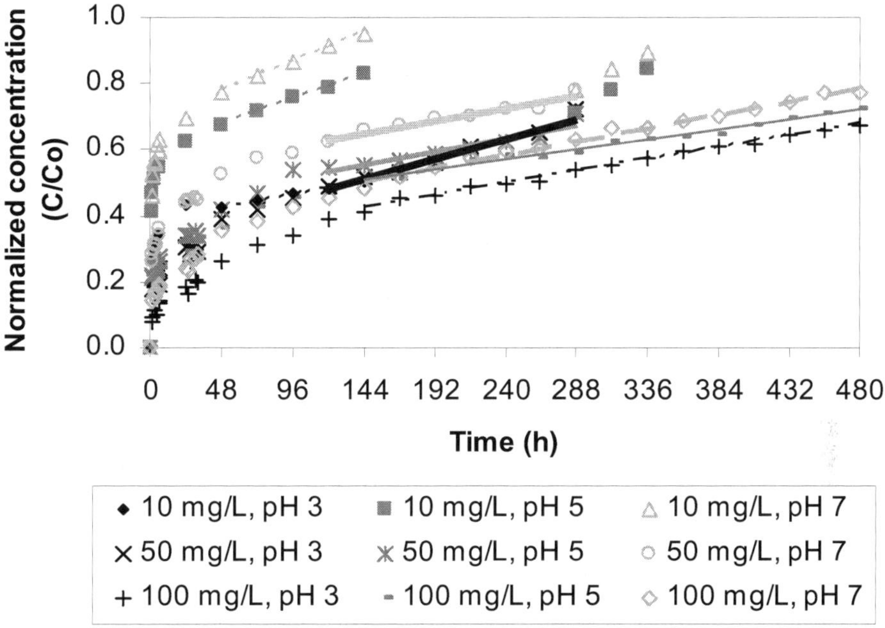

Figure 1: Normalized concentration (C/Co) vs. time.

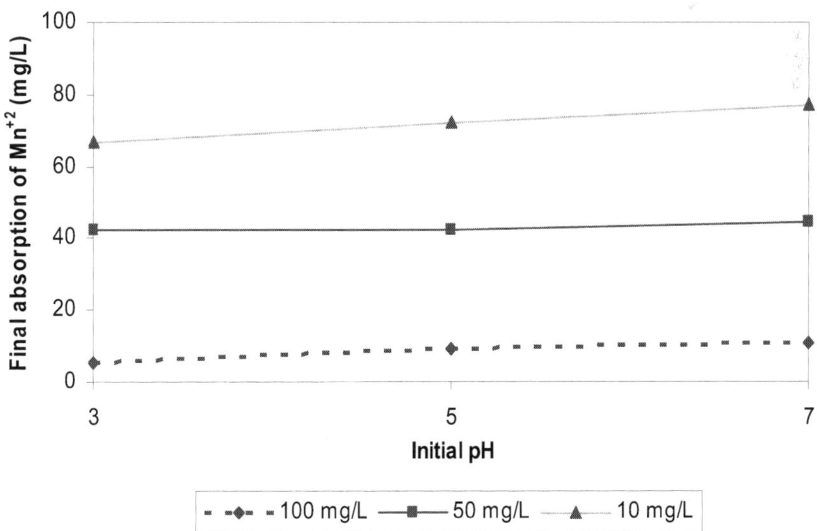

Figure 2: Final absorption of Mn^{2+} (mg/L) depending on the initial pH.

liquid phase. This phenomenon was not appreciated in the other initial concentrations studied, may be due to the low residual concentration at the end of the slow phase when it was on 10mg/L and to a contact time not sufficient in the case of higher concentration, 100 mg/L.

The Mn concentrations in each section within *E. crassippes* are shown in table 2, along with the percentage that they represented, related with the water hyacinth translocation capacity under each experimental conditions studied.

In the data presented in table 2, it was noteworthy that the concentration inside the plant was a function of the C_0 as well as the pH_0, increasing the metal concentration in the plant with any of these two parameters.

It is also shown in table 2 that the water hyacinth had a good capacity for Mn translocation. Even with the lowest concentration studied, independent of pH, the root was the section that reached lower metal concentration, perhaps because the contact time was enough to achieve a good translocation of Mn within the plant. While at 50 mg/L and 100 mg/L, the order of the Mn concentration was R > P > L, in similar percentages to those reported for other metals such as Cd, Pb, and Sr [10].

Table 2: Concentration of Mn within the water hyacinth, mg Mn/kg (dw) and %.

Initial conditions		**R**		**P**		**L**	
C_0, mg/L	**pH_0**	**mg/kg (dw)**	**%**	**mg/kg (dw)**	**%**	**mg/kg (dw)**	**%**
	3	**906**	**29**	**1224**	**40**	**960**	**31**
10	**5**	**1110**	**28**	**1244**	**32**	**1564**	**40**
	7	**1451**	**27**	**2151**	**40**	**1784**	**33**
	3	**4235**	**48**	**3433**	**39**	**1219**	**14**
50	**5**	**4755**	**48**	**3565**	**36**	**1486**	**15**
	7	**5966**	**52**	**3761**	**33**	**1704**	**15**
	3	**10016**	**53**	**6663**	**35**	**2216**	**12**
100	**5**	**10826**	**53**	**6942**	**34**	**2742**	**13**
	7	**16488**	**65**	**6642**	**26**	**2084**	**8**

Subsequently, the bioconcentration factor (BCF) was determined according to the following equation:

$$BCF = \frac{\text{metal concentration in plant} \ \left(\frac{mg}{kg}\right)}{\text{residual metal concentration in water} \ \left(\frac{mg}{L}\right)}$$

Table 3 showed that the BAF decreased when increasing the metal initial concentration as well as generally accepted: by increasing the metal concentration in water, the amount absorbed (table 2) increased but the bioconcentration factor decreased [2, 15, 17]. In table 3, the importance of the initial pH could also been seen, because the BCF was clearly superior at any of the initial concentrations under study, when that parameter was initially adjusted to the neutrality.

Thus, taking into account the concentration values within the plant (table 2) and the BCF data obtained (table 3), it can be concluded that water hyacinth is a major bio-accumulator of Mn, which can be used in the remediation of wastewater contaminated with this metal, especially at pH 7.

Table 3: Bioconcentration factors of Mn in water hyacinth, $\frac{mg/kg}{mg/L}$.

Initial Conditions		R	P	L
C_0, mg/L	pH_0			
10	3	168	227	178
	5	585	655	824
	7	2638	3911	3244
50	3	534	433	154
	5	622	467	195
	7	1126	710	322
100	3	335	223	74
	5	390	250	99
	7	799	323	101

Finally, in order to assess the possible reuse of the contaminated vegetal material in nanotechnology, it was analyzed the presence of metallic aggregates within *E. crassipes*, and its size.

As shown in fig. 3, in samples of roots and leaves from the bioassays with 100 mg Mn^{2+}/L and pH_0 5, it was detected the presence of Mn through EELS (fig. 3a) and it was possible to verify the existence of elementary aggregates in the samples analyzed (bright points in fig. 3b). Magnifying the image in fig.the sizes of the metal aggregates localized (fig. 3c) could be estimated: the two largest, present almost in the center of the image, measured about 20 nm diameter, two other particles (both located on the right side of the image, one at the top and one about half) were of size around 10 nm, and the rest of the aggregates had a size of about 5 nm.

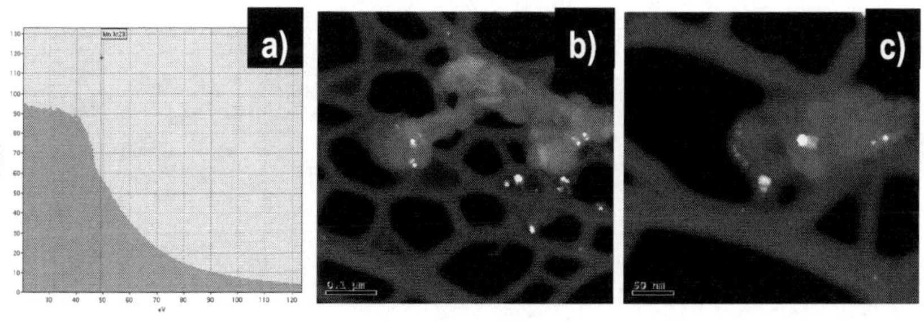

Figure 3: Root and leaf images by EELS and TEM, from bioassay at 100 mg/L and pH_o 5.

Furthermore, the results of the EELS and TEM analysis of samples of petiole from the bioassay at 50 mg/L and pH_o 7 are shown in fig. 4. The image 4a showed the presence of Mn, whereas in figs. 4b and 4c, the images obtained with TEM using clear field and dark field with a high-angle annular detector (HAADF) were shown, in both images an amorphous structure like a sponge can be seen, but in the dark field (fig. 4c) it was easier to distinguish the contrast due to the presence of metallic aggregates located at the top and at the bottom of the structure.

Thus, the formation of nanoparticles could be due to a state of located reduction state rather than to the interaction of the molecules with the metal, than which could imply the formation of microenvironments in those that the conditions around the metals were sufficient to reduce the metal and to derived in aggregates of Mn^0.

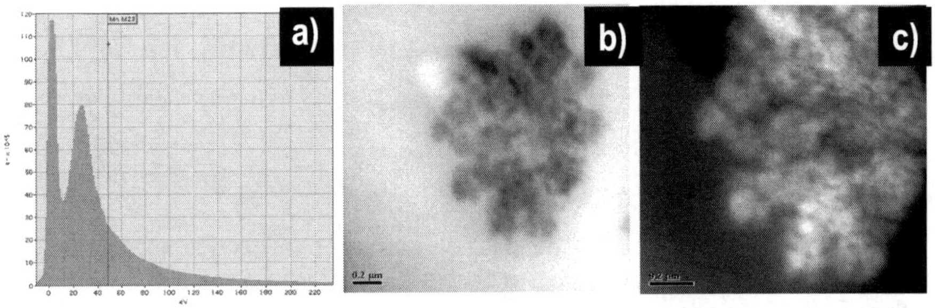

Figure 4: Petiole images by EELS and TEM, from bioassay at 50 mg/L and pH_o 7.

These results confirmed the recently opened bridge between biotechnology and science of science, getting an alternative economically and scientifically viable for the disposal of the metal contaminated vegetal material, so it could be considered in an environmental friendliness sustainable exploitation of the water hyacinth, so to change its name as the worst aquatic weed worldwide to nanotechnology raw material.

4 Conclusions

After analyzing the results presented in this work, it could be possible to reach the following conclusions:

- The water hyacinth did not neutralize the solution that was in contact with it, the pH during all the tests tended to that of the carbon dioxide /water system equilibrium.

- The EC was not a good parameter to control the metal absorption using distilled water.*E. crassippes*was able to eliminate Mn effectively over a wide range of concentrations and initial pH values.
- The final absorption percentage of Mn showed positive dependence with pH for all initial concentrations studied.
- There were 2 phases of absorption: an initial phase of rapid absorption, followed by a slow one, almost linear. With the intermediate concentration (50 mg/L), it was possible to observe a cyclical behavior of the absorption kinetics.
- By increasing the initial metal concentration in water, the metal concentration in the plant (mg/kg) increased but the BCF decreased.
- Mn is well translocated within the water hyacinth. The metal translocation was dependent on both pH and concentration, increasing with both parameters.
- The presence of Mn in nanometric size aggregates could be revealed, suggesting that *E. crassipes* can be seen as a raw material in the field of nanotechnology.

Acknowledgements

The authors wish to thank the National Council for Science and Technology of Mexico (CONACyT) and the Autonomous University of the State of Mexico (UAEMéx) for making this research possible through funding of Projects 91184 and 2506/2007U. Also, we want to thank the IBT-UNAM and the Universidad Michoacana de San Nicolás de Hidalgo for their technical support.

References

[1] Water hyacinth control and possible uses. Technical brief or Intermediate Technology Development Group, www.Hdg.org

[2] Cordes, K.B., Mehra, A., Farago, M.E. & Banerjee D.K., Uptake of Cd, Cu, Ni and Zn by the water hyacinth, Eichhnornia crassipes (Mart.) solms from pulverized fuel ash (PFA) leachates and slurries. *Environmental Geochemistry and Health,* **22**, pp. 297–316, 2000.

[3] Mehra, A., Farago, M.E. & Banerjeee, D.K., A study of Eichhornia crassipes growing in the overbank and floodplain soils of the River Yamuna in Delhi, India. *Environmental Monitoring and Assessment,* **60**, pp. 25-45, 2000.

[4] Fritioff, A., Kautsky, L. & Greger, M., Influence of temperature and salinity on heavy metal uptake by submerged plants. *Environmental Pollution,* **133**, pp. 265–274, 2005.

[5] Mirkin, C.A., Letsinger, R.L., Mucic, R.C. & Storhoff, J.J., A DNA-based method for rationally assembling nanoparticles into macroscopic materials. *Nature,* **382**, 607-606, 1996.

[6] Mirkin, C.A. & Taton, T.A., Materials chemistry: semiconductors meet biology. *Nature,* **405,** 626-627, 2000
[7] Boyce Thompson Institute, www.geocities.com/pbarrosvanc/tema12.html.
[8] Rosas, I., Báez, A., Belmont, R. & Gómez, S., Eichhornia crassipes como un indicador de la presencia de cadmio. *Revista geofísica, Instituto Panamericano de Geografía e Historia*, **12**, pp. 43-54, 1980.
[9] Kay, S.H., Haller W.T. & Garrard L.A., Effects of heavy metals on water hyacinths (Eichhornia crassipes (Mart.) Solms). *Aquatic Toxicology*, **5**, pp. 117–128, 1984.
[10] Mazen, A.M.A. & El Maghraby, O.M.O., Accumulation of cadmiun, lead y strontium and a role of calcium oxalate in water hyacinth tolerance. *Biologia Plantarum,* **40(3)**, pp. 411-417, 1998.
[11] Soltan, M.E. & Rashed, M.N., Laboratory study on the survival of water hyacinth under several conditions of heavy metal concentrations. *Advances in Environmental Research,* **7**, pp. 321–334, 2003.
[12] Akcin, G., Güldede, N. & Saltabas, O., Zinc removal in strongly basic solutions by water hyacinth. *Environmental Science and Health,* **A28(8)**, pp. 1727-1735, 1993.
[13] Vesk, P.A., Nockolds, C.E. & Allaway, W.G., Metal localization in water hyacinths roots from an urban wetland. *Plant, Cell and Environment* **22(2)**, pp. 149-158, 1999.
[14] O'Keeffe, D.H., Hardy, J.K. & Rao R.A., Cadmium uptake by the water hyacinth: effects of solution factors. *Environmental Pollution,* **34A**, pp. 133–147, 1984.
[15] Hardy, J.K. & Raber N.B., Zinc uptake by the water hyacinth: effects of solution factor. *Chemosphere,* **14(9)**, pp. 1155–1166, 1985.
[16] Wolverton, B.C. & McDonald, R.C., Bioaccumulation and detection of trace levels of cadmium in aquatic systems by Eichhornia crassipes. *Environmental Health Perspectives,* **27**, pp. 161-164, 1978.
[17] Wang, W. & Lewis, M.A., (eds). *Metal accumulation by aquatic macrophytes*, Plants for Environmental Studies: Lewis, Boca Raton, 1997.

Section 7
Wastewater treatment and management

Fate and removal of pharmaceuticals and personal care products (PPCPs) in a conventional activated sludge treatment process

S. Suarez, F. Omil & J. M. Lema
School of Engineering, University of Santiago de Compostela, Spain

Abstract

The fate and behaviour of 16 Pharmaceutical and Personal Care Products (PPCPs), including fragrances, hormones and medicines with different physico-chemical properties, was assessed during a conventional biological wastewater treatment process in a denitrifying/nitrifying pilot plant. The reactor was fed with a synthetic mixture that reproduced the chemical characteristics of a medium charged urban wastewater with an average composition of 500 mg/L of COD, 40 mg/L of $N\text{-}NH_4^+$ and 8 mg/L of $P\text{-}PO_4^{-3}$. Selected PPCPs were spiked to this feed at concentrations between 10 and 40 μg/L. The pilot plant has been operated continuously for three years under stable conditions. The occurrence of PPCPs along the different compartments of the pilot plant was determined in a first step by measuring their concentrations in the liquid phase. This was further complemented with detailed mass balances in which the most relevant removal mechanisms during biological treatment have been included (volatilisation, sorption and degradation).

The lowest removal efficiencies were obtained for carbamazepine, diazepam and diclofenac, whereas the highest transformations (>80%) were achieved for fragrances, fluoxetine, ibuprofen, naproxen and natural estrogens. Sorption has shown to play an important role in the biotransformation of musk compounds, which had previously shown not to be easily biodegraded, probably by enhancing their retention inside the pilot plant. The removal of the fragrance celestolide was additionally influenced by volatilisation in the aerobic, which supposed up to 45% of its overall elimination.

WIT Transactions on Ecology and the Environment, Vol 135,
www.witpress.com, ISSN 1743-3541 (on-line)
doi:10.2495/WP100221

The SRT of the plant had only an effect on the transformation degree of compounds with a significant sorption potential. The positive effect of operating at warmer temperatures was only significant for two antibiotics, for which an increase in their transformation degree of around 30% was determined.

Keywords: activated sludge, fragrances, hormones, influence of operational conditions, removal mechanisms, pharmaceuticals.

1 Introduction

Most of the current large-scale Sewage Treatment Plants (STPs) are based on Conventional Activated Sludge (CAS) reactors, operated at a Hydraulic Retention Time (HRT) of 4-14 h, which achieve very efficient removal of organic matter. Moreover, in the last two decades, important progresses regarding the simultaneous elimination of organic matter and nutrients have been achieved, in some cases driven by stricter legal requirements. More recently, when trying to improve the quality of water, the main focus shifted from conventional pollutants (organic matter, solids and nutrients) to more specific xenobiotic compounds, some of which were detected at the low μg/L level and therefore described as micropollutants, which include Pharmaceuticals and Personal Care Products (PPCPs). These observations showed that STPs receive a large number of different trace polluting compounds, although these installations have not been specifically designed to remove them. The reported overall removal rates of PPCPs in full-scale STPs vary strongly and clearly show that their elimination is often incomplete [1]. As a consequence, a significant fraction is discharged within the final effluent into the aquatic environment or sorbed onto primary and secondary sludge, whose deposition on land can be another significant pathway of releasing these substances into the environment.

The removal efficiency achieved for a given micropollutant in a STP will depend on several factors, such as the physico-chemical properties of the compound, the technology used and the operational conditions considered. Parameters such as Hydraulic Retention Time (HRT), Sludge Retention Time (SRT), redox conditions and temperature are thought to affect the removal of PPCPs. The HRT represents the mean time that the liquid phase remains within the treatment process. It was shown to affect elimination of ibuprofen and ketoprofen [2] in a way that lower removal was observed for shorter HRTs. The SRT determines the mean residence time of microorganisms in the reactor, consequently only organisms which are able to reproduce themselves during this time can be retained and enriched in the system. According to this definition, high SRTs allow the enrichment of slowly growing bacteria and consequently the establishment of a more diverse biocoenosis with broader physiological capabilities. For several PPCPs a positive effect on their removal has been observed when working at higher SRT and a critical value for this parameter of 10 d was identified [3]. Regarding redox conditions and temperature, differences in the removal efficiencies for some PPCPs have been reported [4, 5].

The vast majority of data published in the field of removal of PPCPs from wastewaters refer to full-scale STPs, where only the raw influent and final effluent are sampled, in order to measure soluble concentrations of the considered PPCPs. Therefore, only the overall removal efficiency including primary and secondary treatment can be determined, without distinguishing between sorption, volatilization or transformation [1]. The aim of the present work was to perform a detailed study of the fate and behaviour of 16 PPCPs in a pilot plant that represents the most common technology used in full-scale STPs. The reactor was fed with a synthetic medium in order to maintain a complete control of the system and to avoid the complexity of real wastewater, such as the presence of conjugates, metabolites or colloidal solids that could interfere with the reliable quantification of the considered substances in the influent. An extensive sampling including the different streams of the system was carried out so as to evaluate the influence of the different redox conditions (anoxic and aerobic) on the transformation of selected micropollutants. Additionally, the effect of temperature and SRT on the performance of the system was analyzed.

2 Material and methods

The reactor used is a conventional activated sludge plant divided into a first anoxic and a second aerobic compartment, supplied with a secondary sedimentation tank (fig. 1). The total useful volume of the reactor is 30 L, of which 40% correspond to the anoxic fraction and the rest to the aerobic compartment.

The pilot plant was inoculated with 2 g VSS /L taken from full-scale STP based on the Biodenipho® process. The reactor was fed with a synthetic mixture that reproduced the chemical characteristics of a medium charged urban wastewater with an average composition of 500 mg/L of COD, 40 mg/L of N-NH_4^+ and 8 mg/L of P-PO_4^{-3}. After a start-up period of 3.5 months, the set of PPCPs listed in table 1 was spiked to the feed at concentrations in the range of 10 to 40 μg/L.

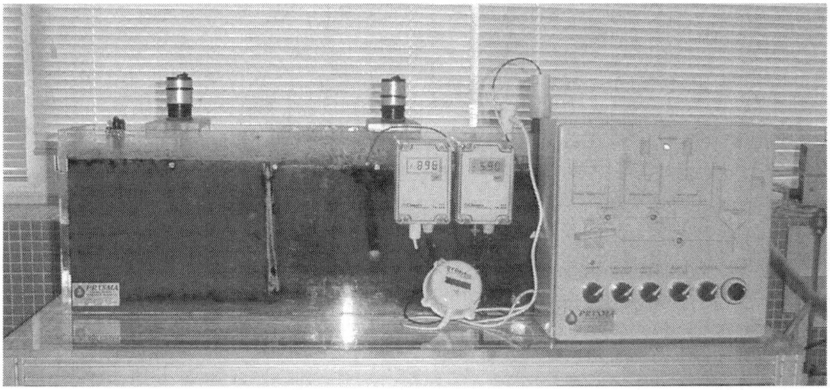

Figure 1: Conventional activated sludge pilot plant.

Table 1: Concentration of PPCPs in the feed.

Compound	C (μg/L)	Compound	C (μg/L)
Anti-depressants:		*Antibiotics:*	
Fluoxetine (FLX)	20	Trimethoprim (TMP)	
Citalopram (CTL)		Roxithromycin (ROX)	10
Estrogens:			
β-Estradiol (E2)	10	Sulfamethoxazole (SMX)	
α-Ethynylestradiol (EE2)		Erythromycin (ERY)	
Anti-inflammatories:		*Musks:*	
Ibuprofen (IBP)	10	Galaxolide (HHCB)	40
Naproxen (NPX)		Tonalide (AHTN)	
Diclofenac (DCF)		Celestolide (ADBI)	
Anti-epileptic:	20	*Tranquilliser:*	20
Carbamazepine (CBZ)		Diazepam (DZP)	

The pilot plant has been operated continuously during 3 years. The operational parameters HRT (1 d) and composition of the feed were not modified during the whole process. Temperature was not controlled and varied therefore according to the ambient temperature. Consequently, two operation periods could be differentiated, corresponding to moderate temperatures (14-18°C) and warmer values (18-23°C). In the case of SRT, the following three ranges were distinguished: <20 d, 20-40 d and >40 d. In some cases this variation was due to natural fluctuations of VSS concentrations inside the pilot plant and in the final effluent, whereas in other cases it was manipulated through purges.

Samples were collected from the feed, anoxic and aerobic compartments, external recirculation (from the settler to the inlet) and final effluent. These samples were analysed weekly for conventional operation parameters (COD, solids, nitrogen). Regarding the concentration of PPCPs, two samples per month were taken during the first four months and afterwards the sampling frequency was reduced to one campaign every 1-2 months. Analyses of PPCPs were performed after collecting 2-L samples in glass or aluminium bottles, which were immediately prefiltered (AP4004705, Millipore) and supplemented with a pinch of sodium azide (~0.3 g/L). For the analysis of PPCPs sample extraction based on Solid Phase Extraction (SPE) or Solid Phase MicroExtraction (SPME) was used as pre-concentration technique prior to their quantitative determination.

Liquid or Gas Chromatography coupled to Mass Spectrometry (LC-MS or GC-MS, respectively) was used for the final quantification [6-9].

3 Results and discussion

3.1 Pilot plant operation

Conventional operation parameters of the pilot plant remained stable, being the pH of the reactor around 8 without any adjustment and the dissolved oxygen concentration in the aerobic compartment always higher than 2.5 mg/L. Sludge concentration inside the reactor increased from the 2 g VSS/L inoculated at the start-up to 4 g VSS/L during the first 140 d, remaining at this value until day 590. After a period of around 130 d of decrease in sludge concentration due to a significant loss of biomass within the effluent combined with a slower sludge growth rate, it stabilised again at 2 g VSS/L. The nitrogen mass balance is correctly accomplished in the reactor (fig. 2) and very high removal efficiencies were achieved for both, COD and nitrogen (fig. 3).

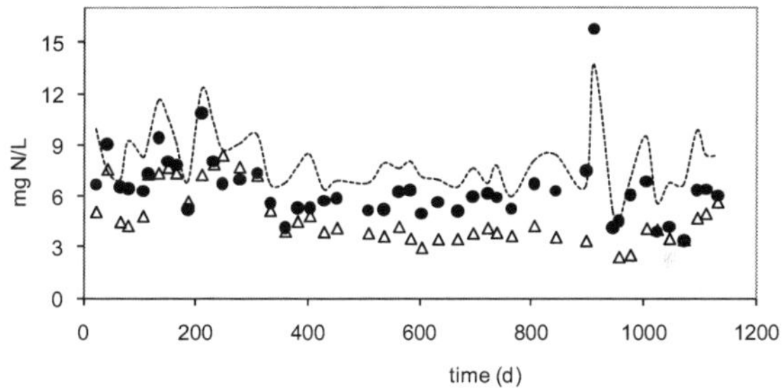

Figure 2: Concentration of nitrogen as ammonium in the anoxic tank experimentally measured (•) and calculated according to dilution (---), as well as nitrate in the aerobic compartment (Δ).

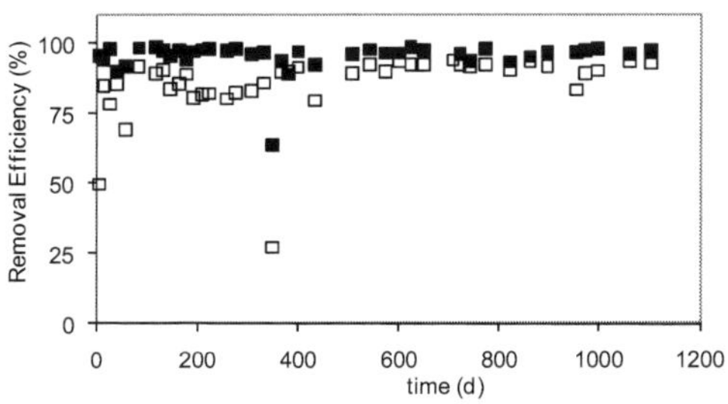

Figure 3: Nitrogen (□) and COD (■) removal efficiencies.

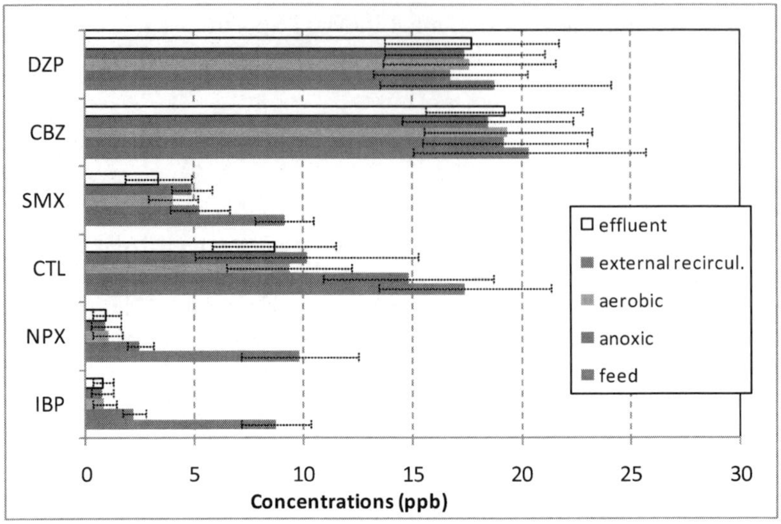

Figure 4: Concentration profile of selected PPCPs inside the pilot plant.

3.2 Fate of PPCPs along the pilot plant

The monitoring of the selected PPCPs led to a total of 20 samplings performed over five different points inside the pilot plant. These data have been summarized for some compounds in fig. 4 in which mean concentrations for the whole set of data were considered.

In fig. 4 different behaviours of PPCPs can be observed. On one hand, carbamazepine and diazepam were not significantly removed during the treatment as shown by the flat concentration profiles in fig. 4. In the case of CBZ the low removal percentage achieved after treatment in the pilot plant was consistent with previously reported data, which contributed to propose it as marker for anthropogenic influences on the aquatic environment by Clara *et al* [10]. Data about the removal of DZP during biological treatment in full-scale STPs are normally not provided due to the fact that concentrations of this substance are frequently below the limits of detection in the influent and effluent of the studied STP [11, 12]. Considering the sorption potential (log K_d 1.3) and biological degradation kinetics ($k_{biol} < 0.1$) of this compound, neither sorption nor degradation is expected to be significant, which was actually confirmed in the studied pilot plant. However, there are also some PPCPs, such as IBP and NPX which are removed very effectively during the treatment (removals > 85%). These very high removal rates for IBP have been previously observed in activated sludge full-scale STPs of Germany, Spain, England or Japan [13-16]. The performance of the pilot plant regarding NPX removal was slightly better than formerly reported in literature [11, 15, 16]. Finally, CTL and SMX are compounds with intermediate removal efficiencies (40-60%). While CTL is usually not included in previous works dealing with the removal of PPCPs during sewage treatment, for SMX, eliminations in the range of -138 up to 70% have been found in previous works [11, 17]. The variability in these data could be a consequence of working with real wastewaters which can contain not only the parent compound but also metabolites, as for example *N*^{4}-acetylsulfamethoxazole, which can subsequently be transformed back to SMX during biological treatment.

3.3 Mass balances of PPCPs

The removal processes considered in the mass balances of selected PPCPs include biological degradation, sorption and volatilisation. The relative contribution of each mechanism to the overall removal depends on the specific properties of each compound, mainly the tendency to be biologically degraded (commonly represented as biological degradation rate constant: k_{biol}), the sorption coefficient (K_d) and the Henry's coefficient (H), respectively. The total mass flows in the different streams and compartments of the pilot plant have been represented in fig. 5 for selected PPCPs, including the dissolved, sorbed and volatilised fraction of each compound.

In fig.5, CBZ represents the mass flow of a recalcitrant compound, for which the internal (i.e. from aerobic to anoxic tank) and external recirculation increases the amount of CBZ that enters the anoxic reactor with respect to the feed. The flat profile between the anoxic inlet and aerobic outlet is a clear indication that no removal of CBZ occurs inside the plant. In the case of IBP, very high removal efficiency is obtained (>90%), but transformation only occurred in the aerobic tank of the plant. This dependency of IBP degradation on redox conditions has previously been reported by Zwiener *et al* [18]. The behaviour of FLX is similar to that of IBP, that is a compound with a negligible sorption tendency although removed to a significant extent by biological degradation (removals > 70%), although in this case a partial removal already occurs in the anoxic compartment. Additionally, a slightly positive effect was observed when working at warmer temperatures, with an increase in the removal efficiency of around 10% (SRT < 20 d in fig. 5(c) and (d)).

In the case of the fragrances HHCB and ADBI, transformations of around 90% were reached. The contribution of volatilisation was significant on the total mass flow of ADBI that left the aerobic tank (fig. 5(f)), being its contribution to the overall removal of this compound in the range of 15-45%. The high contribution of sorption onto sludge on the total mass flow of musk compounds HHCB and ADBI in the different streams of the pilot plant is clearly shown in fig. 5(e) and (f), which is related to their strong lipophilic character. Kupper *et al* [19] reported degradation efficiencies for these fragrances of 40-50%. Additionally, Bester [20] confirmed a partial biological transformation of HHCB by the detection of one metabolite, HHCB-lactone. There are also other authors who associate the removal of fragrances to sorption rather than to biodegradation [21].

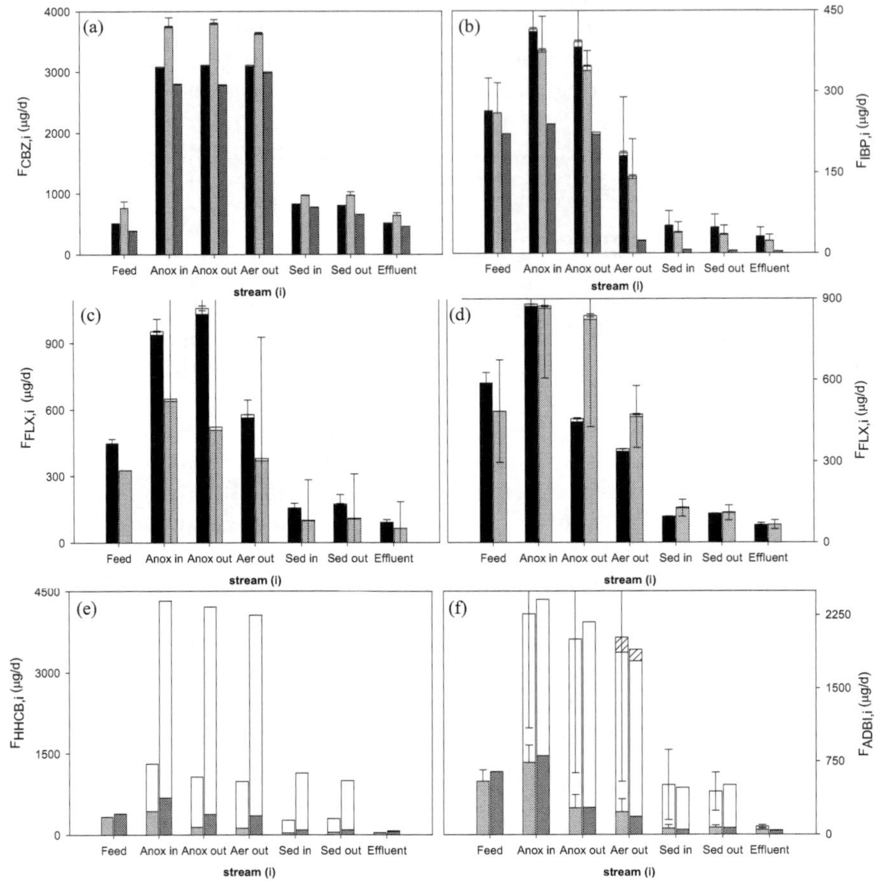

Figure 5: Total mass flow of selected PPCPs in the different streams of the pilot plant working at a SRT of <20 d (■), 20-40 d (■) and >40 d (■). The contribution of sorption (□) and volatilisation (▨) is indicated. For FLX data from operation at low (C) and high (D) temperatures are included.

3.4 Effects of operational parameters on the removal of PPCPs

Some operational parameters of the pilot plant, such as HRT, composition of the synthetic feed and dissolved oxygen level, have been maintained constant during the whole process, whereas temperature, SRT and the concentration of sludge varied. Accordingly their influence on the process could be evaluated.

√ The SRT of the plant had only an effect on the transformation degree of compounds with a significant sorption potential (estrogens and SMX), presumably because it enhances the retention of the compound inside the plant and consequently its availability for biological degradation.

√ The positive effect of warm temperature comparing to moderate ones, was observed for E1+E2, FLX, ERY and SMX, being the influence highest for the last compound (~30%) and lowest for the first one (~5%), that is inversely proportional to the biological degradation rate.

√ The removal of NPX in the studied pilot plant has been increasing from 30% up to 99% during the first 300 days of operation, indicating a possible acclimation of bacteria to this pharmaceutical. On the other hand, after day 600, a clear correlation between sludge concentration in the pilot plant and the efficiency in the elimination of NPX was observed.

4 Conclusions

The performance of the pilot plant with regard to its removal efficiency for a set of 16 PPCPs, including musk compounds, hormones and pharmaceuticals of different therapeutic classes, varied strongly from one compound to another. The following groups of compounds can be distinguished on the basis of its behaviour:

√ Very persistent compounds as CBZ, DZP and DCF, which were not significantly removed during the treatment.

√ Compounds with good biodegradability, such as FLX, NPX and IBP, for which high removal of the parent compound was observed, being sorption to the sludge negligible.

√ Musk fragrances, which were efficiently transformed inside the pilot plant, presumably due to their enhanced retention by sorption onto sludge.

√ Compounds moderately removed, such as CTL, SMX, TMP and ERY, for which the transformation degree in the pilot plant was at least 40%, according to their moderate sorption and biodegradation potential.

Some operational parameters, such as temperature, SRT and sludge characteristics have shown to influence the degree of removal achieved in the pilot plant.

Acknowledgements

This work was supported by Spanish Ministry of Education and Science (MICROFARM project: CTQ2007-66265/PPQ, NOVEDAR_Consolider project: CSD2007-00055 and Research Fellowship). The authors thank Oliver Gans (Austrian Federal Environment Agency) for the analyses of antibiotics.

References

[1] Suarez, S., Carballa, M., Omil, F. & Lema, J.M., How are pharmaceutical and personal care products (PPCPs) removed from urban wastewaters? *Rev Environ Sci Biotechnol,* **7**, pp. 125–138, 2008.

[2] Tauxe-Wuersch, A., de Alencastro, L.F., Grandjean, D. & Tarradellas, J. Occurrence of several acidic drugs in sewage treatment plants in

Switzerland and risk assessment. *Water Research*, **39 (9),** pp. 1761-1772, 2005.

[3] Clara, M., Kreuzinger, N., Strenn, B., Gans, O. & Kroiss, H. The solids retention time - a suitable design parameter to evaluate the capacity of wastewater treatment plants to remove micropollutants. *Water Research*, **39 (1),** pp. 97-106, 2005.

[4] Ternes, T.A., Stumpf, M., Mueller, J., Haberer, K., Wilken, R.D. & Servos, M. Behavior and occurrence of estrogens in municipal sewage treatment plants - I. Investigations in Germany, Canada and Brazil. *Science of the Total Environment*, **225 (1-2),** pp. 81-90, 1999.

[5] Joss, A., Andersen, H., Ternes, T., Richle, P.R. & Siegrist, H. Removal of estrogens in municipal wastewater treatment under aerobic and anaerobic conditions: Consequences for plant optimization. *Environmental Science & Technology*, **38 (11),** pp. 3047-3055, 2004.

[6] García-Jares, C., Llompart, M., Polo, M., Salgado, C., Macías, S., Cela, R. Optimisation of a solid-phase microextraction method for synthetic musk compounds in water. Journal of Chromatography A., 963, 277-285(2002).

[7] Lamas, J.P., Salgado-Petinal, C., García-Jares, C., Llompart, M., Cela, R. & Gómez, M. Solid-phase microextraction–gas chromatography–mass spectrometry for the analysis of selective serotonin reuptake inhibitors in environmental water. *Journal of Chromatography A.,* **1046 (1-2),** pp. 241-247, 2004.

[8] Quintana, J.B., Carpinteiro, J., Rodriguez, I., Lorenzo, R.A., Carro, A.M. & Cela, R. Determination of natural and synthetic estrogens in water by gas chromatography with mass spectrometric detection. *Journal of Chromatography A.,* **1024 (1-2),** pp. 177-185, 2004.

[9] Rodríguez, I., Quintana, J.B., Carpinteiro, J., Carro, A.M., Lorenzo, R.A. & Cela, R. Determination of acidic drugs in sewage water by gas chromatography-mass spectrometry as tert-butyldimethylsilyl derivates. *Journal of Chromatography A.,* **985,** pp. 265-274, 2003.

[10] Clara, M., Strenn, B. & Kreuzinger, N. Carbamazepine as a possible anthropogenic marker in the aquatic environment: investigations on the behaviour of Carbamazepine in wastewater treatment and during groundwater infiltration. *Water Research,* **38 (4),** pp. 947-954, 2004.

[11] Carballa, M., Omil, F., Lema, J.M., Llompart, M., Garcia-Jares, C., Rodriguez, I., Gomez, M. & Ternes, T. Behavior of pharmaceuticals, cosmetics and hormones in a sewage treatment plant. *Water Research,* **38 (12),** pp. 2918-2926, 2004.

[12] Clara, M., Strenn, B., Gans, O., Martinez, E., Kreuzinger, N. & Kroiss, H. Removal of selected pharmaceuticals, fragrances and endocrine disrupting compounds in a membrane bioreactor and conventional wastewater treatment plants. *Water Research,* **39 (19),** pp. 4797-4807, 2005.

[13] Gomez, M.J., Martinez Bueno, M.J., Lacorte, S., Fernandez-Alba, A.R. & Agüera, A. Pilot survey monitoring pharmaceuticals and related compounds in a sewage treatment plant located on the Mediterranean coast. *Chemosphere,* **66 (6),** pp. 993-1002, 2007.

[14] Jones, O.A.H., Voulvoulis, N. & Lester, J.N. The occurrence and removal of selected pharmaceutical compounds in a sewage treatment works utilising activated sludge treatment. *Environmental Pollution*, **145**, pp. 738-744, 2007.

[15] Nakada, N., Tanishima, T., Shinohara, H., Kiri, K. & Takada, H. Pharmaceutical chemicals and endocrine disrupters in municipal wastewater in Tokyo and their removal during activated sludge treatment. *Water Research*, **40 (17),** pp. 3297-3303, 2006.

[16] Ternes, T.A. Occurrence of drugs in German sewage treatment plants and rivers. *Water Research*, **32 (11),** pp. 3245-3260, 1998.

[17] Gobel, A., McArdell, C.S., Joss, A., Siegrist, H. & Giger, W. Fate of sulfonamides, macrolides, and trimethoprim in different wastewater treatment technologies. *Science of the Total Environment*, **372 (2-3),** pp. 361-371, 2007.

[18] Zwiener, C., Glauner, T. & Frimmel, F.H. Biodegradation of pharmaceutical residues investigated by SPE-GC/ITD-MS and on-line derivatization. *Journal of High Resolution Chromatography*, **23 (7-8),** pp. 474-478, 2000.

[19] Kupper, T., Plagellat, C., Braendli, R.C., de Alencastro, L.F., Grandjean, D. & Tarradellas, J. Fate and removal of polycyclic musks, UV filters and biocides during wastewater treatment. *Water Research,* **40 (14),** pp. 2603-2612, 2006.

[20] Bester, K. Retention characteristics and balance assessment for two polycyclic musk fragrances (HHCB and AHTN) in a typical German sewage treatment plant. *Chemosphere*, **57 (8),** pp. 863-870, 2004.

[21] Joss, A., Keller, E., Alder, A.C., Gobel, A., McArdell, C.S., Ternes, T. & Siegrist, H. Removal of pharmaceuticals and fragrances in biological wastewater treatment. *Water Research,* **39**, pp. 3139-3152, 2005.

The use of limestone, lime and MnO_2 in the removal of soluble manganese from acid mine drainage

A. Aguiar, G. Xavier & A. Ladeira
CDTN, Center for the Development of Nuclear Technology, Brazil

Abstract

Acid mine drainage (AMD) is one of the main environmental issues faced by the mining industry. The acid mine water generally contains metals above the permissible discharging levels. Manganese is particularly present in this effluent and its removal is notoriously difficult due to its high solubility over a wide range of pH. While most of the metals precipitate at a pH below neutrality, the pH necessary for manganese precipitation is very high, above 10. Most systems that effectively remove this element from mine waters use the oxidation of Mn (II) followed by precipitation at an elevated pH. Precipitation consumes a great amount of lime, which implies high operational cost. Besides, the process generates a large amount of sludge containing metals which have to be disposed of. The objective of this study is to optimize the removal of manganese by using a laboratory prepared acid solution and an acid effluent from Poços de Caldas uranium mine (Brazil), in order to achieve the Mn permitted level for discharging (<1mg/L) and to reduce the amount of sludge generated. The precipitation process has been studied using lime, limestone and a non-conventional catalyst/adsorbent (MnO_2 residue). The results obtained showed that both lime and limestone are effective in the removal of Mn in a pH higher than 10. However, there is a slight difference between the two reagents and lime shows a better performance. The volume of precipitate generated by the addition of lime was 50% smaller than that obtained when limestone was used. The use of the non-conventional material made the removal of almost 100% of the Mn possible as the concentration of manganese was reduced from 140mg/L to <1mg/L at a pH near neutrality (6.8 to 7.2). The final effluent complies with the recommended value for manganese discharge.

Keywords: manganese, acid mine drainage, precipitation, manganese oxide.

WIT Transactions on Ecology and the Environment, Vol 135, © 2010 WIT Press
www.witpress.com, ISSN 1743-3541 (on-line)
doi:10.2495/WP100231

1 Introduction

Manganese is a typical contaminant from mining effluents, as it is one of the constituents of gold, uranium and zinc ores. The contamination of surface water with manganese is increasing expressively due to mining activities. Specifically in acid mine drainage, manganese is particularly present in high concentrations much higher than the permissible levels for discharging. Acid mine drainage (AMD) has been considered one of the main environmental problems faced by the mining industry and it may contain 5 to 10mg/L of Mn but values higher than 50mg/L are not uncommon. The concentration depends on the amount of Mn-containing minerals and on the process used to treat these minerals [1]. The maximum admissible concentration of manganese in liquid effluents, according to Brazilian legislation is 1.0mg/L. There are pragmatic problems in attaining such low concentration [2] especially when the manganese is in the form of Mn (II), typical case of mine water. In that oxidation state, manganese is much more stable at higher pH than the oxidized form (Mn (IV)). Most systems that effectively remove manganese from mine waters use the oxidation of the Mn (II) species at an elevated pH followed by precipitation as sulphite or manganese oxides [2, 3].

It is well known that for decades most water contaminants have been removed by chemical precipitation. Chemical treatment can provide reasonable remediation treatment for AMD, however, it has been contested, both technically and economically, due to its high operational and environmental costs incurred from the formation of large amounts of sludge which need to be disposed of [4, 5]. In general, the removal of Mn (II) is carried out through the addition of lime, which demands an excessive consumption of reagent as a result of high pH values required for its precipitation [3]. Even so, after the precipitation, the solution free of manganese cannot be discarded, as it generally has a pH above 10.0 [4]. Another difficulty related to the process of precipitation is the time-consuming kinetics in the formation of stable precipitates. Oxidizing reactions are usually kinetically too slow, which means a problem for large flows, as currently found in acid mine drainage (>50mg/L) [6, 7]. The practice of precipitation through the formation of hydroxides, although widely distributed and relatively simple, can be further optimized through studies aiming at achieving a faster, more practical and more effective removal.

Mn removal can be catalyzed by the use of some solids, including manganese oxide [6]. That is because many manganese oxides exhibit redox properties, which implies in high reactivity with respect to the sorption phenomena [8]. The U.S Environmental Protection Agency recognizes many commercially available Mn oxides as oxidizing and adsorptive media in industrial applications. Nowadays, the great challenge of the industry in the removal of manganese, especially when present in acidic effluents, is the low limit permitted for discharging (<1mg/L). This fact stimulates the development of new researches, aiming at the definition of technically and environmentally feasible processes.

This work presents a comparative study of manganese removal from acid solutions, including an acid mine effluent, by precipitation using lime and limestone. The researches also assessed the influence of a non-conventional material (MnO_2) on promoting an efficient removal of manganese at pH lower than 10 which may reduce reagent consumption and, consequently, the amount of precipitate. MnO_2 was expected to act as a catalyst increasing the kinetics of the oxidation and also as an adsorbent of the Mn^{+2}.

2 Methodology

2.1 Reagents

A laboratory solution with Mn concentration around 150mg/L was prepared by dissolution of analytical grade $MnSO_4.H_2O$. An acid mine effluent containing 144mg/L of Mn was supplied by Brazilian Nuclear Industry (INB). Samples of commercial lime and limestone reduced to size <0.044mm were used as precipitants. The manganese oxide (MnO_2) was a residue supplied by the zinc industry. The MnO_2 specific surface area was 46.0 (m^2/g) and was determined by BET method multiple point technique (Quantachrome Corp., NOVA-2200).

2.2 Batch experiments

Three distinct procedures were carried out in order to remove the manganese from the solutions:

Precipitation experiments: The volume of the laboratory solution was set at 200mL and the pH range was from 4.2 to 11, adjusted with lime or limestone. The suspension was magnetically shaken for 30 minutes after pH stabilization at room temperature (25°C). Subsequently, the suspension was filtered and the filtrate was analyzed for Mn through flame Atomic Absorption Spectroscopy (model AA240FS, Varian). The dissolved oxygen was monitored with the use of an oximeter (model DM-22, Digimed) during the process.

Batch Adsorption Experiments: The laboratory solution volume was fixed at 300mL with Mn content of 150mg/L. Different amounts of MnO_2 (0.75 to 3.00g dry basis) were added to the solution. The pH range adopted was set in the range of 6.8 to 7.2. Samples were collected every 2 hours, filtered and analyzed for Mn according to procedure 1. The Eh of the system was monitored using a platinum electrode.

Isotherms: The isotherms were obtained with the laboratory solution and the acid mine effluent. The pH of the Mn solutions was first adjusted to the pre-determined values (6.8-7.2), which leads to a decrease of the Mn concentration to values around 100mg/L, and then the solutions were filtered. Subsequently, different amounts of MnO_2 (0.2 to 1.5g dry basis) were added to 150 ml of the solution. The pH was adjusted to 6.8-7.2 when necessary with the use of 1.0M NaOH and 0.5M Na_2CO_3 for experiments with lime and limestone, respectively. The suspension was magnetically shaken for 24 hours at room temperature (T=25°C). Afterwards, the suspension was filtered and the filtrate was assayed for manganese through Atomic Adsorption Spectroscopy (model AA240FS, Varian).

3 Results and discussion

3.1 Influence of the pH in the precipitation

As can be notice in figure 1 the manganese removal is very low at pH ranging between 4.0 and 8.0. For pH above 8.5, the removal increases significantly and reaches the value set by the protection standards (e.g. 1.0mg/L) at pH 9.1 for limestone and pH 10.0 for lime. According to literature [6, 9, 10] the formation of stable manganese oxides is obtained only at pH above 7.5 and the importance of the pH increase in the removal of Mn is extensively discussed. Figure 1 also shows that the curve for limestone stands below the curve for lime in almost the whole range of pH. It means that limestone shows a removal efficiency slightly higher than that presented by lime.

The mass of limestone used to attain the Mn limit recommended for discharge is greater than the mass of lime. This implies in the generation of an additional volume of residue in the treatment with limestone, which incurs an extra expense, namely disposal and/or storage. In industrial procedures, the viability of the use of limestone depends on the form it is used, which could be in fixed bed or column. This application has not been assessed yet.

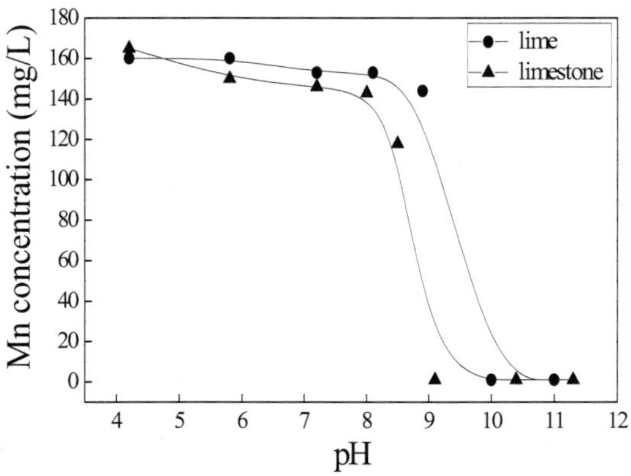

Figure 1: Mn concentration as a function of pH, without MnO_2 addition; T=25°C.

3.2 The influence of dissolved oxygen

Figure 2 shows the correlation between the concentration of dissolved oxygen and the concentration of Mn^{+2} in the system. It is observed that the experiments with lime show an inverse dependence between these two parameters: the higher the rate of manganese removal, the smaller the concentration of dissolved oxygen. This can be explained by the fact that stable forms of solid manganese in alkaline solutions are constituted of oxides (Mn_3O_4 (Hausmannite), Mn_2O_3

(Manganite or Feitknechtite), MnO_2 (Pyrolusite) and Mn_7O_{13} (Birnessite)) and for their formation the dissolved oxygen is consumed.

Another possible explanation for the low concentration of dissolved oxygen in the experiments with lime is the formation of a highly viscous sludge, which hinders the exchange of gases between the air and the solution. As the experiment was conducted without aeration, this could have been responsible for the diminution in the concentration of dissolved oxygen in the solution.

On the other hand, the experiments with limestone show no clear correlation between manganese removal and the concentration of dissolved oxygen in the system. In such case, it can be considered that the formation of stable Mn solid forms consume carbonate and not dissolved oxygen. This statement is supported by the fact that the values of dissolved oxygen observed during the experiments are similar to the solubility of oxygen in aerated water (e.g. 7-8mg/L). The values of dissolved oxygen in the system are close to 8mg/L even in high percentages of Mn removal. This result differs from the one obtained with lime and suggests that the dissolved oxygen do not participate in the formation of stable Mn solid forms in the systems with limestone. An example of manganese carbonate is Rhodochrosite ($MnCO_3$) that forms in systems with low concentrations of O_2 and high concentrations of CO_2 [11], despite the fact that the system under study does not present low concentrations of O_2.

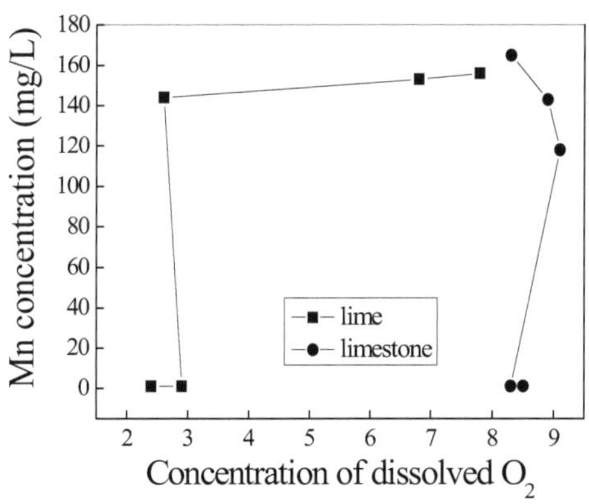

Figure 2: Influence of dissolved oxygen on manganese removal. Experiments carried out without aeration, T=25°C.

3.3 The influence of MnO_2

The removal of manganese may be catalyzed by solids, including manganese oxides (autocatalysis) [6]. Moreover, the presence of small amounts of solids through the recycling of the precipitates or due to turbidity may enhance the rate of oxidation from Mn^{+2} to Mn^{+4}. The oxidation accelerates the precipitation process as the latter is favoured by the presence of the oxidized form of manganese. Figure 3 shows the data for manganese precipitation with lime at pH between 6.8 and 7.2 by adding different amounts of the MnO_2 residue. It is observed that the larger the quantity of the MnO_2 added to the system, the more effective the manganese removal is. Furthermore, for amounts greater than 2.25g and reaction time of 6 hours, the final solution reaches the maximum admissible concentration of manganese in liquid effluents. By comparing these results with the ones from figure 1, one can conclude that at the pH range 6.8-7.2 and without the addition of MnO_2, the removal of Mn was lesser than 10% and the final concentration of manganese was extremely high (above 140mg/L) whether at the same pH range and with the addition of 2.25g of MnO_2 the removal was very effective and the Mn final concentration was smaller than 1mg/L.

Figure 4 shows the influence of the use of MnO_2 residue in Mn removal by using limestone to set the pH in a range of 6.8-7.2. Similar to the experiment

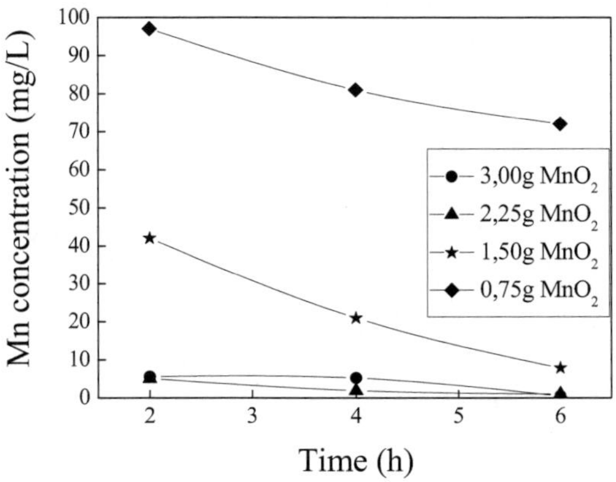

Figure 3: Influence of MnO_2 in the removal of Mn, pH adjusted with lime (6.8-7.2), aeration and T=25°C.

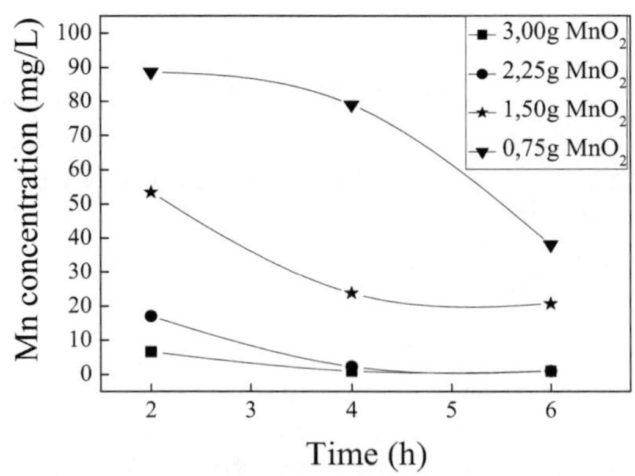

Figure 4: Influence of MnO_2 in the removal of Mn, pH adjusted with limestone (6.8-7.2), aeration and T=25°C.

with lime, the addition of MnO_2 increased the rate of Mn removal. Amounts greater than 2.25g of MnO_2 and 6 hours of reaction time imply that the final solution complies with the legal requirement for Mn concentration in liquid effluents. It was demonstrated that the Mn removal at pH near the neutrality is only possible, in the systems under investigation, with the addition of MnO_2.

It is also observed from figure 3 and 4 that the metal removal increases over time. In the experiments with limestone, the effluent reached the maximum admissible concentration for Mn after a reaction time of 6 hours. The only exception is observed with the trial with 3.00g of MnO_2, which reached the maximum permitted concentration within a reaction time of 4 hours. The results obtained with lime are similar to the ones obtained with limestone, however by using lime, the rates of Mn removal are initially higher and Mn concentrations lower than 10mg/L are more easily obtained.

3.4 Adsorption isotherms

Figure 5 shows the adsorption isotherms for the laboratory solution and the acid mine effluent using the MnO_2 residue as adsorbent/catalyst. The pH was adjusted with lime or limestone to 6.8-7.2 [11]. The solid lines represent the adjustment of

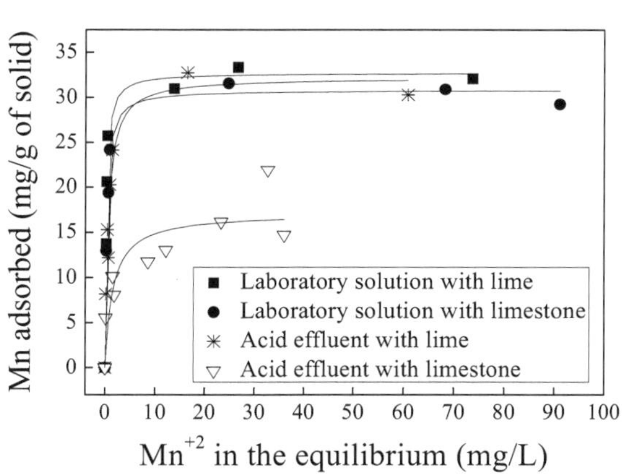

Figure 5: Equilibrium isotherms for Mn adsorption with MnO_2. The solid lines represent the adjustment of the experimental data to Langmuir's equation. T=25°C.

the experimental data to the Langmuir's equation. The clear definition of the plateaus, observed in figure 5, suggests that the manganese removal occurs by adsorption on the solid surface and not by precipitation. Langmuir's equation is well accepted to describe the adsorption phenomenon and is presented below:

$$Q = \frac{Q\text{max} \times k \times Ce}{1 + k \times Ce} \tag{1}$$

where,

Q = the amount of manganese adsorbed by the solid MnO_2 (mg/g of solid); Qmax = maximum adsorption capacity (mg/g); Ce= equilibrium concentration of Mn in the solution (mg/L); k= constant of affinity between the metal and the adsorbent.

Maximum adsorption capacities (Qmax) were determined by data extrapolation using Langmuir's equation. The Qmax for each experimental condition is shown in table 1. The results were very similar, except for the experiment with acid effluent and pH adjusted with limestone, in which the adsorption plateau was not clearly formed.

Table 1: Values for K and Qmax (mg/g) determined by fitting Langmuir's equation.

Sample	Qmax (mg/g)	k
Laboratory solution with lime	32,73	4,405
Laboratory solution with limestone	30,83	3,252
Acid effluent with lime	32,21	1,678
Acid effluent with limestone	17,13	0,643

The values of Qmax were between 17mg/g and 32mg/g. Literature reports Qmax values around 110mg/g for Mn adsorption on the mineral scolecite [12]. However, the initial concentration of the solutions used in the aforementioned study was up to 5000mg/L. The use of more concentrated solutions implies in higher loading capacity of the adsorbent. Moreover different experimental conditions make the comparison of the results difficult.

The acid effluent is an acid mine water with pH around 2.7 and an initial concentration of sulphate of 2.4g/L. In the presence of sulphate, the maximum adsorption capacity of the MnO_2 residue was expected to be smaller, due to the possibility of precipitation of calcium sulphate on the surface of the solid. However, the maximum capacity obtained for the acid effluent having its pH adjusted with lime is close to the maximum capacity obtained for the laboratory solutions. On the other hand, the Qmax for the acid effluent having its pH adjusted with limestone (17mg/g) is inferior to that obtained for the other solutions, indicating some sort of interference of other ions of the effluent. This statement is supported by the decrease in the affinity constant (K). So far, there is no explanation for this lower capacity.

4 Conclusion

The present study highlights the importance of the pH in the achievement of an effective Mn precipitation which means a final effluent within the recommended values for discharging. The necessity of alkaline pH in the formation of solid stable forms of the metal is then confirmed.

By using a non-conventional adsorbent/catalyst (MnO_2 residue) it was possible to attain the recommended value for manganese discharge, i.e. 1.0mg/L, at pH range of 6.8-7.2, whereas when this non-conventional absorbent/catalyst was not used, a pH next to 10 was required. Mn removal at pH next to neutrality reduces the volume of precipitate up to 50%, which means a smaller volume of residue for disposal.

The maximum adsorption capacities (Qmax) of the MnO_2 residue were calculated by fitting the Langmuir's equation and were between 17mg/g and 32mg/g.

Acknowledgements

The authors would like to thank FAPEMIG and CNPq for financial support.

References

[1] Johnson, D.B. & Hallberg, K.B., Acid mine remediation options: a review. *Sci. Total Environ*, **338**, pp. 3-14, 2005.

[2] Mariner, R., Johnson, D.B. & Hallberg, K.B., Characterisation of an attenuation for the remediation of Mn (II) contaminated waters. *Hydrometallurgy*, **94**, pp. 100-104, 2008.

[3] Bamforth, S.M., Manning, D.A.C., Singleton, I., Younger, P.L. & Johnson K.L., Manganese removal from mine waters – investigating the occurrence and importance of manganese carbonates. *Applied geochemistry*, **21**, pp. 1274-1287, 2006.

[4] Fernandes, H.M., Franklin, M.R. & Veiga, L.H., Acid rock drainage and radiological environmental impacts - A study case of the Uranium mining and milling facilities at Poços de Caldas. *Waste Management*, **18**, pp. 169-181, 1998.

[5] Nascimento, M.R.L., Fatibelllo, O.F.O. & Teixeira, L.A., Recovery of uranium form acid mine drainage waters by ion exchange. *Mineral Processing & Extractive Metall. Re,* pp. 129-142, 2004.

[6] Lovett, R.J. Removal of manganese from acid mine drainage. *J. Environ. Quality*, **26(4),** pp. 1017-1024, 1997.

[7] Aklil, A., Mouflih, M. & Sebti, S. Removal of heavy metal ions from water by using calcined phosphate as a new adsorbent. *J. Hazardous Materials* **A112**, pp. 183-190, 2004.

[8] Dias, A., Sá, R.G., Spitale, M.C., Athayde, M. & Ciminelli, V.S.T., Microwave-hydrothermal synthesis of nanostructured Na-birnessites and phase transformation by arsenic (III) oxidation. *Materials Research Bulletin*, **43**, pp. 1528-1538, 2008.

[9] Goncalves, C.R. & Ladeira, A.C.Q. Remoção de manganês e recuperação de urânio de águas ácidas de mina. *Anais do XXI Encontro Nacional de tratamento de minérios e metalurgia extrativa*, v.2, eds. I.A.S. de Brum & C.H. Sampaio, pp. 57-63, Natal, Brasil, 2005.

[10] Ladeira, A.C.Q. & Goncalves, C.R. Uranium recovery and manganese removal from acid mine drainage. *Proceedings of the 9th International Conference on Water Pollution: modeling, monitoring and management*, v.1. eds. D.P. Rico, C.A. Brebbia & Y.V. Esteve, WIT press, Southampton, pp. 465-474, Alicante, Spain, 2008.

[11] Sikora, F.J., Behrebds, L.L., Brodie, G.A. & Taylor, H.N. Design criteria and required chemistry for removing manganese in acid mine drainage using subsurface flow wetlands. *Water Environment Research*, **5**, pp. 536-544, 2006.

[12] Dal Bosco, S.M., Jimenez, R.S. & Carvalho, W.A., Removal of toxic metals from wastewater by Brazilian natural scolecite. *Journal of Colloid and Interface Science*, **281**, pp. 424-431, 2005.

Removal of lead from aqueous solutions using Saudi activated bentonite

S. Al-Shahrani
Department of Chemical Engineering and Material Science, King Abdulaziz University, Saudi Arabia

Abstract

Raw Saudi bentonite which was obtained from the Khulays bentonite deposit, 95 km north of Jeddah, was ground to -325 mesh (44 μm). Samples of ground bentonite were activated using sulfuric acid in order to increase its adsorptive capacity. The removal characteristics of lead ions from aqueous solution using Saudi activated bentonite were investigated under various operating variables such as shaking time, solution pH and initial metal concentration. The results showed that the sorption of lead ions on Saudi activated clay was very fast and the equilibrium was reached after only 10 min. They also showed that increasing the initial lead concentration decreased lead removal percentage due to the saturation of clay with lead ions. Furthermore, the adsorption of lead increases from 20 to 100% with an increase in solution pH from 1 to 5.0 and then 100% lead was removed when solution pH was from 5 to 11. The adsorption isotherm data were well fitted well with Langmuir better than Freundlich models. Lead adsorption onto Saudi activated bentonite was well represented by the pseudo-second-order kinetic model.

Keywords: lead removal, Saudi activated bentonite, Langmuir isotherm, Freundlich isotherm, Adsorption kinetics.

1 Introduction

Metal industries such as electroplating, metal finishing, metallurgical, tannery, chemical manufacturing, mining and battery manufacturing use large quantities of water in their processes. Wastewater streams from these processes contain toxic substances, heavy metal, acids, alkalis, and other substances. Removal of

WIT Transactions on Ecology and the Environment, Vol 135, © 2010 WIT Press
www.witpress.com, ISSN 1743-3541 (on-line)
doi:10.2495/WP100241

heavy metals such as lead from aqueous solution is necessary because it can be readily adsorbed by marine animals and directly enters the human food chains, thus presenting a high health risk to consumers [1–3]. Lead contamination is known as a worldwide problem due to its long term and widespread use. It is known to damage the kidney, liver and reproductive system, basic cellular processes and brain functions. The toxic symptoms are anemia, insomnia, headache, dizziness, irritability, weakness of muscles, hallucination and renal damages [4, 5].

Montmorillonite mineral, which is known commercially as bentonite or calcium bentonite, is characterized by the presence of exchangeable cations (Na, Ca, and Mg). Montmorillonite can be activated to produce active clay and increase its adsorptive capacity [6, 7]. Bentonite can be utilized as a cost effective sorbets material for the removal of lead from industrial waste water in order to minimize its processing cost and solve its disposal problems in an environmentally sustainable way [8].

The solution pH plays an important role in lead adsorption by bentonite. Mishra and Patel [9] investigated the adsorption of lead on bentonite at different pH. They found no lead was removed at pH lower than 3. It was observed that a gradual increase in adsorption with increase in pH from 3 to 6 and the maximum adsorption was at pH 6. Furthermore, lead adsorption increased gradually with increase in pH from 6 to 10 may be due to the formation of the precipitate of $Pb(OH)_2$. pH 6 was considered as optimum condition for the removal lead from waste water.

Adsorption of lead ions by clay has been the subject of several studies [10–14]. The most important factors which may affect sorption of lead were pH, lead concentration, mixing time and clay dosage.

In this research, removal of lead from aqueous solution will be investigated using Saudi activated bentonite.

2 Materials and methods

2.1 Preparation of Saudi activated bentonite

Saudi bentonite is not active in its natural state. It needs some treatment to modify its adsorptive property. Table 1 shows the chemical composition of Saudi Activated bentonite.

Natural bentonite was activated using sulphuric acid leaching. The apparatus and chemical treatment method with H_2SO_4 which was carried out in this study are similar to what have been described previously [15]. The acid treatment was conducted using a Pyrex glass batch reactor with boiler-reflux condenser and a magnetic stirrer/hot plate. Acid activation was done using the optimum operating variables such as, 45% by weight acid concentration, the temperature of boiling, 15 minutes reaction time, 200 rpm degree of mixing, grain size of -325 mesh (44μm) and water to clay ratio of 5:1.

Table 1: Chemical composition of Saudi activated bentonite [15].

Compound	Chemical composition (%)
Sio_2	66.2
Al_2O_3	11.71
Fe_2O_2	3.0
TiO_2	1.5
MgO	0.73
CaO	<0.05
K_2O	0.48
Na_2O	0.12
MnO	0.05
SO_3	<0.05
P_2O_5	<0.05
L.O.I.(1000°C)	15.3

2.2 Adsorption process

Adsorption of lead with clay is carried out in a batch reactor. Stock solution (1000 mg/L) of lead was prepared by dissolving the appropriate amounts of analytical reagent grade $Pb(NO_3)_2$ in distilled water. The stock solutions were diluted as required to obtain standard solutions containing 20–100 mg/l of a required heavy metal. Known amount of Saudi activated bentonite was placed in a conical flask containing lead solution of known concentration and pH. The resulting mixture was then mixed continuously using horizontal shaker with water bath (JULABO SW 22) for a given time period. 200 rpm shaking and 25°C temperature were applied in the shaker. Afterward, samples were taken out from the shaker and the clay was separated by filtration using filter paper (Whatman No., 42). The filtrate was analyzed for lead concentration by Atomic absorption spectrophotometer (AAS). The effects of several factors such as pH, concentration of solution, clay mass and contact time on lead removal percentage were investigated.

3 Results and discussions

3.1 Effect of shaking time

The time-dependent behavior of lead adsorption was investigated by varying the contact time between the adsorbate and adsorbent in the range of 1-40 min. The initial concentration of lead was ranging from 20 to 100 ppm, while the dose of clay sample was 1.0 g/50 ml and the solution pH was fixed at 3.1. The data showed that the sorption of lead ions on Saudi activated clay was very fast and the equilibrium was reached after only 10 min (Figure 1). As found before [16, 17] the adsorption process is fast at the beginning of the reaction due to the adsorption of lead on the surface sites of clay. The results show that shaking the mixture of different initial concentration for 10 minutes was sufficient to reach

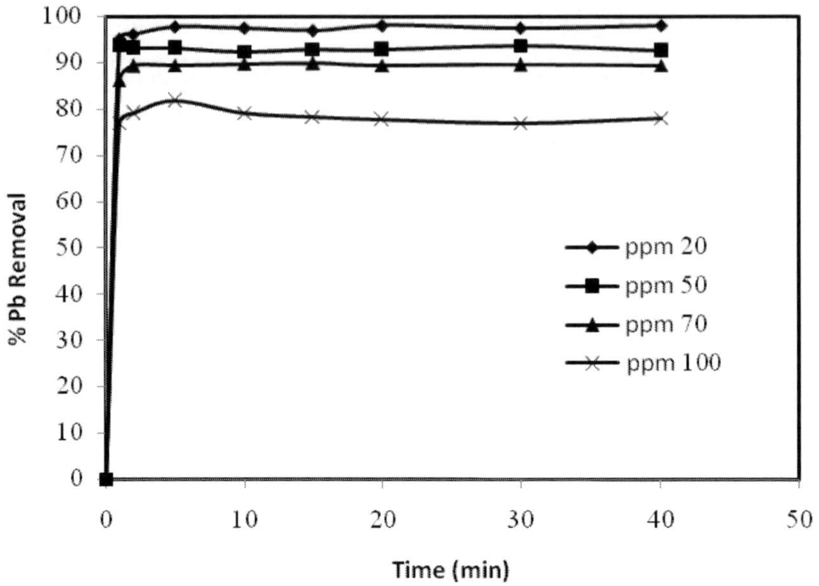

Figure 1: Effect of shaking time on the removal of lead by Saudi activated bentonite. Initial metal concentration varies between 20 and 100 mg/l, clay dosage: 1 g/50 ml, pH = 3.1.

equilibrium. They are also show that increasing the initial lead concentration decreased lead removal percentage due to the saturation of clay with lead ions.

3.2 Effect of pH

The effect of pH on the removal of lead ions by Saudi activated bentonite was investigated. The adsorption of lead was studied in the pH range of 1–11 with a constant clay amount of 1 g/50 ml of lead solution, a shaking time of 10 min and lead concentration of 100 ppm. The pH of the metal ion solution was adjusted by adding required amount of dilute NaOH or HNO3 solution. pH was measured using a pH meter.

The results presented in Figure 2 reveal that the adsorption of lead increases from 20 to about100% with an increase in solution pH from 1 to 5.0 and then about100% lead was removed when solution pH from 5 to 11.

The effect of initial pH on lead removal may be explained as mention before [17], where in the acidic conditions, both adsorbent and adsorbate are positively charged (M^{2+} and H^{+}) and therefore, the net interaction is that of electrostatic repulsion. In addition, the H^{+} ions present in higher concentration in the aqueous medium compete with the positively charged Pb^{2+} ions for the surface adsorbing sites, resulting in a decrease in the removal of lead.

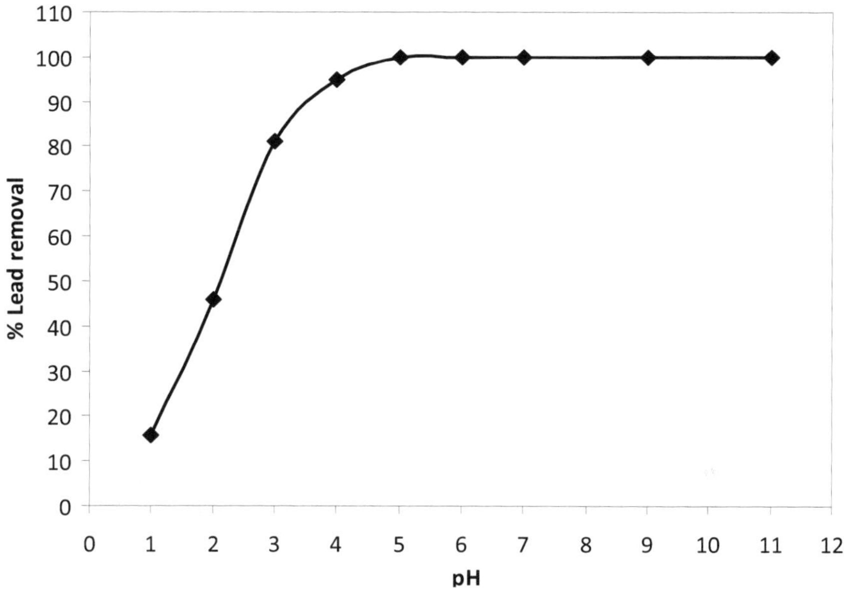

Figure 2: Effect of pH on the removal of lead by Saudi activated bentonite. Initial lead concentrations 100 mg/L, clay dosage: 1 g/50 ml, contact time: 10 min.

3.3 Effect of lead concentration

Figure 3 shows the effect of lead concentration on lead removal when lead concentration varies from 20 to 100 ppm under operating conditions of 10 min shaking time, 1.0 g/50 ml for adsorbent and mixture pH =3.1. The results showed that about 100% of lead was removed when lead concentration was 20 mg/l. Lead removal percentage decreased with increasing lead concentration to about 80% when lead concentration was 100 ppm. Therefore, Saudi activated bentonite removal efficiency needs to be increased by increasing solution pH to 6 to avoid the lead hydroxides precipitation region which it is start at the pH of 6.5 [17].

3.4 Effect of clay dosage

The effect of the amount of Saudi activated bentonite (0.05–0.75 g/100 ml) on lead removal at constant values of initial metal concentration (100 mg/l), contact time (10 min), pH = 6 and temperature (25°C) is shown in Figure 4. The results showed that the removal of lead increased rapidly (98%) until an adsorbent dose of 0.125 g/100 ml and gradually up to 0.25 g/100 ml where about 100% of lead was removed from the solution. However, this result was expected since as the dose of adsorbent increases, the number of adsorbent sites increases. These amounts attach more ions to their surfaces [18]. Similar results were reported where many types of materials were used as adsorbents [19].

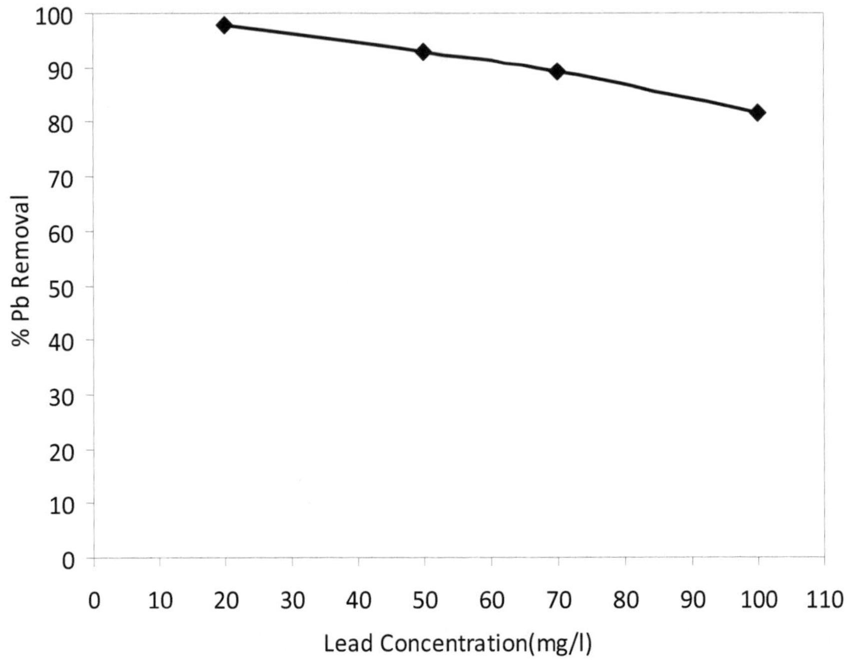

Figure 3: Effect of metal concentration on the removal of lead by Saudi activated bentonite. Clay dosage: 1 g/50mL. pH = 3.1, contact time: 10 minutes.

3.5 Adsorption isotherm

Two different adsorption isotherm models, the Langmuir [20] and Freundlich [21] equations were used to fit the experimental data obtained from this work. These models were tested to determine the maximal capacity of lead removal using Saudi activated bentonite. The quality of the isotherm fit to the experimental data is typically evaluated based on the magnitude of the correlation coefficient for the regression; i.e. the isotherm giving an R^2 value closest to unity is considered to give the best fit. The adsorption isotherms for lead removal were studied using volume of 100 ml at initial metal concentration of 100 mg/l with adsorbent mass ranging from 0.05 to 0.75 g at room temperature (25°C) and solution pH = 6. Afterward, data obtained were fitted to the Langmuir and Freundlich isotherms. The Langmuir equation, in the linear form is written as:

$$\frac{C_e}{q_e} = \frac{1}{bq_{\max}} + \frac{C_e}{q_{\max}} \tag{1}$$

where C_e is the equilibrium concentration of metal ions (mg/l), q_e is the amount of metal ions adsorbed per unit weight of adsorbent (mg/g bentonite), q_{max} is the maximum adsorption capacity (mg/g), and b is the adsorption equilibrium constant (l/mg).

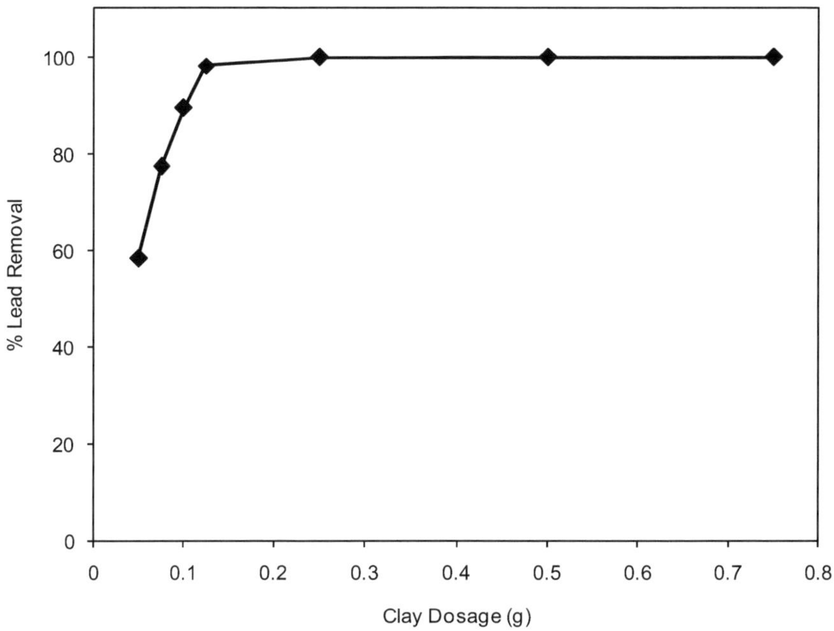

Figure 4: Effect of clay dosage on the removal of lead by Saudi activated bentonite. Initial lead concentrations 100 mg/L, pH = 6, contact time: 10 minutes.

For the Freundlich equation the linear form is written as:

$$\log q_e = \log K + \frac{1}{n}\log C_e \qquad (2)$$

where k and n are the constant characteristics of the system.

The best estimated values of all the equation parameters are summarized in Table 2. The adsorption isotherm data are well fitted with the linearized

Table 2: The Langmuir and Freundlich equation parameters predicted from adsorption isotherm data of lead ions onto Saudi activated bentonite at pH =6, 25°C and initial lead concentration of 100 mg/l.

Isotherm parameters	Value
Langmuir	
q_{max} (mg/g)	116
b (l/mg)	0.989
R^2	0.9932
Freundlich	
K (mg/g)	47.77
n	3.3
R^2	0.8919

Langmuir better than linearized Freundlich equations and give R^2 = 0.989–0.8919 as shown in figures 5 and 6. The maximum adsorption of Saudi activated clay equals to 116 mg/l.

According to Tryball [22], it has been shown using mathematical calculations that n values of between 1 and 10 for the Freundlich isotherm indicate effective adsorption.

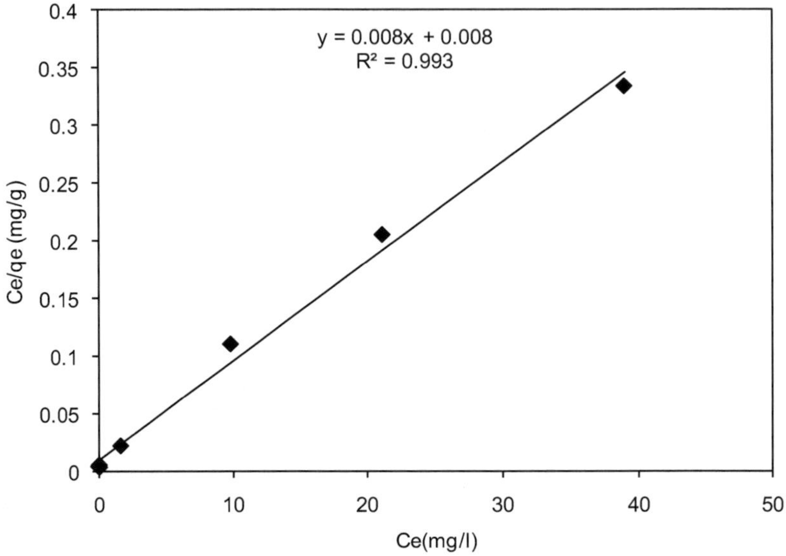

Figure 5: Langmuir plot for the adsorption of lead on Saudi activated bentonite.

3.6 Adsorption kinetic

Kinetics of sorption describing the solute uptake rate is one of the important characteristics defining the efficiency of sorption. The kinetic parameters for the adsorption process were studied on the batch adsorption at room temperature and pH at 3.1. The data were fitted to the first-order Lagergren equation [23],

$$(\log q_e - q_t) = \log q_e - \frac{k_1}{2.303}t \tag{3}$$

where k_1 (min^{-1}) is the first-order rate constant, q_e (mg /g) is the amount of adsorbed metal ions on the Saudi bentonite at equilibrium and q_t (mg /g) is the amount of lead adsorbed at time t (min). The first-order constants can be obtained by plotting log (q_e−q_t) versus time.

The experimental adsorption kinetics data were analyzed by applying the pseudo-second-order kinetics model, which is expressed as:

$$\frac{t}{q_t} = \frac{1}{k_2 q_e^2} + \frac{1}{q_e}t \tag{4}$$

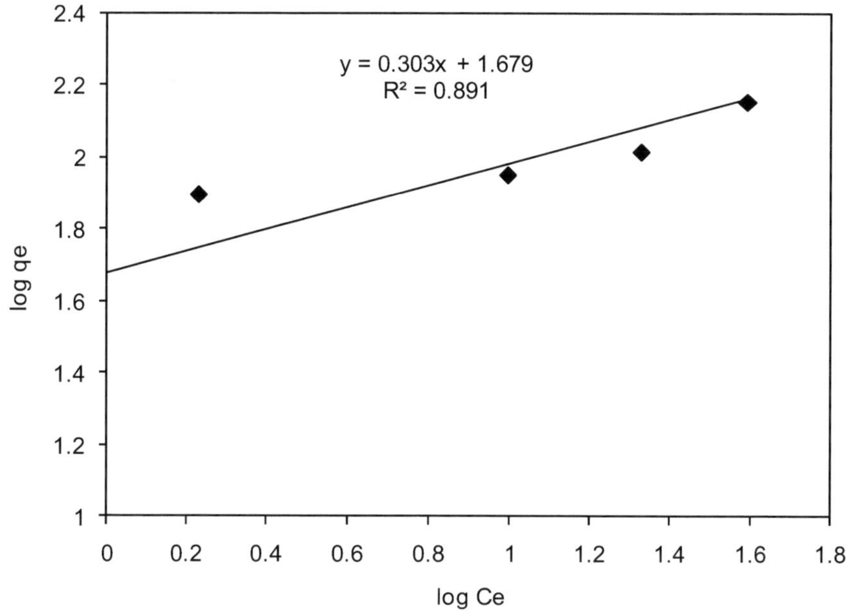

Figure 6: Freundlich plot for the adsorption of lead on Saudi activated bentonite.

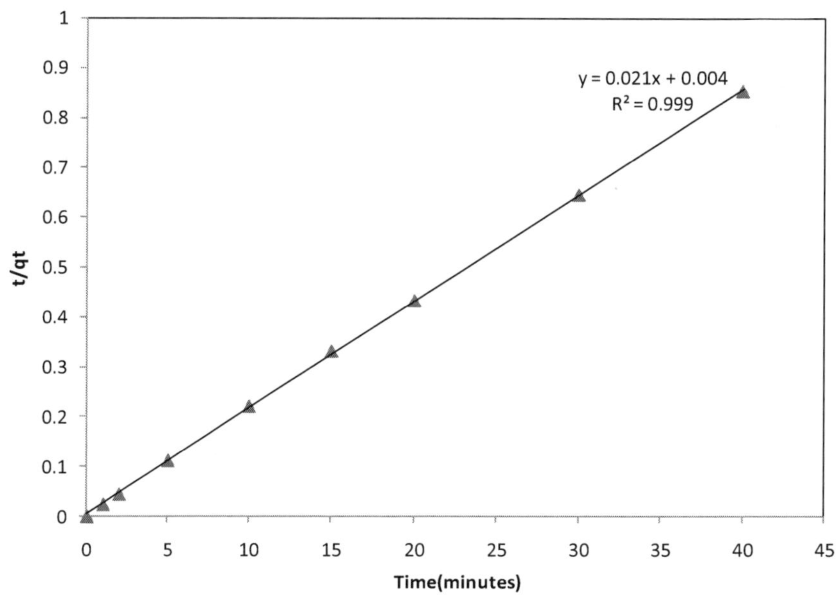

Figure 7: Plot of second order model for lead adsorption by Saudi activated bentonite.

where k_2 (g/(mg/min)) is the pseudo-second-order kinetics constant, q_e (mg /g) is the amount of adsorbed metal ions on the Saudi bentonite at equilibrium and q_t (mg /g) is the amount of lead adsorbed at time t (min). The fit of this model was checked by each linear plot of (t/q_t) versus t as shown in Figure 7.

By comparing to the regression coefficients for each expression, a good agreement of the experimental data with the second order kinetic model was observed. The correlation coefficient for the second order kinetic model is 0.999. Thus, a plot of t/q_t against t should give a linear relationship with a slope of $1/q_e$. These results agree with same results reported before [10, 23].

4 Conclusions

- The sorption of lead ions on Saudi activated clay was very fast and the equilibrium was reached after only 10 min. The results show that shaking the mixture of different initial concentration for 10 minutes was sufficient to reach equilibrium. They are also show that increasing the initial lead concentration decreased lead removal percentage due to the saturation of clay with lead ions.
- Removal of lead ions by Saudi activated bentonite is effected by mixture pH. The adsorption of lead increases from 20 to 100% with an increase in solution pH from 1 to 5.0 and then 100% lead was removed when solution pH increases from 5 to 11.
- Lead removal decreases with increasing lead concentration when lead concentration increases from 20 to 100 ppm.
- The removal of lead increases as the dose of Saudi activated bentonite increases. This is due to increase in the number of adsorbent sites.
- The adsorption isotherm data were well fitted with the Langmuir model better than Freundlich model.
- A batch adsorption kinetic experiment revealed that lead adsorption onto Saudi activated bentonite was well represented by the pseudo-second-order kinetic model. It can be concluded that film diffusion and intraparticle diffusion are simultaneously operating in the whole adsorption process.
- Saudi activated bentonite can be considered as a promising adsorbent for the removal of lead from aqueous solutions.

Acknowledgements

I would like to thank the Deanship of Scientific Research, King Abdulaziz University, for its financial support to accomplish this research. The investigator would like to thank Prof. Yahia Al-Hamid, the chairman of Department of Chemical Engineering King Abdulaziz University, for his continual support and invaluable suggestion during this research.

References

[1] Lai, C.H., Lo, S.L., Lin, C.F., 1994, Evaluating an iron coated sand for removing copper from water. *Wat. Sci. Technol.* 30, 175.

[2] Tonni Agustiono Kurniawan, Gilbert Y.S. Chana, Wai-Hung Loa, Sandhya Babel, Physico–chemical treatment techniques for wastewater laden with heavy metals. *Chemical Engineering Journal* 118 (2006) 83–98.

[3] P.N. Cheremisinoff, Handbook of Water and Wastewater Treatment Technology, Marcel Dekker Inc., New York, 1995.

[4] David M. Ayres, Allen P. Davis, Paul M. Gietka, Removing Heavy Metals from waste water. Engineering Research Center Report, University of Maryland, 1994.

[5] US EPA, Cost and benefits of reducing lead in gasoline. Draft final report, Office of policy analysis, US EPA 230-03-84-005, Washington DC,(1984).

[6] Laurnt, D, Atlas of Industrial Mineral. Ministry of Petroleum and Mineral Resources Publication. KSA, 1993.

[7] Brown, G, The X-Ray Identification and Crystal structures of Clay Minerals. Alden & Mowbray, Oxford, 1997.

[8] S.P. Singh, L.Q. Ma, M.J. Hendry, Characterization of aqueous lead removal by phosphatic clay: Equilibrium and kinetic studies. *Journal of Hazardous Materials* B136 (2006) 654–662.

[9] P.C. Mishra, R.K. Patel, Removal of lead and zinc ions from water by low cost adsorbents. *Journal of Hazardous Materials* 168 (2009) 319–325.

[10] H. Chen, and A. Wang, Kinetic and isothermal studies of lead ion adsorption onto palygorskite clay. *Journal of Colloid and Interface Science.* 307 (2007) 309–316.

[11] I. Chaari, et al., Lead removal from aqueous solutions by a Tunisian smectitic clay, *J. Hazard. Mater.* (2008).

[12] Vassilis J. Inglezakis, Marinos A. Stylianou, Despoina Gkantzou, Maria D. Loizidou, Removal of Pb(II) from aqueous solutions by using clinoptilolite and bentonite as adsorbents. *Desalination* 210 (2007) 248–256.

[13] Y. El Mouzdahir Y. El Mouzdahir, A. Elmchaouri, R. Mahboub, A. ElAnssari , A. Gil, S.A. Korili, M.A. Vicente, Interaction of stevensite with Cd^{2+} and Pb^{2+} in aqueous dispersions *Applied Clay Science* 35 (2007) 47–58.

[14] Murari Prasad, Sona Saxena, Attenuation of divalent toxic metal ions using natural sericitic pyrophyllite. *Journal of Environmental Management*, Volume 88, Issue 4, September 2008, Pages 1273-1279

[15] Al-shahrani, Saad, The feasibility of producing activated clay from a local raw material by sulfuric acid leaching. Msc dissertation, Chemical and Material Engineering Department King Abdulaziz University Jeddah, KSA., 1998.

[16] R. Naseem and S. S. Tahir, Removal of Pb(II) from Aqueous/Acidic Solution by using Bentonite as Adsorbent. *Wat. Res.* Vol. 35, No. 16, pp. 3982–3986, 2001

[17] Islem Chaari, Emna Fakhfakh, Salima Chakroun, Jalel Bouzid, Nesrine Boujelben, Mongi Feki, Fernando Rocha, and Fakher Jamoussi, Lead removal from aqueous solutions by a Tunisian smectitic clay. *Journal of Hazardous Materials*, Volume 156, Issues 1-3, 15 August 2008, Pages 545-551
[18] O. Abollino, M. Acetob, M. Malandrinoa, C. Sarzaninia, E. Mentasti, Adsorption of heavy metals on Na-montmorillonite. Effect of pH and organic substances, *Water Res*. 37 (2002) 1619–1627.
[19] J.H. Potgieter, S.S. Potgieter-Vermaak, P.D. Kalibantonga, Heavy metals removal from solution by palygorskite clay, Miner. Eng. 19 (2006) 463–470.
[20] Langmuir L, The adsorption of gass, mica and platinium. J Am Chem Soc 1918; 40:1361.
[21] Freundlich H. Colloid and capillary chemistry. London: Metheum, 1926, p. 883.
[22] Tryball, R.E., 1980. Mass Transfers Operations, third ed. McGraw, New York.
[23] Lagergren S, About the theory of so-called adsorption of soluble substances. Kungliga Svenska Vetenskapsakademiens. Handlingar, Band 1898; 24:1–39.

High Density Sludge process applied to metal-containing effluent

P. Suvio[1], D. Sapsford[1], A. J. Griffiths[1], K. Williams[1], J. D. Davies[2] & S. Maynard[2]
[1]*School of Engineering, Cardiff University, UK*
[2]*Corus Strip Products, UK*

Abstract

Large volumes of water are utilised during the operation of complex integrated steelworks. This usually results in the discharge of liquid effluents containing low concentrations of metals in solution and also the need to dispose of metal-rich sludges resulting from the treatment of these effluents. This paper presents the current performance of the final effluent plant at a large steelworks in South Wales, UK, focusing on the discharge effluent quality. In order to better understand the metal removal mechanisms a series of laboratory studies have addressed the precipitation of single metals and the co-precipitation of Zn with a range of other metals, including Fe and Cu, using sodium hydroxide as the alkali. Experimental data is compared to the predicted values using the geochemical model, PHREEQCi 2. Large sludge volume reductions were found for effluent treatment using the so-called High Density Sludge (HDS) process, which was operated on a pilot scale. HDS was achieved by mixing recirculating sludge and the feed water prior to adding the alkali, creating a so-called two stage neutralisation process, where recirculated sludge acts as a seed for further metals precipitation. Careful performance monitoring of the HDS process pilot plant was carried out in order to ensure suitable conditions for the formation of HDS. By the end of the trial HDS with high initial settling characteristics was being formed.

Keywords: zinc, hydroxide precipitation, high pH precipitation, steelworks, High Density Sludge, settling velocity.

WIT Transactions on Ecology and the Environment, Vol 135, © 2010 WIT Press
www.witpress.com, ISSN 1743-3541 (on-line)
doi:10.2495/WP100251

1 Introduction

The treatment of aqueous effluents to meet metal concentrations that are acceptable in final discharges to receiving waters is commonplace in a wide range of industrial applications, from mines to large chemical complexes. The current study was undertaken in conjunction with a large integrated steel manufacturing complex in South Wales, UK. Several effluent treatment systems are used at the site, handling flows from individual process facilities. As a means of capturing any residual metal species before final discharge, a conventional alkali precipitation and settlement system, using type II (flocculant) settling [1], has been in operation since 1999. This has been very successful in limiting concentrations of target metals in solution. However, in common with most plants of this type, especially when sodium hydroxide is the alkali of choice, the settled sludge concentrations are very low with maximum concentrations of 5% solids (w/v) [2] in exceptional cases and further dewatering by filter press can be problematic. In many similar situations the so-called High Density Sludge (HDS) process, where precipitates are recycled to mix with the incoming feed, has been found to be highly effective at enhancing sludge dewatering behaviour [3]. This paper examines the existing plant performance and presents the results of a pilot-scale study of effluent treatment by the HDS II process [4], where the incoming effluent water is mixed with recirculated sludge, prior to adding the alkali reagent and air.

2 Experimental

2.1 Full-scale plant performance studies

In order to establish the current level of metal concentration in the final discharge entering and leaving the conventional alkali precipitation and settlement system in use at the steelworks, full-scale plant studies were carried out over a 7 month period, examining the levels of target metals in solution, including Zn, Fe, Ni, Cu, Cr, Pb and Cd. The system is using sodium hydroxide (NaOH), which is mixed together with the incoming feed and Cationic Polyacrylamide by mixing motors in the first chamber of the system. The sample for the incoming water was taken from the mixing chamber, which has a pH controlling system in place, aiming to keep the pH as close as possible to, but above, 8.5. The sample for the outlet water was taken from the clean water weir at the end of the sedimentation system. Prior to testing the samples were filtered, followed by an analysis on a Perkin Elmer Optima DV5300 Inductively Couple Plasma (ICP) atomic emission spectrometer.

2.2 Laboratory study of precipitation and co-precipitation

Several studies have been carried out looking at the precipitation of metals out of solution. In order to determine how the target metals precipitate either on their own or together with other metals in solution, a co-precipitation study using base titration with sodium hydroxide (NaOH) as an alkali was carried out. In this laboratory study 100mg/l of metal concentration for each metal was added to a litre of tap water. The volume of sodium hydroxide (NaOH) required to increase the pH to stable 7, 8, 9, 10 and 11 was measured. Furthermore, a sample was taken at each pH for analysis of the remaining soluble metal concentration on a Perkin Elmer Optima DV2100 Inductively Couple Plasma (ICP) atomic emission spectrometer. The pH meter used in the study was a Hanna HI208. Previous work on removing metal complexes from solution as metal hydroxides has showed that the actual solubility of metal ions and the minimum pH for maximum precipitation varies according to the mix of metal ions present in the solution [5].

2.3 Modelling of expected solution properties

PHREEQCi Version 2 [6, 7] is a free ware computer program written in the C-programming language that is designed to perform a wide variety of low-temperature aqueous geochemical calculations, based on an ion-association aqueous model.

2.4 Pilot-scale HDS plant

A continuous trial with a HDS process pilot plant (Figure 1) was carried out during the two months leading up to 16.02.2010 in order to confirm that HDS could be produced using sodium hydroxide as the alkali reagent. Zn, Cu, Fe, Ni and Mn were added to tap water to give the desired concentrations. The tap water was dosed with the appropriate amounts of Mg and Ca to mimic the background concentrations found at the steelworks (see Table 2).

2.4.1 Plant description

The plant consists of a $1m^3$ Intermediate Bulk Container (IBC) feed water storage tank that was manually stirred daily to ensure the contents were fully mixed. A peristaltic feed pump (Watson Marlow SciQ 323) was used, aiming for a nominal incoming feed flow rate of 10l/h (30 minute retention time in the first two reactors). The plant feed was mixed with the recycled sludge in the 5l Stage I reaction tank (cylindrical with a radius of 0.075m and a height of 0.28m). A top mounted mixer (a Heidolph RZR 2041 operated at speed setting 1 at 302rpm) was installed in the Stage I reactor to keep the solids in suspension.

The overflow from the Stage I reactor was fed into the similarly sized Stage II reactor, where the pH was raised by the addition (NaOH). The alkali was supplied from a storage vessel and dosed to the Stage II reactor via an integral controller/metering pump (Hanna Instruments BL 7916). A pH controller was used to maintain the Stage II reactor pH appropriate for the metals in the feed water. The alkali dosing rate was controlled via a pH probe located at the outlet of the Stage II reactor. The pH controller was calibrated twice a week and the pH measurement was also checked using a Hanna Instruments HI208 portable pH meter. The alkali was stored in a 25l storage vessel and the stock solution was made up to a strength of 5% by dissolving 50 g of sodium hydroxide pellets

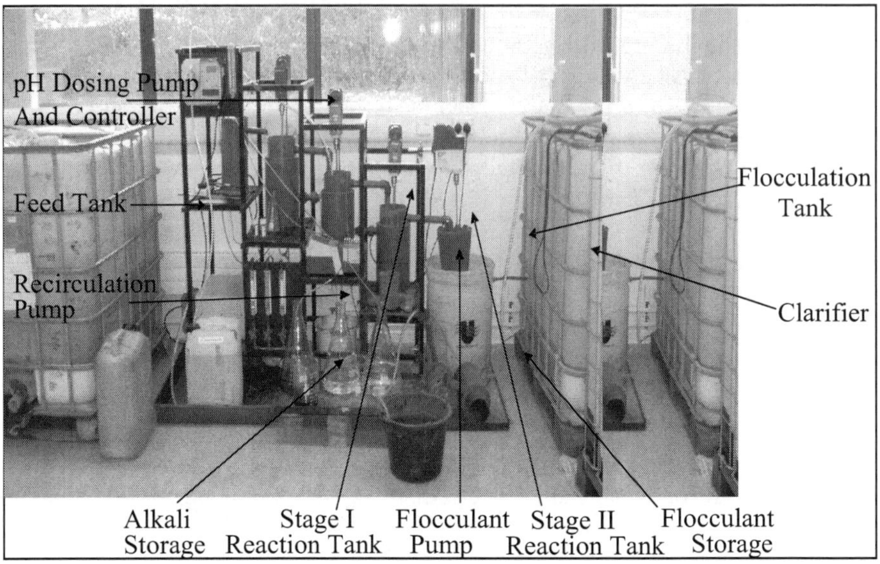

Figure 1: Photograph of the High Density Sludge pilot plant.

per 1 litre of hot tap water, which ensured that all the pellets dissolved. Air was introduced via a diffuser ring at a rate of 10l/h to ensure the full oxidation of the metals. A top mounted mixer (a Heidolph RZR 2041 operated on speed setting 1 at 200rpm) was used to increase oxygen transfer rate and ensure adequate mixing.

The treated water from the Stage II reactor overflowed into the floc-mixing tank (volume 2.5l, cylindrical with a radius of 0.075m and a height of 0.14m). Flocculant was added to the feed line from the Stage II reactor. A slow-speed flocculation mixer (a Heidolph RZR 2041 operated on speed setting 1 at 141rpm) was used for solids and flocculant mixing. The flocculated mixture then flowed via gravity into the clarifier. Anionic flocculant (Superfloc 110) was made up at a concentration of 0.05% w/v (i.e. 0.5g of active flocculant/l). The dosing pump used was a FA Hughes (DCL) peristaltic pump. Solids/liquid separation was achieved in a clarifier/thickener cone, with a maximum surface area of approximately 0.0176m^2. Thickened solids from the clarifier were either recirculated to the Stage I reactor or removed from the system by peristaltic Watson Marlow 604U pump. A slow-speed mixer (a Heidolph RZR 2041 operated on speed setting 1 at 70 rpm) was used to assist solids/liquid separation. Treated water was discharged from the system by overflowing the clarifier unit.

2.5 Pilot plant performance monitoring

The performance of the pilot plant was assessed by monitoring several aspects, including the volume of feed water, flocculant and alkali added daily, as well as the pHs throughout the plant. The volume of sludge returned to the Stage I reactor was regularly monitored and the rate of removal of sludge was controlled. Further, regular samples were taken throughout the plant in order to carry out analysis for soluble metal concentration levels on Perkin Elmer Optima DV2100 Inductively Couple Plasma (ICP) optical emission spectrometer. Daily settling tests were carried out and solids concentration of the sludge was measured by filtering samples and measuring their weight against the sample volume (%w/v) after drying in oven at 105°C for 2 hours. From the settling rate data collected the initial linear position of the curves were plotted against the respective solids concentrations to monitor how sedimentation changes with the age of the sludge.

3 Results and discussion

3.1 Full-scale plant performance

Results for soluble metal concentrations before (In) and after (Out) being treated by the steelwork's final discharge precipitation and settlement system can be seen in Table 1 below. The steelworks are working under consent limits for monthly average soluble metal concentrations in final discharge, including:

- Soluble Zn 2.5 mg/l monthly average
- Soluble Pb 0.3 mg/l monthly average
- Soluble Cr 0.2 mg/l monthly average

As can be seen in Table 1, the steelworks final discharge precipitation and settlement system is working efficiently in removing soluble metals with near or over 90% removal performance for Fe, Zn, Pb and Cr and the average concentrations leaving the steelworks final discharge precipitation and settlement system are well under the consent limit.

3.2 Laboratory-scale studies for precipitation and co-precipitation with expected solution properties

Precipitation and co-precipitation experiments, especially looking for performance of zinc were carried out and compared against predicted

Table 1: Soluble metal removal performance of the steelworks final discharge precipitation and settlement system*.

	In mg/l			Out mg/l		Removal
Metal	Max	Min	Ave	Max	Ave	%
Fe	102.00	2.40	20.60	10.80	1.88	90.86
Zn	29.00	0.30	6.11	2.90	0.74	87.87
Pb	9.40	0.00	0.93	0.05	0.05	94.20
Cr	0.10	0.00	0.03	0.00	0.00	87.50
Ni	0.50	0.00	0.04	0.02	0.02	50.00
Cu	0.90	0.00	0.07	0.05	0.05	23.53

*Cd results and minimum soluble metal concentration levels are 0.00 for all sections.

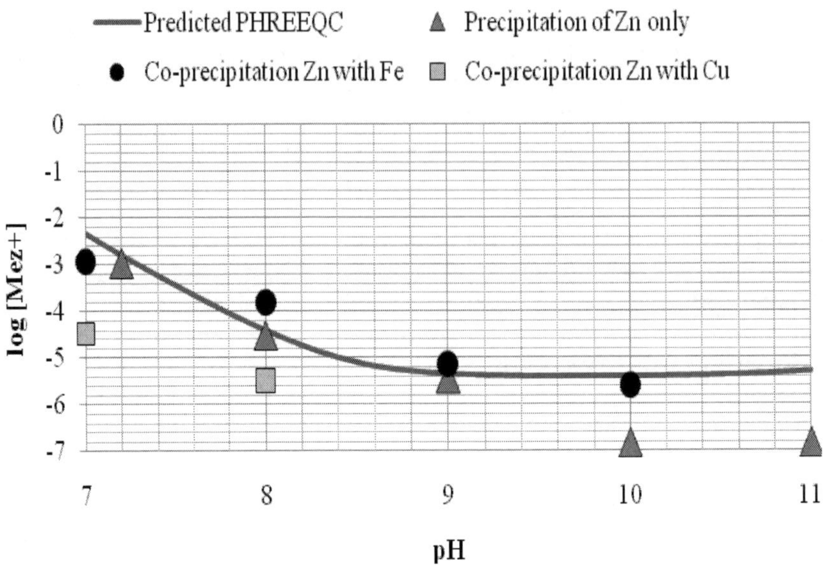

Figure 2: Comparison between predicted (PHREEQC) and experimental soluble metal concentrations at various pHs.

PHREEQCi 2 values (Figure 2). As a simple element, the zinc precipitation results follow predicted PHREEQC values very closely at pH's 7 to 9 and give ever lower concentration values at pH's 10 and 11. When co-precipitating with copper, zinc precipitation results were close to the theoretical, but were found to be higher than the predicted values at pH's 8 and 9. At pH 11, there was no zinc left in the solution. When co-precipitating with Fe^{2+}, zinc precipitation values were well under the predicted values at pH's 7 and 8. At pH's 9 to 11, there was no Zn left in the solution when co-precipitating with soluble Fe^{2+}.

Around pH 8.5, these results resemble closely the performance of the steelworks final discharge precipitation and settlement system for zinc left in solution.

3.3 Pilot plant data

Plant performance analysis can be carried out looking at several aspects of the performance during the trial. The results for soluble concentrations before (In) and after (Out) being treated by the HDS pilot plant can be seen in Table 2. The results show that the HDS pilot plant was able to not only create sludge with good settling qualities (See Figure 8), but also discharge water with very low concentrations of soluble metal.

The variation of pH with time in the feed and throughout the HDS pilot plant streams can be seen in Figure 3. Throughout the trial, the feed pH is hovering around 7, with the Stage I Reactor and discharge pH being around pH 8.

Table 2: Soluble concentrations in the HDS pilot plant feed and discharge water.

Constituent	In mg/l			Out mg/l		
	Min	Max	Ave	Min	Max	Ave
Zn	24.5	76.8	42.7	0.0	5.2	1.4
Cu	0.5	3.4	1.3	0.0	0.4	0.0
Ni	4.3	8.0	6.0	0.0	3.4	1.0
Fe	0.0	0.2	0.0	0.0	0.0	0.0
Mn	12.8	26.2	16.1	0.0	11.3	6.1
Mg	17.5	36.8	22.3	5.2	27.9	20.3
Ca	200.7	470.9	339.2	87.4	391.7	290.6

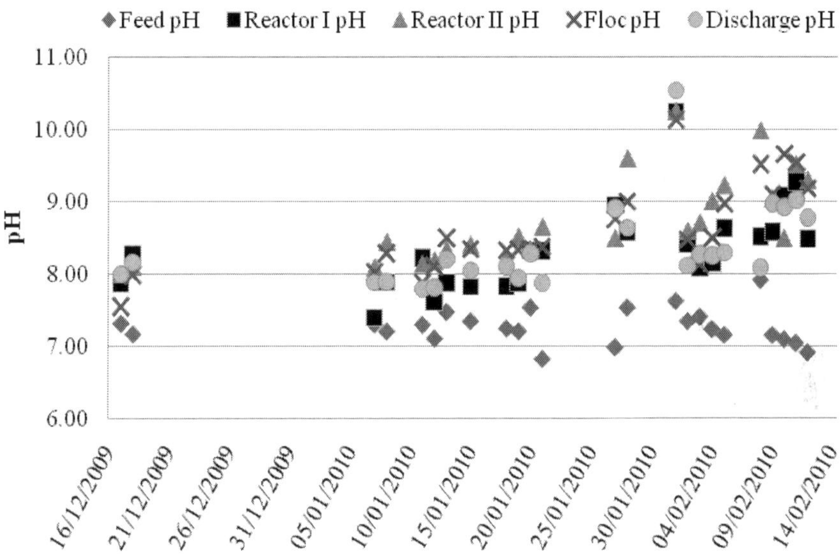

Figure 3: pH variation with time in feed and throughout the HDS pilot plant streams.

Towards the end of the trial, the operating pH was raised slightly to ensure all the metals were being removed from solution. This, together with the higher amount of sludge return at the end of the trial, also affected the amount of soluble manganese left in the solution as can be seen in Figure 4.

The performance of removing soluble Zn (Figure 5) during the trial was also getting better towards the end and lower discharge concentrations of soluble Zn were found at the end of the trial, even though the concentration of the Zn on the feed water was raised from 50mg/l to 100mg/l.

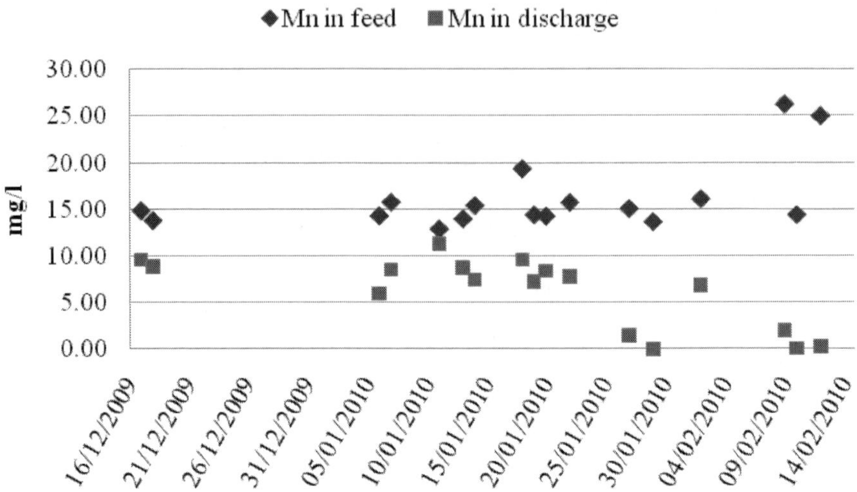

Figure 4: Soluble Mn concentrations with time in the HDS pilot plant feed and discharge water.

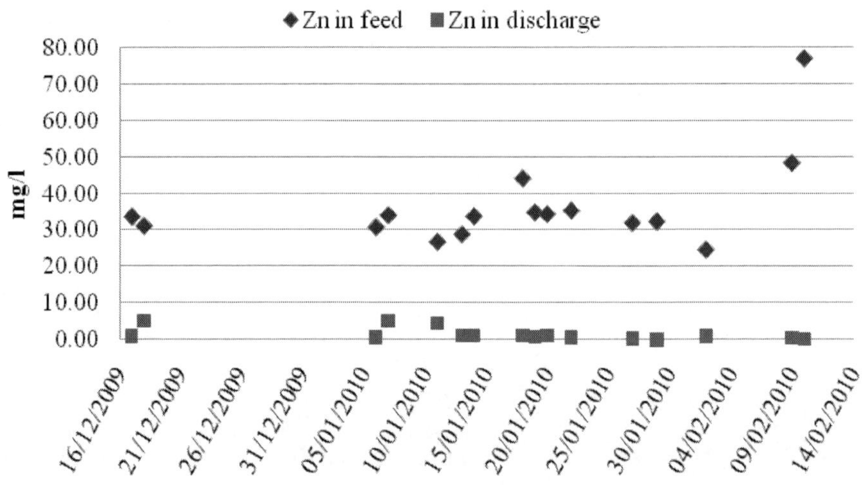

Figure 5: Soluble Zn concentrations with time in the HDS pilot plant feed and discharge water.

Looking at the Mn and Zn removals in more detail it is interesting to note that the soluble Mn levels in the discharge remained close to about 8mg/l up to around 20/01/2010 when the operating pH of the Stage II Reactor was raised close to 9. The effect of increasing amounts of recirculating sludge seemed to be minimal. Once the pH was raised the Mn concentration in the discharge decreased accordingly. For Zn, however, there appears to be a stabilisation of

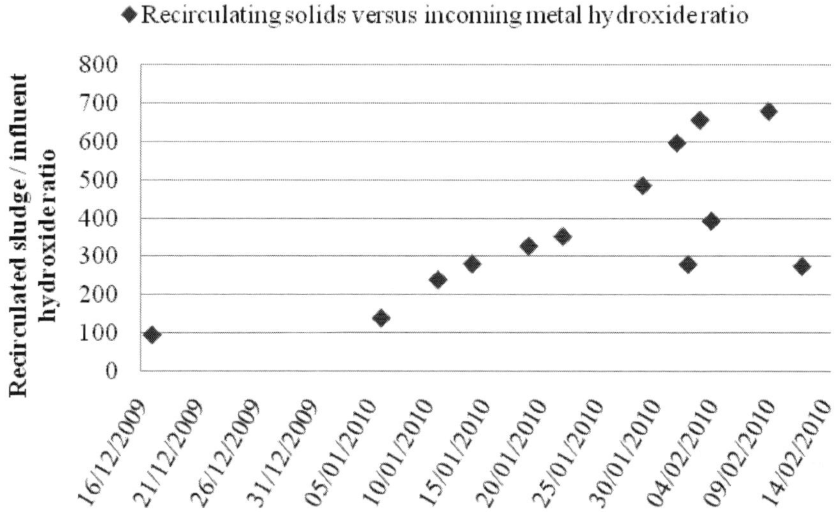

Figure 6: Recirculating solids to incoming metal hydroxide ratio during the HDS pilot plant trial.

the soluble concentration in the pilot plant effluent well before the pH was increased.

Figure 6 shows the metal hydroxide mass recirculation ratio during the trial. The ratio varied between 95.1:1 and 680.4:1 during the whole trial. The largest ratios occurred at the later stages of the trial when the returning sludge concentrations were higher.

As can be seen in Figure 7, the flocculant dosage rate during the trial was between 2 and just under 6 mg of flocculant per litre of feed water treated. The best settling characteristics were found with a flocculant rate of around 2.5-3 mg of flocculant per litre of feed water treated.

There was a very dramatic improvement in the settling rates as the trial progressed as can be seen in Figure 8, which displays the settling rates as a function of the corresponding solid concentrations for discrete ages of the sludges as reflected by the volumes treated. The ability of a floc to settle depends on its size and density. As the HDS process develops the floc density increased and an enhanced settling characteristic is obtained. The data clearly shows the development of floc density (HDS formation) as more and more precipitation occurs. In total, a volume of about 6000 litres was treated over two months of the trial, culminating in exceptionally high settling rates in excess of 22m/h at a relatively high solids concentration of 5%(w/v). For very young sludges it was not possible to achieve high settling velocities even at solids concentrations as low as 0.1%(w/v). Towards the end of the trial it was demonstrated by following the whole settling curve for 24 hours that settled solids concentration in excess of 70%(w/v) were achieved compared to the literature values of less than 5%(w/v) for conventional sludge on the large.

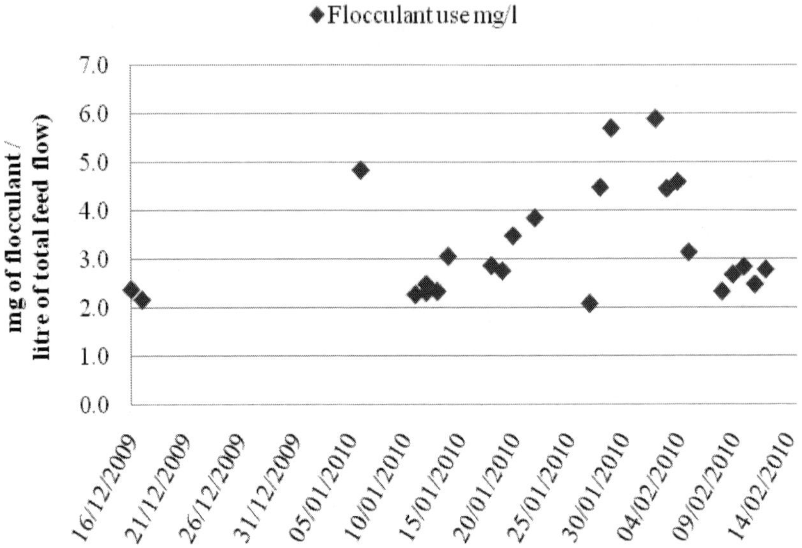

Figure 7: Flocculant dosage rate during the HDS pilot plant trial.

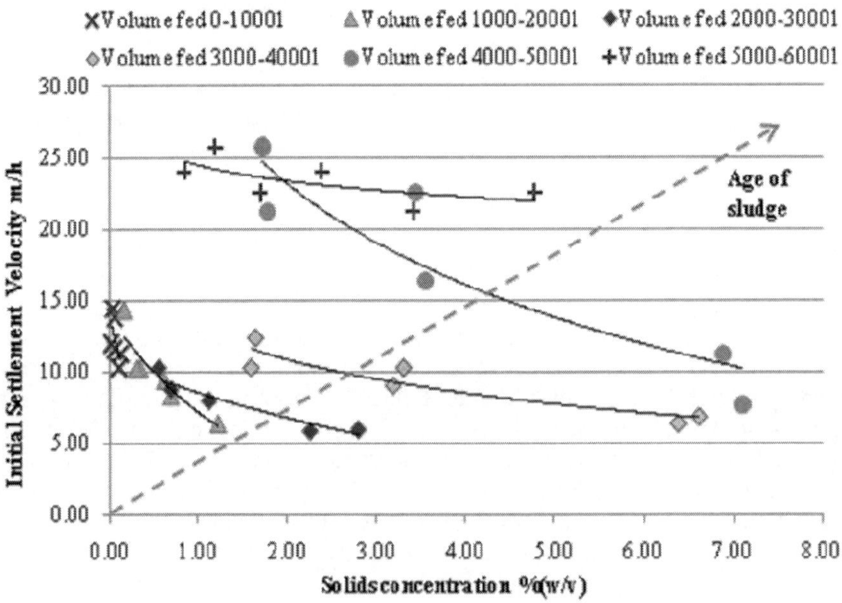

Figure 8: Initial settling rates as functions of volume fed to the HDS pilot plant (data represents approximately two months of operation).

4 Conclusions

The steelworks current final discharge precipitation and settlement system is performing well, but the voluminous sludge produced is problematic to deal with.

PHREEQCi 2 theoretical and the experimental soluble metal concentrations match quite well through the various pHs.

Co-precipitation of Zn with Fe is shown to produce residual soluble metal concentrations much lower than expected as a single species.

HDS forms very readily with this particular steelworks feed, giving precipitates with excellent settling characteristics.

References

[1] Gray N. F., *Water Technology: An Introduction for Environmental Scientists and Engineers*, 2nd Edition, Elsevier Butterworth Heinemann: Oxford, pp. 535-571, 2005.

[2] Dempsey, B.A. and Jeon, B-H., *Characteristic of sludge produced from passive treatment of mine drainage*, Geochemistry: Exploration, Environment, Analysis, Vol. 1, pp. 89-94, 2001.

[3] Bullen C. J., *An Investigation into the Mechanisms Controlling the Formation of High Density Sludge*, PhD. Thesis, University of Wales, Cardiff, 2006.

[4] Coulton, R. H., Bullen, C. J. and Hallet, C., *The Design and Optimisation of Active Minewater Treatment Plants'*, Land Contamination and Reclamation, Vol 11, No 2, pp 273-279, 2003.

[5] Swindley, S.P., *Control of effluent in steel production,* Engineering Doctorate (EngD) Dissertation, University of Wales, Cardiff, 1999.

[6] Parkhurst, D.L., *PHREEQC – a computer program speciation, batch-reactions, one-dimensional transport, and inverse geochemical calculations,* US Geological Survey Water resources Investigations, Report 95-4227, 1995.

[7] Parkhurst, D.L., Appelo, C.A.J., *User's guide to PHREEQC (version 2) – a computer program for speciation, batch-reactions, one-dimensional transport, and inverse geochemical calculation,* US Geological Survey Water resources Investigations, Report 99-4259, 1999.

Modelling of an Upflow Anaerobic Sludge Blanket reactor

R. Rodriguez[1,2] & L. Moreno[2]
[1]*Universidad Nacional de Ingeniería (UNI), Nicaragua*
[2]*Royal Institute of Technology, Sweden*

Abstract

A model describing both physical and biological processes in Upflow Anaerobic Sludge Blanket (UASB) reactors was developed. The main objective of the modelling was to take into account the transient growth of the microorganisms from the start-up of the reactor until a steady state is reached. In addition, the model considers the degradation of the substrate and its reaction with the biomass, which is present in the form of spherical granules of different sizes. For the degradation of the substrate within the granule, the mass transport through the stagnant film around the granule and the intra-particle diffusion are accounted for, together with the specific reaction rate.

In the model, the growth of the biomass follows the Contois kinetics. The amount of biomass reaches a steady state after weeks or months. Biomass is generated when substrate is degraded and a given fraction of biomass disappears per unit of time, which is determined by the decay constant. The value of this constant is taken from the literature. The model also considers that a fraction of the biomass may be carried out by the water flow.

The model was solved by using COMSOL Multiphysics. Data from the literature was used in order to illustrate the processes occurring in the UASB reactor. Modelling can be a useful tool for the design and optimization of UASB reactors.

Keywords: UASB, COMSOL multiphysics, Contois model.

1 Introduction

Biological anaerobic treatment is extensively used to treat wastewater containing high levels of organic matter. Many environmental studies focus on the

WIT Transactions on Ecology and the Environment, Vol 135, © 2010 WIT Press
www.witpress.com, ISSN 1743-3541 (on-line)
doi:10.2495/WP100261

degradation of wastewater in anaerobic reactors due to the fact that this type of processes not only degrades the organic matter, but also produces fuel in the form of biogas and generates low quantities of sludge, making the reactor's maintenance less frequent. One of these reactors is the Upflow Anaerobic Sludge Blanket (UASB) reactor.

The Contois model (Contois [1]) describes the growth of microorganisms with time. This model was developed about 20 years after the Monod model. The principal difference between these two models is that the Monod model considers that the concentration of limiting substrate (when the specific growth rate is half maximum, commonly written as K_S) is independent of population density, whereas the Contois model says that K_S is a function of population density (Contois [1]). Several researchers have compared the Contois vs. the Monod models. However, they have not reached any clear conclusion about which model is better. Bhattacharya and Khai [2] report that the Contois model is better for modelling cow dung digesters. According to Kryłów and Tal-Figiel [3], in order to describe the kinetics of individual stages of processes (substrate degradation, VFA behaviour, microbial growth) for low strength municipal wastewater, it is more convenient to use the Monod model. Hu et al. [4] studied the kinetics of anaerobic digestion using synthetic ice-cream wastewater and they reported that the Contois equation is more adequate than the Monod equation for describing the kinetics process.

Despite the fact that UASB reactors have been used for more than 20 years, few models describing the processes exist. The main processes occurring in the UASB reactors are degradation of substrate, microorganism growth and production of biogas. Singhal et al. [5] developed a model describing the fluid flow of UASB reactors; they divided the reactor in two zones (volume zone and bypassing zone) instead of the three typical zones (sludge bed, sludge blanket and settling zones) to reduce the amount of variables of the model. They injected lithium tracer in the UASB reactor influent and monitored it in the reactor's effluent at different times of the reactor operation. They concluded that the model is suitable for explaining the physical transitions occurring in the zones of UASB reactors. Kalyuzhnyi et al. [6] developed a model focusing on sludge dynamics along the reactor height. The model is one-dimensional dispersed plug flow. It was tested using literature data and as a result, Kalyuzhnyi et al. [6] reported that the model sometimes overestimated or underestimated data from literature in accordance with the working conditions of the model. Korsak et al. [7] developed a model describing processes occurring in UASB reactors. They included the kinetics reactions based on the Monod model.

In this paper, a model describing both physical and biological processes in an Upflow Anaerobic Sludge Blanket (UASB) reactor is developed. The model takes into account the transient growth of the microorganisms and substrate degradation from the start-up of the reactor until a steady state is reached. In addition, the model considers the degradation of the substrate within the biomass, which is present in the form of spherical granules of different sizes. In the granule, the mass transport through the stagnant film around the granule and the intra-particle diffusion are considered accounted for, together with the specific reaction rate. It is assumed that the kinetics follow the Contois model.

The modelling of UASB reactors processes could be a helpful tool for designing and monitoring reactors used in treatment plants.

2 The model

The proposed model explains the substrate behaviour and microorganism growth along the reactor height. The model is one dimensional (all processes depend on the vertical axis). The model kinetics follow the Contois model (equation (1)),

$$\mu = \mu_{\max} \frac{S}{B \cdot X + S} \tag{1}$$

where μ is the specific growth rate (h^{-1}), $\mu_{\max}$ is the maximum specific growth rate (h^{-1}), S is the substrate concentration (Kg/m^3), B is the kinetic parameter [KgCOD (KgVSS)$^{-1}$] and X is microorganism concentration (Kg/m^3).

The model also includes the decay rate of microorganisms that could be caused by toxic metabolism (Kryłów and Tal-Figiel [3]), breaking of the cellular membrane (Kalyuzhnyi et al. [8]) and predation (Bitton [9]). The model takes into account the fraction of microorganisms (biomass) present in the reactor effluent. García et al. [10] studied two identical UASB reactors. They focused on the wash out fraction. They added polymers to one of the reactors and reported that this addition improved the degree of biomass retention inside the reactor.

In order to model the main processes occurring in the UASB reactor, some simplifications or assumptions are made:

-The processes depend on the height of the reactor (one dimensional model)
-Substrate is comprised of a single biodegradable substrate.
-The particles or granules have a spherical shape.
-There is no temperature change inside the reactor.
-Kinetics follows the Contois model

Applying the assumptions described above, the differential equation describing the concentration of substrate in the reactor may be written as

$$\frac{\partial S}{\partial t} = D \frac{\partial^2 S}{\partial Z^2} - vel \frac{\partial S}{\partial Z} - \left(\frac{\mu_{\max}}{Y} \frac{X}{B \cdot X + S} \right) \cdot S \tag{2}$$

$$\frac{\partial X}{\partial t} = (D * w) \frac{\partial^2 X}{\partial Z^2} - [(vel * w)] \frac{\partial X}{\partial Z} + \left(\mu_{\max} \frac{S}{B \cdot X + S} - K_d \right) \cdot X \tag{3}$$

where S is the substrate concentration [Kg m^{-3}], X is the biomass concentration [Kg m^{-3}], D is the dispersion coefficient [m^2 h^{-1}], vel is the upflow velocity [m h^{-1}], Y is the yield coefficient [KgVSS (KgCOD)$^{-1}$], w is the washout fraction which represents a percentage of microorganisms present in the effluent, t is time [h] and Z is the vertical axis, refers to the height of the reactor.

In both equations (2) and (3) the first term on the left hand side is the accumulation term, the first term on the right hand side is the dispersion term, the second one is the advective or transport term and the third one is the reaction term.

In a UASB reactor, the microorganisms are found in the reactor forming granules or particles. The reaction takes place mainly within the particles and depending on the reaction rate, they grow with time. The growth rate is high in zones with high activity and substrate degradation and small in the region with low activity.

The location of the granules in the reactor is determined mainly by the size of the granules and the up-flow velocity of the water; i.e. by the action of the gravity and the friction forces particle-fluid. Gas generation is also very important, since the gas bubbles adhere to the particles and lift them. The location of the particles is continuously varying, therefore only an average location along the vertical axis may be defined for the particles in function of the size; large size granules are found in locations close to the bottom.

2.1 The reaction within the granule

For the interaction between substrate degradation, microorganism growth and particle (granule) size, a model was developed which allows determining the kinetics constant defined in equations (7) and (8).

In the model, it is assumed that the reaction of degradation of the substrate takes place within the granules, in presence of microorganisms. The model includes the transport of substrate by diffusion through the stagnant water film around the granules, the diffusion from the surface into the granule where the degradation reaction takes place, and finally the degradation reaction. In summary, we can talk about three resistances in series. Due to the biological reaction, microorganisms grow and the granule is increased in size. Since the variations in substrate concentration within the granules are small with time, a quasi steady state is obtained for the concentration of substrate within the granule. Therefore, the mass balance for the substrate in the granule may be written for the steady state.

$$\frac{1}{r^2}\frac{d}{dr}\left(r^2\frac{dS_p}{dr}\right)=\frac{k}{D_A}S_p \tag{4}$$

where D_A is the effective diffusion coefficient of substrate within the granule [m^2h^{-1}], S_p is the concentration of substrate within the granule (particle), [kgm^{-3}]; k is the substrate volumetric conversion rate, [h^{-1}]; and r is the radial distance from the centre of the particle, [m]. The equation is solved for a substrate concentration S [kgm^{-3}] in the bulk liquid, and for a mass transfer coefficient k_m [m h^{-1}] for the substrate through the stagnant water film around the granule. Once the concentration profile for the substrate in the granule is known, the mass flow of substrate into the particle, W_R [kgh^{-1}], may be obtained [5]:

$$W_R = 4\pi R^2 k_m S \cdot \left(\frac{D_A(\phi \cosh\phi - \sinh\phi)}{D_A(\phi \cosh\phi - \sinh\phi) + Rk_m \sinh\phi} \right) \tag{5}$$

where $\phi = R\sqrt{k/D_A}$ is the Thiele modulus (dimensionless value) and R is the granule radius [m]. The term W_R corresponds to the amount of substrate that is degraded within each granule. Therefore, the amount of substrate degraded per unit volume reactor may be calculated considering φ_P, which is the fraction of total volume occupied by the granules [m^3 granule $(m^3 reactor)^{-1}$]. Once the substrate degradation rate is known, the kinetics constant (K) in equations (7) and (8) may be determined,

$$K = 3k_m \frac{\varphi_P}{R} \left[\frac{D_A(\phi \cosh\phi - \sinh\phi)}{D_A(\phi \cosh\phi - \sinh\phi) + Rk_m \sinh\phi} \right] \tag{6}$$

The model for the substrate and microorganisms concentration in the UASB reactor is transient and the location of the granules along of the reactor height is determined by their size. However, the reaction within the granules is solved for the steady state. The governing equation for substrate degradation and microorganism growth may be rewritten in function of the kinetic constant and they are shown in equations (7) and (8), respectively.

$$\frac{\partial S}{\partial t} = D \frac{\partial^2 S}{\partial Z^2} - vel \frac{\partial S}{\partial Z} - (K \cdot S) \tag{7}$$

$$\frac{\partial X}{\partial t} = (D * w) \frac{\partial^2 X}{\partial Z^2} - [(vel * w)] \frac{\partial X}{\partial Z} + (K \cdot Y \cdot S - K_d \cdot X) \tag{8}$$

In equation (8) the decay constant (K_d) is included. This constant represents the microorganism death and its unit is [h^{-1}]. The parameter w in equation (8) represents the fraction of microorganisms that wash out by the upflowing water. It is important to indicate that the amount of biomass is included in the Kinetics constant through the value of φ_P.

2.2 Model data

Sponza and Uluköy [11] compared three models (Monod, Modified Stover-Kincannon and Grau second order) in order to determine which one is suitable for predicting the substrate removal kinetics of a UASB reactor. They used synthetic wastewater containing glucose and 2,4 dichlorophenol (2,4 DCP) as substrate for feeding the reactor. As a result, Sponza and Uluköy reported Monod kinetic model was more suitable than Stover-Kincannon and Grau second order to describe the reactor's behaviour. In this paper we used data reported by Sponza and Uluköy [11], but we applied Contois model in place of Monod model to test our model. Table 1 shows the parameters used in the model. It is assumed that the substrate concentration both in input flow and initial is the same (3 Kg/m^3). However, in the input flow, there is no microorganism concentration but the initial microorganism concentration in the UASB reactor is 5E-3 Kg/m^3. The kinetic parameter B was determined plotting the SRT/(1+SRT*Kd) vs. X/S where SRT is the sludge retention time (Sponza and Işik [12]).

The experimental UASB reactor has 1 m of length and a diameter of 6 cm, the effective volume is 2.5 litres and the UASB reactor was inoculated with partial granules collected from a UASB reactor treating wastewater of Pakmaya Baker Factory in Izmir but the input fluid flow for the experiments was synthetic wastewater. The reactor worked under mesophilic conditions (Sponza and Uluköy [11]).

Both microorganism and substrate performance was set for a period of 648 hours (HRT varying 20-10 h), by Sponza and Uluköy [11]. To know the behaviour of substrate, chemical oxygen demand (COD) data was used as substrate. The washout fraction was assumed as well as the Péclet number. In Figures 1 and 2, the x-axis represents the reactor height (m) and the y-axis is the substrate and microorganism concentration (Kg/m^3) respectively. Figure 3, the x-axis is the reactor's height (m) and the y-axis is the granule's radius (m).

Table 1: Parameters used in the model.

Parameter	Value	Units
Péclet number	50	Dimensionless
Hydraulic retention time*	9	h
Upflow velocity*	0.098	m^3/m^2.h
Dispersion coefficient	1.96 E-3	m^2 h^{-1}
Maximum specific growth rate*	8.75E-3	h^{-1}
Yield*	0.00078	KgVSS(KgCOD)$^{-1}$
Kinetic parameter B*	0.1053	KgCOD (KgVSS)$^{-1}$
Decay constant*	3.875E-5	h^{-1}
Washout fraction	0.0002	Fraction microorganism

* Taken from Sponza and Uluköy [11].

3 Results and discussion

The equations of the model were solved using COMSOL Multiphysics 3.5. The model response for substrate degradation and microorganism growth are shown in Figures 1 and 2. Figure 3 shows the radius of the granule as a function of the height in the reactor.

Regarding the substrate behaviour, the input substrate concentration is 3 Kg/m^3, but due to dispersion, Figure 1 shows a value slightly less (2.9 Kg/m^3). The substrate concentration decreases continuously with the height, reaching a value below 1.5 Kg/m^3 at the reactor top. The steady state is reached after a long time of operation (more than 650 h).

In Figure 2, the microorganism concentration reaches its maximum at the reactor bottom, where the maximum substrate degradation rate takes place. The concentration of microorganisms decreases with the height due to the less concentration of substrate. An increase of biomass concentration at the reactor bottom is remarkable compared to that of other heights of the reactor. The behaviour of the reactor is very sensible to the parameters chosen in the simulations. In this case, the concentration of the microorganisms in the reactor is lower for the combination of parameters used.

Sponza and Uluköy [11] reported a COD (substrate) reduction of 83% with an operation time of 27 days. In our model, the COD is consumed at the reactor's bottom where the major concentration of biomass is present. With respect to Figure 2, Sponza and Uluköy did not report values of the microorganism

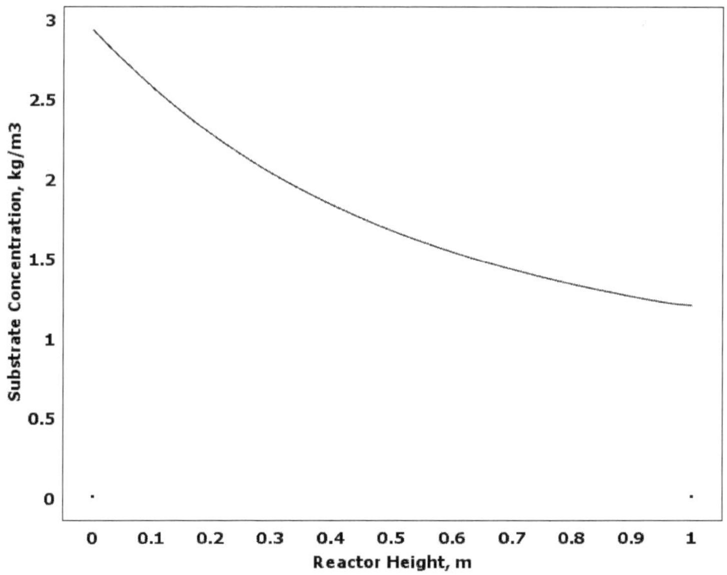

Figure 1: Model response for the substrate concentration profile.

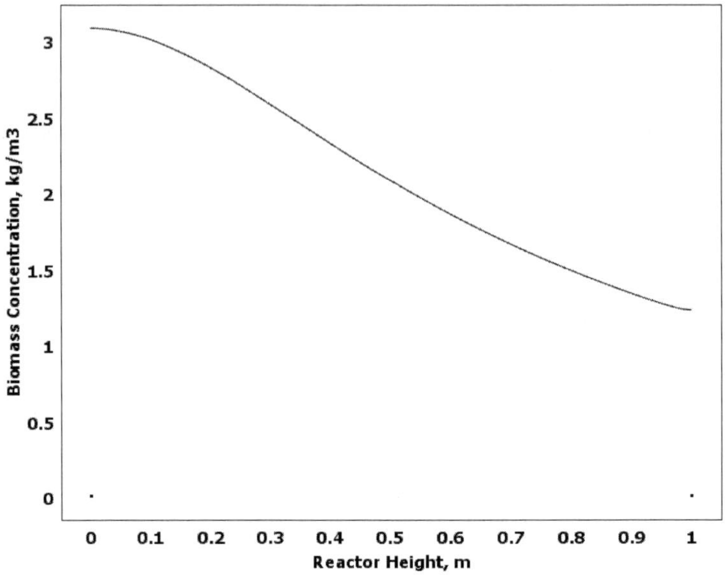

Figure 2: Model response for the microorganism concentration profile.

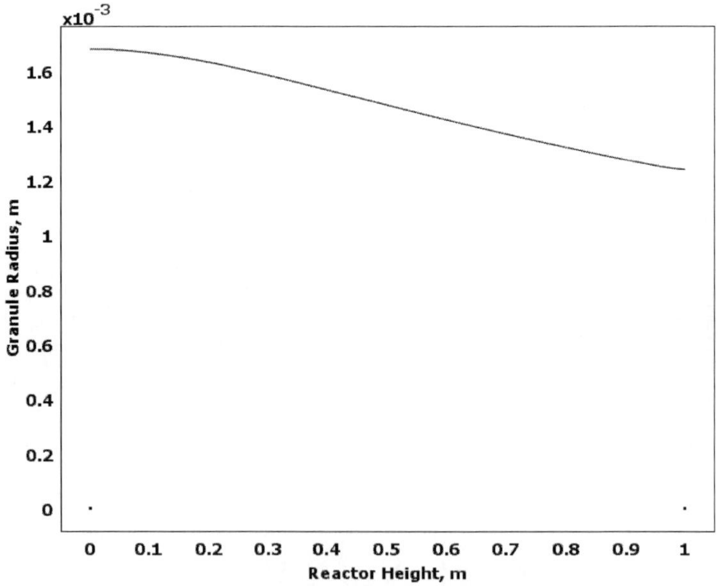

Figure 3: Model response for the granule radius.

behaviour; their work was focused on the comparison of three kinetics models. Initial biomass concentration was 5E-3 Kg/m^3, but after long operation time, biomass grows and reaches 3.53 Kg/m^3. The major growing of biomass is found at the reactor's bottom.

Figure 3 shows the profile of granules sizes along the reactor's height. The response model shows granule diameters ranging between 2.5 – 3.2 mm, after a long operation time.

In our model, we have it transient but with time it reaches steady state conditions. An important aspect in this modelling is the high sensibility of the model response to the value used for the different parameters. A study of the influence of the most important parameters on the response of the model is currently in process.

4 Conclusion

The proposed model describes the degradation of the substrate (as COD) and the growth of the microorganisms (biomass) in the reactor. The model has been successfully used to simulate the UASB reactor performance studied by Sponza and Uluköy [11]. Two kinetics models, Contois and Monod were used to study the results and no significant differences were found between these kinetics. These kinetics reactions include the growth and death of the microorganisms. Since the reaction takes place within the granules a model was developed which takes into account diffusion resistance in the granule and in the film around the granule. The granules in the reactor are distributed according their sizes, the large ones trend to locate at the bottom of the reactor.

The results of the model are quite sensitive to some model parameters used in the simulations; e.g. the relative values between the decay constant, K_d, and the growth rate. If the decay constant is too large, the microorganisms will decrease in the reactor, decreasing the degradation of the substrate.

This type of model may be a useful tool in the development and design of UASB reactors. It allows study the consequences on the reactor performance of different types of substrates, inlet substrate concentrations, flowrates, and different kind of biomasses.

Acknowledgement

This work is supported by the Swedish International Development Cooperation Agency (SIDA/SAREC).

References

[1] Contois D. Kinetics of Bacterial Growth: *Relationship between Population Density and Specific Growth Rate of Continuous Cultures*. Microbial. 21, 40-50, 1959.

[2] Bhattacharya S. and Khai P. *Kinetics of Anaerobic Cowdung Digestion.* Energy Vol. 12, No. 6, pp. 497-500, 1987.
[3] Kryłów M. and Tal-Figiel B. *Kinetics of Anaerobic Processes.* Proceedings of a Polish-Swedish seminar, 2003.
[4] Hu W., Thayanithy K. and Forster C. *A Kinetic Study of the Anaerobic Digestion of Ice-cream Wastewater.* Process Biochemistry 37, 965–971, 2002.
[5] Singhal A., James Gomes, Praveen V. and Ramachandran B. *Axial Dispersion Model for Upflow Anaerobic Sludge Blanket Reactors.* Biotechnol. Prog. 14, 645-648, 1998.
[6] Kalyuzhnyi S. Fedorovich V. and Lens P. *Novel Dispersed Plug Flow Model for UASB Reactors Focusing on Sludge Dynamics. Proceedings of 9th international conference on anaerobic digestion.* Antwerpen pp 123-128, 2001.
[7] Korsak L., Cabezas A. and Moreno L. *A Model of Kinetics and Mass Transfer in an UASB reactor.* Proceedings of 4th international conference on environmental science and technology, 2008.
[8] Kalyuzhnyi S. Fedorovich V. and Lens P. *Dispersed plug Flow Model for Upflow Anaerobic Sludge Bed Reactors with Focus on Granular Sludge Dynamics.* Microbiol Biotecnol pp. 221-237, 2006.
[9] Bitton G. Wastewater Microbiology. 3rd Ed. Wiley Editor, 2005.
[10] García H., Rico C., García P. and Rico J. *Flocculants Effect in Biomass Retention in a UASB Reactor Treating Dairy Manure.* Bioresource Technology 99, 6028–6036, 2008.
[11] Sponza D. and Uluköy A. *Kinetic of Carbonaceous Substrate in an Upflow Anaerobic Sludge Blanket (UASB) Reactor Treating 2,4 Dichlorophenol (2,4 DCP.* Journal of Environmental Management 86, 121–131, (2008).
[12] Sponza D. and Işik M. *Substrate Removal Kinetics in an Upflow Anaerobic Sludge Blanket Reactor Decolorising Simulated Textile Wastewater.* Process Biochemistry 40, 1189–1198, 2005.

On-line monitorization in a decarbonator-settling tank for water treatment

A. Tijero, A. Moral, J. Tijero, A. Blanco & C. Negro
Department of Chemical Engineering,
Universidad Complutense de Madrid, Spain

Abstract

The objective of this work is the on-line control of the solid-liquid separation process in a water treatment plant using a focused beam reflectance measurement sensor (FBRM) in order to improve the coagulation, precipitation and flocculation of this plant. A previous study of the system is carried out in order to analyze the possible influence of the external variables on the floc mean chord size though of the monitorization of the flocculation tank. The formation of particle aggregates is the product of the addition of an inorganic flocculant ($FeCl_3$) with a subsequent treatment with an organic synthetic flocculant (anionic polyacrilamide) that determine the formed floc properties. The FBRM study of the variations of the mean chord size realised after the changes in the system show an improvement in flocculant diffusion and a greater phase contact. The final objective is to improve the calcium carbonate precipitation conditions, to reduce the membrane fouling and avoid the deposit formation in the plant surface. The trials carried out a characterization of the waters before and after changes in the system in order to establish whether the final properties of the water fulfil the parameters required for its posterior use. The results obtained show that the measurement system FBRM is an on-line sensor that is good enough to be used as a control and prediction method in the water treatment industries. Furthermore, the relations between the data obtained by the sensor and the properties of the water after the depuration allow control of the outlet water.

Keywords: flocculation, coagulation, FBRM control on-line, river water treatment.

doi:10.2495/WP100271

1 Introduction

For the best distribution of electric power, new Combined Cycle Power Plants (CCPP) are installed near consumer centres. Due to their strategic location, in many power plants the water supply is from a river polluted by wastewater and by municipal and industrial wastewater treatment plants' effluents.

The object of this work is the optimization of the water pre-treatment plant in a CCPP, where raw water is supplied by a river that has a basin with dense population and industrialization areas, and to obtain by this pre-treatment water with the adequate characteristics needed for use in the refrigeration towers.

The pre-treatment process can be designed as "enhanced coagulation and enhanced precipitative softening", which refers to the process of improving the removal of natural organic matter (OM) compared with other conventional coagulation of softening water treatment processes.

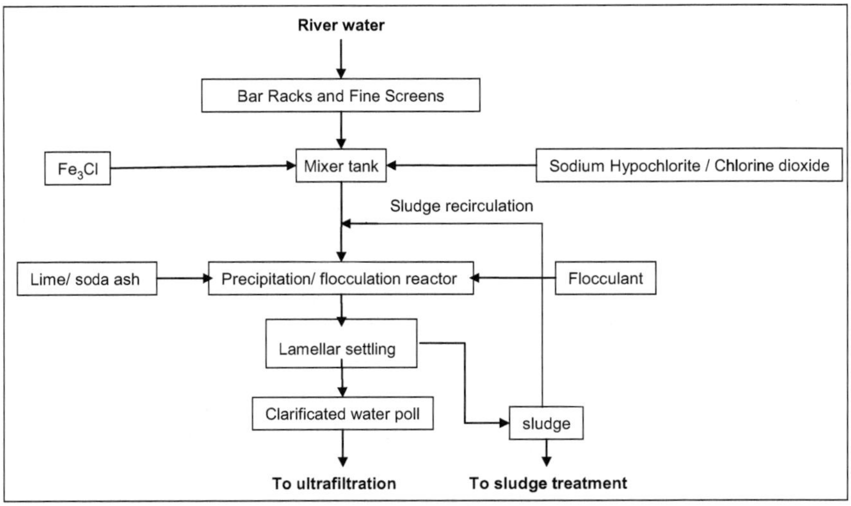

Figure 1: Plant block diagram.

In the preliminary step, river water is carried by a channel with a bar rack and screen and pumped to the raw water river basin.

The first step of the physic-chemical treatment is coagulation with iron trichloride in a mixer tank for organic matter (OM) and removal, and addition of chlorine/chlorine dioxide for OM oxidation and disinfection.

When added to water, iron salts dissociate and react to yield hydrolysis products. As first evidenced by Mattson in 1928, these hydrolysis products form the coagulant species that remove turbidity and colour during water treatment [1]. The hydrolysis of $FeCl_3$ proceeds with monomers, dimmers, double corner trimers, and polycations that consist in the arrangement of 24 Fe atoms in a B-FeOOH local structure [2]. Further neutralization of the Fe solution induces the association of Fe_{24} polycations within fractal clusters [3]. Unlike river waters, municipal sewage provides a system with abundant organic matter and complexing anions that should extend the concentration range in which hydrolyzing coagulant species interact with the raw water content [4]. The removal of this organic matter from the bulk water can occur by direct precipitation or by adsorption of the organic matter onto metal hydroxides [4].

The second step is precipitative softening carried out in a continuous stirred-tank reactor (CSTR), with hydroxide ions (as $Ca(OH)_2$) addition for the carbonate speciation to CO_3^{-2} which facilitates the precipitation of $CaCO_3$. This process generally takes place at a pH of 10. Removal of magnesium hardness is caused by the precipitation of $Mg(OH)_2$ which generally takes place at a pH of 10.5 or higher. Depending on the characteristics of the source water and the relative concentrations of calcium and magnesium ions and carbonate and noncarbonate ions, the amount of lime and soda ash required for precipitative softening may vary. Rivers with strong seasonal variations of flow with water composition of magnesium salt in summer can require a necessary soda ash addition for magnesium precipitation.

In the third stage the flocculation process is carried out. Flocculant is added to the precipitation/flocculation reactor. In this case we have a dual system defined by a complex flocculation, which is by a flocculation combined by a previous coagulation mechanism. These traditional dual systems involve the use of two different polymers, a high molecular weight anionic polymer (anionic polyacrilamide) and a low or middle molecular weight high density cationic polymer, or a metallic cation ($FeCl_3$). This mechanism involves the adsorption of the cationic polymers or metallic cations into the negative colloidal surface, forming cationic patches, and the interaction between the high molecular weight polymer and positive charges or previous formed patches form a bridging between the colloidal particles. The bridging mechanism between the colloidal particles by an electrostatic-patch is show in figure 2.

The result of this kind of mechanism is flocs that combine the resulting properties of the flocs formed by a patch and bridging flocculations. The flocs are quite resistance and a partial re-flocculation readily occurs under low shear conditions. However, the efficiency of these systems depends on the order of the

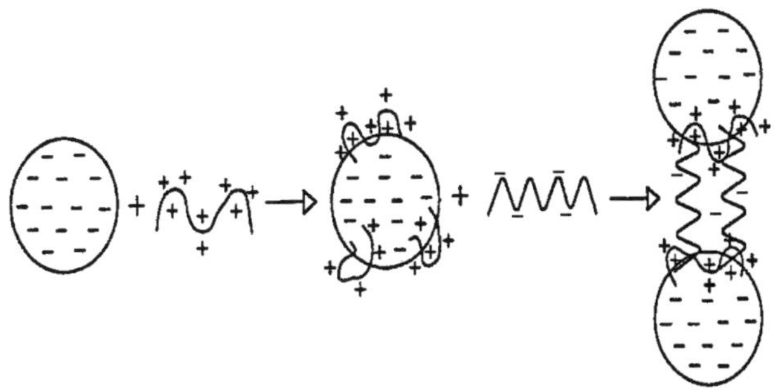

Figure 2: Bridging mechanism by electrostatic patch.

additions of the components and of the time interval between additions [5]. In our industrial level application the depuration entails an ultra-filtration process and due to that the previous particle flocculation is vital because it increases membrane lifetime and the process efficiency.

The flocs formation into precipitation/flocculation reactor was studied by monitoring of the flocs size with a FBRM sensor. These measures, in real time, are related to the process variables. The control set-points of these process variables for chemicals dosification is established from supply river water characteristic.

This paper tries to evaluate the possibility to carry out modifications in the pre-treatment plant to reduce to the minimum the employment of chemicals as well as to solve the problems of the inlay and the adherence. These modifications can be in the treatment process or installation revamping

Finally, the objective is, to study the flocs formation and its evolution along successive plant steps. A preliminary study shows what main phenomena occur inside the precipitation/flocculation stirred tank; for this motive the selected localization for flocculation probe is immersed in the turbulence zone.

The fundaments of measurement with the FBMR sensor will be explained in 2.1. The results of the flocs' measurements were compared with the variables that affect the coagulation-flocculation.

Supply water is also a characterized and the outlet water from the settling pond, before and after of the system modifications, with the objective to obtain enough data to evaluate if the final characteristics of the water fulfil the specified parameters for use as cooling tower water, bearing in mind the seasonal variation of the river water characteristic for several time periods.

2 Experimental

2.1 FBRM probe description and location

The methodology to monitor flocculation is based on using a focused beam reflectance measurement system, FBRM M500P, manufactured by Lasentec Mettler Toledo, Seattle, USA. The FBRM technique provides the chord length distribution of the particles in the suspension in real time. This distribution depends on the shape, size and concentration of particles [6].

The FBRM measurement principle, as shown in figure 3, is based on backward light scattering. A laser beam is coupled to an immersible probe via an optical fibre. This laser beam is deviated from the probe's central axis and focussed into a disperse medium with an optical conduit. When this laser beam intersects with a particle, light scattering occurs. A certain fraction of the light is scattered back into the system. This back scattered light is coupled via a beam splitter to a second fibre and conduced to a detector [7]. The rotational velocity of the laser is constant. The time span in which back scattering is detected is therefore directly proportional to the path length of the laser on the particle. It is assumed that the particle velocity is small compared to the laser rotational velocity. The length of the laser path on the particle is therefore proportional to

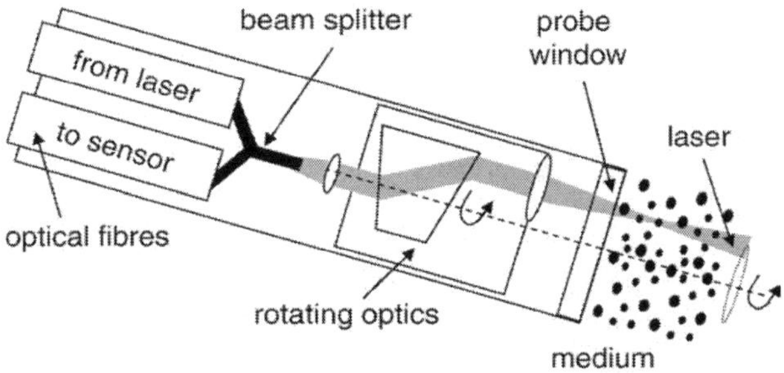

Figure 3: Schematic drawing of a FBRM probe.

Figure 4: Location of the FBRM.

the time span in which scattering is detected. This path length is called a chord length.

Depending on the laser position, different chord lengths are measured even for a single particle. Those chord lengths are generally different from any characteristic particle length [8].

The detail methodology to study flocculation processes and floc properties and the principles of the measurements are described in previous papers [9, 10].

The right selection of the probe location with respect to the full flow sheet of the process and the orientation inside the tank is essential to obtain accurate data. It was decided to install the FBRM after coagulant addition in the flocculation tank. The location of the FBRM is show in the figure 4.

The FBRM probe is immersed into the precipitation/flocculation tank, fixed on the bridge of the agitator, as shown in figure 3. This zone inside of the flocculation tank has height agitation, without bubbles, to obtain representative data from the system and to keep adherents particles from the lens off the FBRM. This tank is a crucial zone of the depuration process because inside it takes place the precipitation of calcium carbonate and magnesium carbonate/

magnesium hydroxide, the growth of precipitate particles for recirculation of part of the sludge and the addition of polyelectrolyte to the water previously coagulated to produce a complex flocculation.

2.2 Recording time optimization

In the selection of the recording time of particle size distribution we have to take into account that the shorter the measuring time, the greater the sensibility will be, although, at the same time any natural "noise" from the measurement will be recorded. Moreover, the management of very short measuring times entails having to work with a great amount of information. All the data presented correspond to measurements carried out inside the flocculation tank with the purpose of optimizing the measuring time of this study system. The recording time optimization data is show in the figure 5.

In recording times of 30 minutes the graphic have lines with few peaks and this show that three lines are good to obtain a global development, however we lose important information for shorter intervals. For times lower than 10 minutes (5 and 2 minutes) the results obtained are very difficult to interpret because the natural noise from the measurement meddles significantly with the great amount

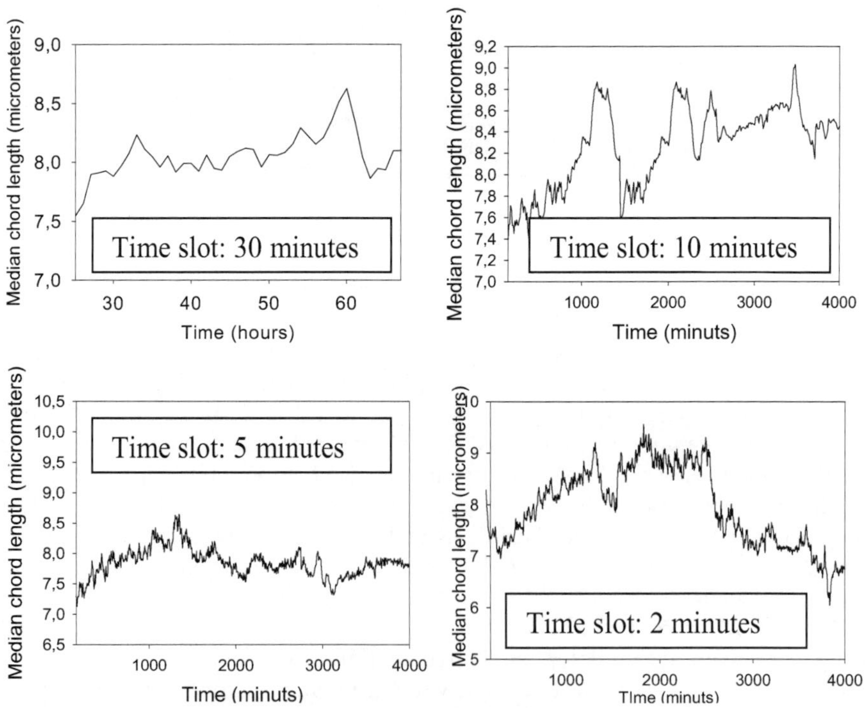

Figure 5: Recording time optimization data.

of information obtained. Best time interval measurement of the particle size distribution is 10 minutes. This interval permits to obtain a very stable measurement and delayer time is low compared with change rate in the system.

3 Results and discussion

In the industrial pre-treatment plant operation, all process variables are controlled, also are analysed and monitoring numerous parameters. In this work the selected process parameters for optimization are:

Point and amount of polyelectrolyte dosified
Sludge recirculation
Flocs size as long of cord

Treatment efficiency is giving for:

Turbidity, NTU
Total organic carbon, TOC
Hardness as total and calcium hardness, in mg/L of Ca CO_3

3.1 Water characterization

The analytical parameters systematically determined in the pre-treatment plant are:

Calcium, magnesium, bicarbonates, carbonates, hydroxides, chlorides, sulphates, nitrates, phosphates, fluorides, nitrites, pH, conductivity, turbidity, TOC, COD, anionic detergents, ammonia, silica solved and total, chlorine free , chlorine demand, solids (dissolved, suspended and total), iron, manganese and aluminium

The water samples obtained in different points of the plant present a common characteristic, their anionic charge. Water cationic demands are determined in different points of process.

The flocculant used during this work realization is of medium molecular weight, and moderately anionic. The results obtained from these samples show a very low variability (figure 6).

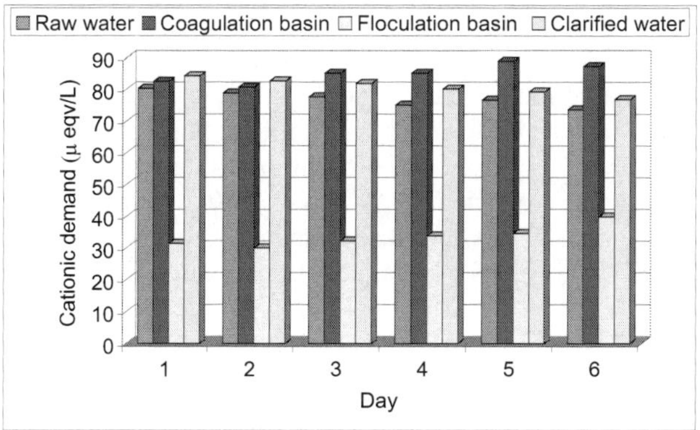

Figure 6: Cationic demands of process water.

3.2 Correlating the effect of flocculant dosages with sensor data

Several plant trials were carried out with different polyacrilamide concentrations and without change the rest of the parameters. The on-line measure system FBRM situated in the flocculation tank detected that when there is an increase in the polyacrilamide dose the particle mean chord size increase. The optima flocculant dose needed to obtain a maximum mean chord size was previously calculate by the star-up supplier. However, it is observed that when there is an increase of the dose with respect to the recommended dose the mean chord size increase and this is a probe that there is problem with the dosification or the homogenization.

After a detailed study of the plant, it is detected in the second mix chamber that the dosification of flocculant is carried out in a far away zone of the propeller that permits the homogenization of the system by stirring, and because of that the contact between the flocculant, the environment and the lime is not carried out of adequate form. Because of that it is necessary to make a change in the flocculant addiction pipe placing it in the maximum agitation zone in order to improve the contact between phases.

3.3 Monitoring the effect of system changes

After the changes in the process, the average chord size increases, that is an indication of a better precipitative softening. As well, it is observed that the biggest flocs have more stability, that is, they suffer less size variation with the time the next figure shows that, after the changes in the system, the mean chord size rise from about 7µm values to 14µm that is a betterment of the system.

3.4 Correlating sensor data with water properties

Changes realised in addition point and dosification of flocculant and in sludge recirculation can be appreciated in floc size and floc stability shown in figure 7. This best flocculation is related with selected treatment analytical parameters.

3.4.1 Turbidity

This parameter reflects the efficiency of flocculation. Raw river water has very different turbidity, between 10 and 35 NTU, the graphics shown in figure 8 show the changes in the systems before and after of modifications in the system.

Turbidity reduction efficiency is about 80% previous to a plant's modifications and get to 90% after these changes.

Clarified water NTU is given for lineal equations.

Previous at changes in process:

$$NTUclarified\ water = 0{,}197NTU\ raw\ water. \quad (1)$$

After changes in process:

$$NTUclarified\ water = 0{,}102NTUraw\ water. \quad (2)$$

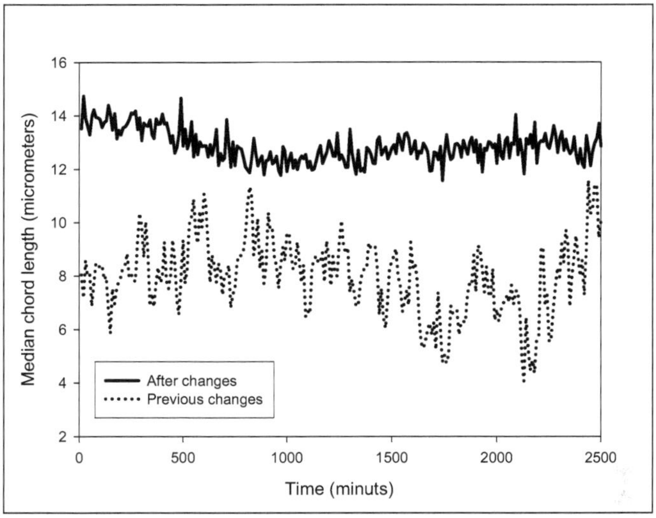

Figure 7: Mean chord size previous and after the changes in the system.

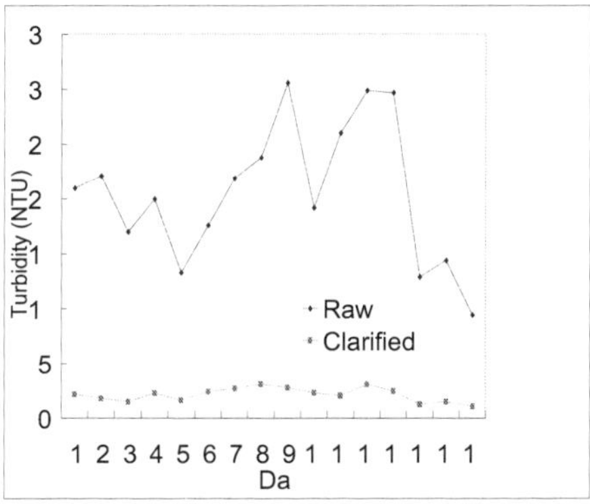

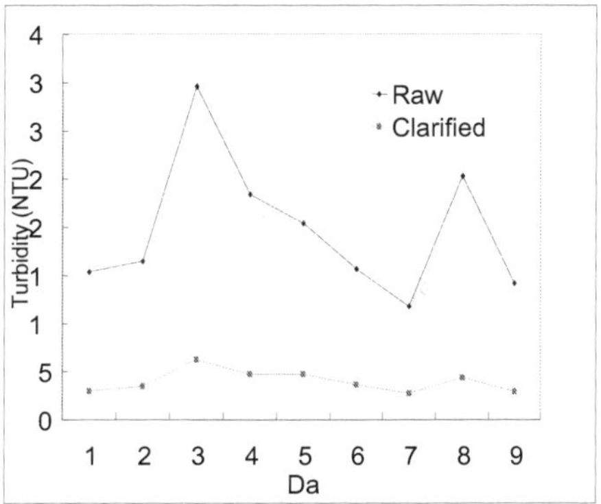

Figure 8: Comparison of turbidity of the raw water with clarified water, before (to the left side) and after (to the right) process modifications.

3.4.2 TOC

Enhanced coagulation is a treatment technique that has the purpose of removing organic matter (OM) to reduce the formation of Disinfection By-product DBPs.

Enhanced softening refers to the process of obtaining improved removal of MO, DBP precursors, by precipitative softening. The analytical parameter adopted is TOC. Figure 9 shows the variability of raw water TOC.

Coagulation and precipitative softening, before treatment modifications, has an efficiency of TOC reduction in clarified water of about 35% of raw water TOC. After modifications TOC reduction is about 50%.

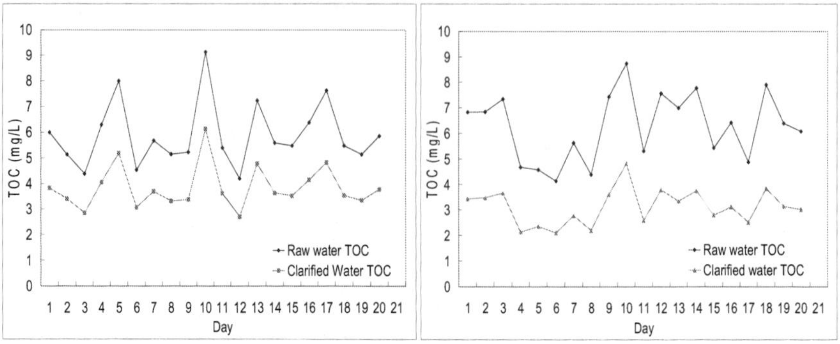

Figure 9: Comparison of the raw water TOC with clarified water TOC, before (to the left side) and after (to the right) process modifications.

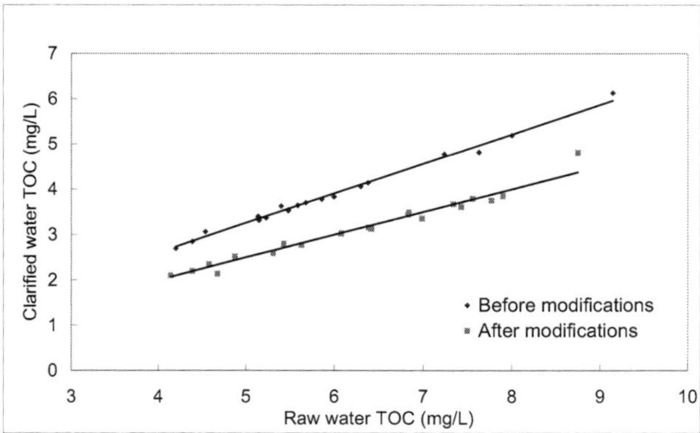

Figure 10: Clarified water TOC vs. raw water TOC.

Clarified water TOC is related to raw water TOC by the lineal equations.
Previous at modifications:

$$TOC_{\text{clarified water}} = 0{,}64TOC_{\text{raw water,}}\ R^2 = 0{,}98 \tag{3}$$

After modifications:

$$TOC_{\text{clarified water}} = 0{,}49TOC_{\text{raw water}}\ ,\ R^2 = 0{,}96 \tag{4}$$

Figure 10 show the clarified water TOC versus raw water TOC. TOC reduction efficiency of treatment before modifications is about 36% of TOC of water inlet. TOC reduction efficiency rise to about 50% after process modifications

3.4.3 Hardness

Influent hardness has high variability, and softening treatment is not always able to get the maximum specified calcium hardness 360 mg/L as calcium carbonate and 480 mg/L as calcium carbonate of specified maximum total hardness.

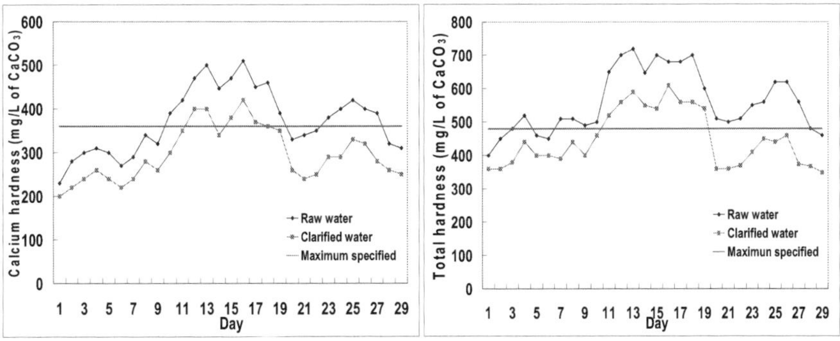

Figure 11: Comparison of calcium hardness (left-hand side) and total hardness (right-hand side) of the raw water with clarified water.

The graphics in figure11 shows that it is suitable to be more efficiency in the water softening. Process changes are operating after the 20th day, but hardness values of clarified water are not sufficiently low. For more hardness reduction it is necessary to make changes in pH, different lime and soda ash addition and, also, the addition of another kind of flocculant, preferment cationic, in order to get softer water.

4 Conclusions

The methodology based on utilization of a focused beam reflectance system of measure FBRM is suitable to supervise process of precipitative softening and flocculation.

Modifications in process are related to floc size and this floc size with the treatment analytical parameters, as turbidity, TOC and hardness.

Changes in the dosage point and dosage amount of flocculant improved the mix and diffusion, the fixation on the floc surface, and the physical-chemical adsorption equilibrium achieving between adsorbed flocculant and the flocculant in solution.

These process modifications are important because the flocculant left in the water, increasing membranes fouling and fixing into the surfaces, forming appreciable velocity growth of incrustation

References

[1] Mattson, S., Cataphoresis and the electrical neutralization of colloidal material. *Journal of Physical Chemical*, 32, pp. 1532-1552, 1928.

[2] Bottero, J.Y., Manceau, A, Villieras F. &, Tchoubar D., Structure and mechanism of formation of FeOOH (Cl) polymers. *Langmuir*, 10, pp. 316-319, 1994.

[3] Tchoubar, D., Bottero, J .Y., Quienne, P, Arnaud, M. Partial hydrolysis of ferric chloride salt. Structural investigation by phonton-correlation spectroscopy and small angle X-ray scattering. Langumir 1991; 7: 398-402.
[4] El Samrani, A. G., Lartiges, B. S., Montargès-Pelletier, E., Kazpard, V., Barrès, O. & Ghanbaja, J., Clarification of municipal sewage with ferric chloride: the nature of coagulant species, *Water research*; 38, pp.756-768, 2004.
[5] Csempesz, F., Enhanced flocculation of colloidal dispersions by polymer mixtures. *Chemical Engineering Journal*, 80, pp 43–49, 2000.
[6] Negro, C., Sánchez, L. M., Fuente, E., Blanco, A. &. Tijero, J., Polyacrylamide induced flocculation of a cement suspension. *Chemical Engineering Science,* 61, pp 2522 – 2532, 2006.
[7] Sparks, R. G., Dobbs, C. L., The use of laser backscatter instrumentation for the online measurement of the particle-size distribution of emulsions, *Particle &. Particle System Characterization,* 10, pp. 279–289, 1993.
[8] Tadayyon, A. & Rohani, S., Determination of particle size distribution by Par-Tec (R) 100: modeling and experimental results, *Particle &. Particle System Characterization,* 15, pp. 127–135, 1998.
[9] Blanco, A., Fuente, E., Negro, C. & Tijero, J., Flocculation monitoring: focused beam reflectance measurement as a measurement tool. *The Canadian Journal of Chemical Engineering,* 80, pp.734–740, 2002.
[10] Blanco, A., Fuente, E., Negro, C. & Tijero, J., Focused beam reflectant measurement as a tool to measure flocculation. *Tappi Journal*, 1, pp.14–20, 2002.

Section 8
Novel techniques for water treatment

Development of a novel high performance cake-less filtration system

T. Mori, H. Satone, T. Hirata & J. Tsubaki
Department of Molecular Design and Engineering, Graduate School of Engineering, Nagoya University, Japan

Abstract

A new cake-less continuous filtration system with a spiral guide rod inserted into a ceramic tube filter and slurry prepared in a well dispersion state has been developed. We tried to filter some difficult to filter materials such as iron oxide and activated sludge. Especially, the algal suspension and activated sludge was not able to be filtrated by a conventional filtration system like filter press. Concentrated slurries of iron oxide and activated sludge were discharged at 35 vol% and 7.55 mass% respectively. The concentrates of every sample retained flowability. In addition, to apply this system to a real process, we tried to calculate the filter tube number from the relationship between the filtration flux and slurry concentration. As a result, the size of this system can be expected to be very compact. Also, in this system, the batch operation was more effective than continuous operation.

Keywords: dispersant, dynamic filtration, purification, solid-liquid separation, suspension, thickening, washing.

1 Introduction

In the cake filtration used to concentrate slurry in many industries, there are various problems such as; the packing fraction of the formed cake is low because of the added flocculants, the cake formed on the filter media has no flowability, and therefore has to be scraped mechanically, which makes the instrument becomes big and complicated.

In order to overcome these problems, we developed a novel filtration system, in which a dispersant is added instead of a flocculant and the concentrate can be collected continuously without any scraper because the concentrate still has

WIT Transactions on Ecology and the Environment, Vol 135, © 2010 WIT Press
www.witpress.com, ISSN 1743-3541 (on-line)
doi:10.2495/WP100281

fluidity [1]. However, in our previous system, it is difficult to scale up due to its structure [2]. Thus, in our previous report [3], the new cake-less filtration system using a ceramic tube filter within the spiral guide rod was developed. In this work, the purpose is to filtrate various hardly filterable materials using the new filtration system and to discuss the scale-up of the system. In this paper, we tried to concentrate various hardly filterable materials such as sericite, iron oxide, algal suspension and activated sludge of the excrement of farm animals. In addition, we tried to calculate the number of the ceramic filter that were necessary to concentrate from the target concentration and throughput. Moreover, we also compared efficiency of the batch operation with that of continuous operation.

2 Experiments

2.1 Filtration unit

The filtration system used in this paper is same as in our previous paper. Slurry in the feed tank was fed into the filter part at a set pressure by pump. The filter media was a ceramic tube. The pore size was 1.5 μm and the inner diameter, outer diameter and length were 9, 12, and 300 mm, respectively. The spiral guide rod was set in the ceramic tube. Material of the core rod of the spiral guide rod was acrylic. On the core rod, a lead wire was wound as a spiral guide. The diameter of the core rod and the lead wire were 6 and 1.5 mm, respectively. The pitch of the spiral guide rod was 10 mm. In order to concentrate the slurry, only the dense slurry was returned to the tank. The filtration pressure and flow rate of circulation were controlled by the pressure regulator and the valve at outlet of the filter and measured by a pressure gauge and flow meter, respectively. The filtration flux was calculated by weighting the filtrate during a period of time.

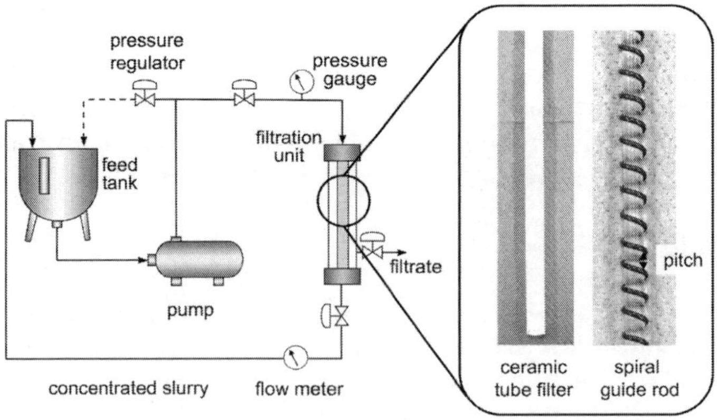

Figure 1: Schematic illustration of the new filtration system.

2.2 Concentration of iron oxide slurry

The iron oxide slurry (density of 5.2g•cm^{-3}) was supplied by the JFE Chemical Co. Ltd. The initial concentration was 8.3 vol% (32 mass%). Now, in the plant, this slurry was concentrated up to 70 mass% (31 vol%) using a filter press.

This iron oxide slurry from ironworks originally contained ammonium sulfate as a flocculant. Therefore, we used the water glass as a dispersant. The additive amount of the dispersant was 2.4 mg•g^{-1}iron oxide. The filtration pressure was 0.4 MPa and the flow rate of the concentrate was 16.8 L•min^{-1}.

2.3 Concentrating of activated sludge of the excrement of farm animals

The initial concentration of the activated sludge of the excrement of farm animals was 1.2 mass%. The state of the slurry was a good dispersion state without the addition of any dispersant. The slurry was concentrated up to 7.5 mass% by sunlight because it was also not able to be filtrated by a conventional filtration system like a filter press. The filtration pressure was 0.38 MPa and the flow rate of the concentrate was 17 L•min^{-1}.

3 Results and discussions

3.1 Concentration of iron oxide slurry

Figure 2 shows the experimental results of the iron oxide slurry. The filtration flux also decreased gradually while the slurry was being concentrated. In this test, the slurry was concentrated up to 35 vol% (74 mass%) which is comparable to a conventional filter press. The pump did not work well at 35 vol%, although the flowability of the concentrate was good as shown in Figure 3. Therefore, it also seems to be possible to concentrate the slurry more highly using a high performance pump.

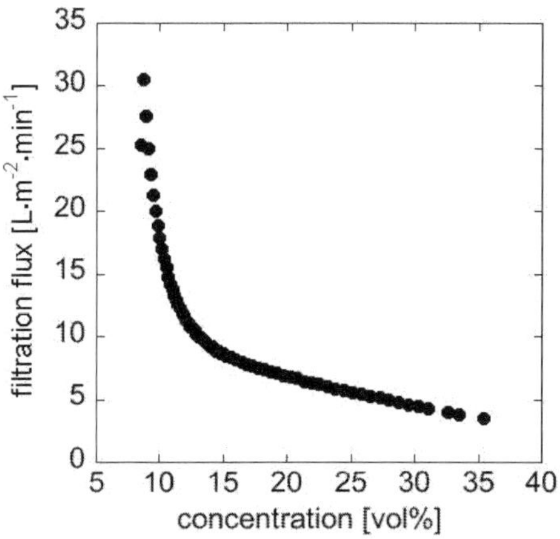

Figure 2: Relationship between the filtration flux and the volumetric concentration of the iron oxide slurry.

Figure 3: Flowability of the concentrated iron oxide slurry.

3.2 Concentrating of activated sludge of the excrement of farm animals

Figure 4 shows the relationship between the filtration flux and the slurry concentration. This slurry also could be filtrated without fouling and the filtration flux did not change although the filtration flux rapidly decreased because the particles fouled the filter just after starting filtration like the algal suspension. In this experiment the filtration test had to stop at 7.55 mass% because of shortage of the feed slurry. However, there was no cake observed on the spiral guide rod after end of concentration test and the dense slurry had flowability as shown in Figure 5. Therefore, it seems to be possible to concentrate more highly.

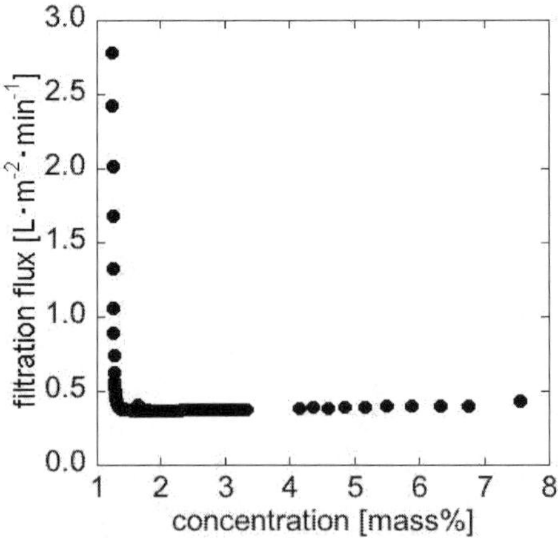

Figure 4: Relationship between the filtration concentrated flux and the slurry concentration of the activated sludge.

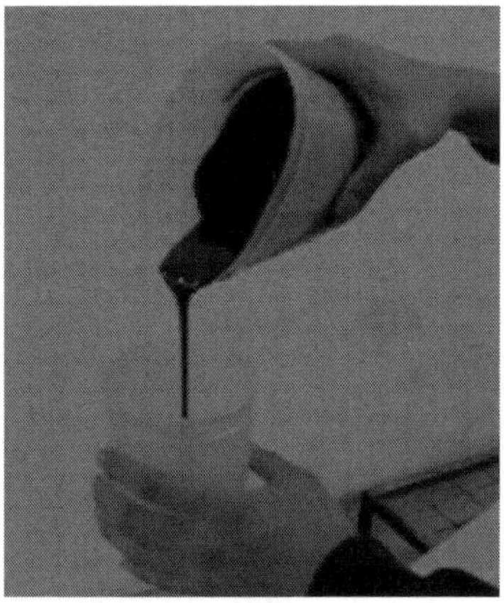

Figure 5: Flowability of the activated sludge.

4 Scale-up of the system

The scale-up of the instrument is necessary to treat a large quantity of samples. In this system, the scale-up is to increase the number of ceramic filters. The number of ceramic filters can be calculated from the required filtration area, and therefore, we tried to determine the required filtration area of each of the cases of batch operation and continuous operation from the target concentration and throughout, and compared the results of each.

4.1 Batch operation

If there is a correlation between the filtration flux and slurry concentration, the required filtration time and filtration area can be calculated as follows.

We considered that the initial (t=0) slurry, of which the volume was V0 [m^3] and the concentration is ϕ_0 [-], was concentrated to V [m^3] and ϕ [-] during t [s]. Afterwards, during dt [s], the volume decreased as $Aq(\phi)\mathrm{d}t$ [m^3]. Here A [m^2] and $q(\phi)$ [m•s^{-1}] are the filtration area and the filtration flux respectively, and the concentration increase was dϕ [-]. Assuming that there were no particles in the filtrate, this process can be described by the following equation from the material balance equation.

$$V_0\phi_0 = V\phi = \left(V - Aq(\phi)\mathrm{d}t\right)\left(\phi + \mathrm{d}\phi\right) \quad (1)$$

From this equation, the next equation can be derived.

$$\mathrm{d}t = \frac{V_0\phi_0}{A\phi^2 q(\phi)}\mathrm{d}\phi \quad (2)$$

This equation is integrated in the condition; (t : 0 → t, ϕ : ϕ_0 → ϕ).

$$t \cdot A = V_0\phi_0 \int_{\phi_0}^{\phi} \frac{\mathrm{d}\phi}{\phi^2 q(\phi)} \quad (3)$$

From this equation, the relationship between the required filtration time and filtration area can be calculated.

4.2 Continuous operation

In the case of continuous operation, it is necessary to keep the concentration in the tank target concentration ϕ_t [-]. In stable operation, the slurry, which has a concentration of ϕ_0 [-], is fed at the flow rate of V_f [$m^3 \cdot s^{-1}$], and the dense slurry, which has a concentration of ϕ_t [-] was discharged at a flow rate of V_t [$m^3 \cdot s^{-1}$]. Then, the filtrate was drained at a flow rate of $Aq(\phi)dt$ [$m^3 \cdot s^{-1}$]. From the material balance equation, these two following equations can be obtained.

$$V_f \phi_0 = V_t \phi_t \tag{4}$$

$$(1-\phi_0)V_f = Aq(\phi_t) + (1-\phi_t)V_t \tag{5}$$

From these two equations, we can calculate the required filtration area for the flow rate of feed slurry.

$$A = \frac{V_f(\phi_t - \phi_0)}{\phi_t q(\phi_t)} \tag{6}$$

In the case where the slurry which has a whole volume of V_s [m^3] is filtrated during t [s], the volume of $\alpha \cdot V_s$ [m^3] ($0 < \alpha < 1$) is filtrated by batch operation, and the remaining volume of $(1-\alpha) \cdot Vs$ [m^3] is filtrated by continuous operation afterwards. The relationship between the required filtration time and filtration area of this case is expressed by following equation.

$$t \cdot A = \alpha \cdot V_s \cdot \phi_0 \int_{\phi_0}^{\phi_t} \frac{d\phi}{\phi^2 q(\phi)} + (1-\alpha) \cdot V_s \cdot \frac{\phi_t - \phi_0}{\phi_t q(\phi_t)} \tag{7}$$

4.3 Determination of the filter tube number of the concentrating system

We considered the concentration of the whole volume of 1 m^3 of each hardly filterable slurry. We tried to calculate the relationship between the required filtration time and filtration area of batch operation and continuous operation from Eqs. (3) and (7), respectively. In the case of continuous operation, α is assumed to be 0.2. However, it is difficult to measure the volumetric concentration of the activated sludge of the excrement of farm animals, so the mass concentration was used instead. c [-] and $q(c)$ [$kg \cdot m^{-2} \cdot s^{-1}$] were used instead of ϕ[-] and $q(\phi)$ [$m \cdot s^{-1}$], respectively. In addition, we considered the concentration of the total weight of 1000 kg.

The correlation between the filtration time and the number of the filters are shown in Figures 6–7. In the case of the slurry that the filtration flux decreased gradually while the slurry was being concentrated, like in the iron oxide slurry, the batch operation was more effective than that of continuous operation. This is because, in case of continuous filtration, it is necessary to always keep the concentration in the tank at the target concentration, and therefore, the filtration flux is always late. However, in case of the batch filtration, we could use all filtration fluxes, including high filtration flux at a low concentration. On the other hand, in the case of the slurry where the filtration flux did not change although the filtration flux rapidly decreased just after starting filtration, like in the activated sludge of the excrement of farm animals, the efficiency of the continuous operation was not different from that of batch operation.

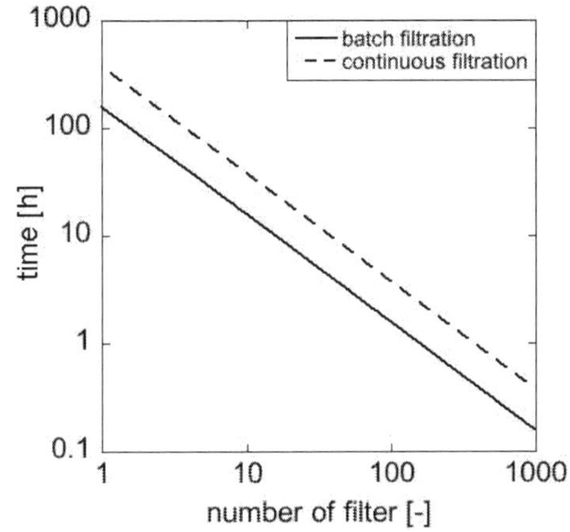

Figure 6: Correlation between the filtration time and the number of ceramic filters of the iron oxide slurry.

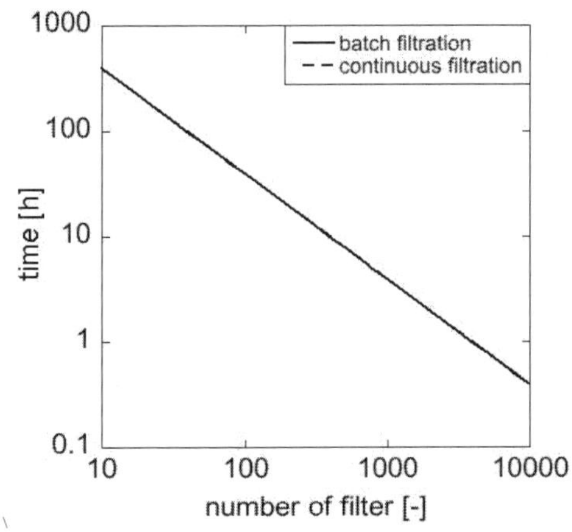

Figure 7: Correlation between the filtration time and the number of ceramic filters of the activated sludge.

The maximum number of filters that were necessary to concentrate to the target concentration during one hour by batch operation in the case of the activated sludge of the excrement of farm animals, was 3,950. 3,950 filters fit into the box in which the length, breadth and height are 1000, 1000 and 500 mm, respectively. Compared with conventional filtration systems like a filter press, it is very compact, even when including the feed tank and pump.

5 Conclusions

In this paper, a kind of cross flow tubular ceramic filter with an inserted spiral guide rod was discussed. We tried to filter some hardly filterable materials such as iron oxide and activated sludge of the excrement of farm animals by the new system. Every sample was able to be concentrated up to enough concentration with flowability and no cake formed on the filter media. In addition, to apply this system to real processes, we were able to determine the filter tube number from the relationship between the filtration flux and slurry concentration. In every sample, the size of this system could be expected to be very compact. Also, in this system, the batch operation was more effective than continuous operation. From these results, this new system can be very useful in various industrial processes.

References

[1] Tsubaki, J, Mori, T., Unenbat, T., Ochirkhuyag, B., 2006. Development of a novel slurry condensation method by applying dispersant instead of flocculant, Journal of Society of Powder Technology Japan 43, 731-736.

[2] Ochirkhuyag, B., Mori, T., Tsubaki, J., Katsuoka, T., Satone, H., Choi, Sugimoto, T., 2008. Development of a High Performance Cake-less Continuous Filtration System, Chemical Engineering Science 63, 5274-5282.

[3] Katsuoka, T., Satone, H., Yamada, H., Mori, T., Tsubaki, J., 2010. Development of a novel high performance cake-less filtration system – Optimization of operating conditions –, submitted to Chemical Engineering Science.

Development of an iron-amended biofilter for removal of arsenic from rural Canadian prairie potable water

A. M. Gottinger[1,2], D. J. Wild[1], D. McMartin[2],
B. Moldovan[3] & D. Wang[4]
[1]*Mainstream Water Solutions Inc., Canada*
[2]*Faculty of Engineering & Applied Science, University of Regina, Canada*
[3]*Cameco Corporation, Canada*
[4]*Centre for Sustainable Infrastructure Research, National Research Council of Canada, Canada*

Abstract

The current study examined the use of a ZVI (zero valent iron)/sand filter for the removal of arsenic (As) from Canadian Prairie ground water sources. Batch isotherm data indicated a favourable reaction represented by the Langmuir isotherm equation with loading capacities of 5000 and 2000 mg As/kg ZVI. Column experiments using arsenate-spiked RO water (50 μg/L) and varying volumetric ratios of ZVI to sand indicated no statistical difference in arsenic-removal performance above a ZVI/sand ratio of 20/80 (%, v/v) with removal efficiencies of greater than 98%. A second column study using two ground water sources with 50/50 and 40/60 ZVI/sand filters achieved 89 – 96% As removal.

A pilot study using a 50/50 ZVI/sand filter integrated into the existing small-scale biological system showed arsenic removal efficiency of approximately 99.7%. By incorporating this ZVI/sand filter into existing biological treatment, it was capable of removing As to concentrations below 0.1 μg/L and reducing the concentrations of other contaminants, such as ammonia, iron and manganese. The overall performance of the pilot system indicates the ZVI/sand filter is a viable option for arsenic removal from drinking water for small communities (populations < 5000).

Keywords: arsenic, ZVI/sand filter, adsorption isotherm, slow sand filtration.

WIT Transactions on Ecology and the Environment, Vol 135, © 2010 WIT Press
www.witpress.com, ISSN 1743-3541 (on-line)
doi:10.2495/WP100291

1 Introduction

Arsenic (As) occurs in natural waters and its concentration depends on the local hydrology, geology and geochemical characteristics of the aquifer [1]. The most common As-containing mineral is arsenopyrite, which upon oxidation, releases arsenic into water [2]. Arsenic can exist in four oxidation states (As^{5+}, As^{3+}, As^{0}, and As^{3-}). In most ground waters, arsenic exists primarily as As^{3+} and As^{5+} with As^{3+} being up to 60 times more toxic and generally more mobile and difficult to remove than As^{5+} [3].

The As concentrations in most Canadian Prairie ground water supplies is typically less than 25 μg/L, therefore, there has not been significant concern until the maximum allowable concentration (MAC) for As was reduced from 25 to 10 μg/L in the new Guidelines for Canadian Drinking Water Quality [4].

Many of the currently available As removal technologies for small to medium-sized water treatment systems require a considerable capital investment. These treatment methods also have associated operating costs such as chemical addition for co-precipitation, replacement of expensive media, or contaminated sludge removal that also requires a skilled operator.

An assessment identified 63 rural, remote and resort communities in the western Canadian provinces that do not meet the new MAC guidelines for As and will require innovative potable water treatment solutions [5]. With the need for treatment identified Mainstream Water Solutions Inc., initiated a project to develop a suitable small-scale As removal technology. The present study consisted of initial research, identification of options, and engineering optimisation to develop a proprietary technology using chemical-free methods of treatment.

Pokhrel *et al.* [6] evaluated various small water treatment configurations for As removal on Saskatchewan ground water including two slow sand filter (SSF) systems. They found that the SSF systems were highly efficient, at up to 96% As removal. SSF systems are re-emerging technologies that are particularly well suited to small communities because they are simplistic in operation, require little process adjustment and technical expertise, and are economical to install and operate. Several biological SSF municipal water treatment systems have been recently installed in the Canadian Prairie provinces.

Under conditions of sufficient Fe (soluble including both ferric and ferrous) content, SSFs can effectively remove As from ground water. A ratio of at least 20:1 Fe to As must exist to promote adsorption and co-precipitation with As [7]. Where Fe is lacking, it may be added through media such as ZVI (zero valent iron) filings. ZVI filings are a proven and effective media for As removal in permeable reactive barriers, in ZVI filings columns or in a mixture with filtration sand in columns.

This research was designed to identify and quantify an optimised chemical-free potable water treatment method using ZVI filings for regulatory As removal from ground water. The specific objectives were: (1) to investigate the use of ZVI filings for As removal, (2) to identify the optimal ratio of ZVI/sand for As removal, (3) to evaluate the designed ZVI filings configuration for As removal from the source water of two local communities, (4) to determine the ZVI filing's loading capacity and, (5) to determine the impacts on performance of a scaled-up pilot unit. These objectives were met through batch adsorption experiments, continuous-flow column studies, and a pilot study.

2 Materials and methods

2.1 Water sources

Raw ground water samples were obtained from the resort villages of Kannata Valley and Buena Vista, SK on Last Mountain Lake, approximately 50 km northwest of the capital city, Regina. The ground water at each community contains Fe:As ratios between 56:1 and 9:1, respectively (table 1). Each represents a particular challenge to removal of As by conventional filtration methods.

Despite a theoretically adequate Fe:As ratio of 56:1, pilot-scale SSF treatment at Kannata Valley indicated insufficient As removal. In particular, the Kannata Valley water matrix contains interfering ions including high concentrations of sulfate and bicarbonate as well as organic matter. As(III) is the predominant chemical species (30 to 34 μg/L).

Removal of As from Buena Vista raw water proved difficult during previous SSF pilot-scale tests and was attributed primarily to an insufficient Fe:As ratio at 9:1. Unlike Kannata Valley, As(V) is the predominant chemical species (12 to 13 μg/L).

Table 1: Summary of raw water quality for communities of Kannata Valley and Buena Vista, SK.

Parameter	Kannata Valley	Buena Vista
	mg/L	
Arsenic	0.0370	0.0170
Iron	2.09	0.160
Manganese	0.130	0.0620
Ammonia-N	3.37	1.03
DOC	5.00	3.00
Total Alkalinity (as $CaCO^3$)	502	358
Bicarbonate	612	423
Chloride	174	13.0
Fluoride	0.200	0.400
Calcium	93.0	23.1
Sodium	531	158
Sulfate	840	87.3
Total Hardness (as $CaCO_3$)	427	100
Turbidity (NTU)	16.8	0.780
pH	7.70	7.80
Conductivity (μS/cm)	3.10×10^3	880

2.2 Adsorbent media

ZVI filings were 20-40 mesh (ASTM) aggregate (0.45-0.55 mm) obtained from Peerless Metals (Detroit, Michigan). BET analysis to determine filings surface area was completed with indeterminable results likely due to pore sizes smaller than 3.5 Å. However, research has indicated that Fe^0 surface area is not a primary factor controlling As adsorption [8].

2.3 Experiments

2.3.1 Batch experiments

10 g of ZVI filings were added to 100 ml of water containing between 1 μg/L to 1.2 g/L As(V). The solutions were agitated on an orbital table shaker (100 rpm) for 48 h after which an aliquot was drawn, filtered with a 0.45 μm GF/C membrane and analysed by GFAAS for residual As concentrations. Adsorption data was fitted to Freundlich and Langmuir isotherms and equation parameters tested for significance at the 95% confidence interval using the student's t-test.

2.3.2 Continuous-flow column studies

There were two phases included in the column studies. Phase I examined the performance of various ratios of ZVI and sand in the columns with RO water, while Phase II determined the performance of columns loaded with two different ratios of ZVI/sand to remove As from Kannata Valley and Buena Vista ground water.

Column experiments were designed to replicate a full-scale SSF operation. The columns comprised triplicate sets of 2 L HDPE Nalgene carboys with a bottom outlet. Biologically active filtration sand (d_{10}: 0.45-055 mm) was obtained from an operating SSF. Time series effluent samples were collected from the column outlets, filtered with a 0.45 μm GF/C membrane, preserved with nitric acid and analysed by ICP-MS. Arsenic speciation tests (ion exchange tube/ICP-MS) were completed on raw, ozonated and treated water.

2.3.2.1 Phase I: RO water RO water spiked with sodium arsenate and a complex nutrient solution was pumped through columns containing 6 ratios of ZVI/biologically-active sand (table 2).

Table 2: Composition of Phase I columns.

Column #	% ZVI by volume	% sand by volume
1 – 3	0	0
4 – 6	0	100
7 – 9	50	50
10 - 12	40	60
13 - 15	30	70
16 – 18	20	80

The columns were assembled by adding a base of gravel (effective size, d_{10}: 3.0-3.5 mm), including both blank and control columns, and then the ZVI/sand media of various ratios (table 2). The total media volume was 1.17 x 10^{-2} m^3 (0.0414 ft^3). The columns were constructed in triplicate with each media configuration.

The nutrient solution (containing inorganic nutrients, such as nitrate and phosphate, to maintain biological sand activity in the column) was added to the influent water throughout the duration of the experiment. As(V) was further added to the nutrient-modified RO influent at a concentration of 50 μg/L. Influent water was prepared no more than 24 h prior to beginning the column experiments.

A column rack was built to support 18 columns that were supplied with the modified RO influent water using a submersible pump into a PVC feedline. Kynar needles valves on the influent and effluent lines regulated the flow rate through each column. The columns were operated in downflow mode with a hydraulic loading rate of 0.023 m/h (0.01 gpm/ft^2).

The columns were allowed to stabilize for 7 days before As-laden water was applied. A time series of effluent samples were collected from each column over the 4-week experimental period.

2.3.2.2 Phase II: Kannata Valley and Buena Vista ground water In Phase II, two separate column apparatuses were built to remove As from ground water samples collected from two rural resort communities near Regina, SK. Again, triplicate columns were investigated at two chosen media ratios effective in Phase I. The ZVI/sand ratios tested were 50/50 and 40/60 and also included triplicate columns at 0/100 (v/v).

Two 66 cm HDPE ozone contacting tanks were used in Phase II influent water preparation. According to Mainstream Water Solutions standard design each tank included a sealed lid fitted with a balanced barometer and catalytic ozone destructor. Ozone was supplied by a 4 g/h corona discharge generator (air fed) and applied via sidestream injection. The raw water was ozonated until the oxidation-reduction potential (ORP) reached approximately 500 mV at which point As(V) should be the predominant species present. The feed rate into each column was regulated using kynar needle valves. The columns were acclimatized for 6 days using nutrient-spiked RO water. On day 7, ozonated Kannata Valley and Buena Vista raw water was introduced. The columns were operated in downflow mode with a hydraulic feed rate of 0.023 m/h.

2.3.3 Pilot study

The pilot study was designed to confirm the laboratory results regarding optimal ZVI/sand media and anticipated As removal efficiency at one community, Kannata Valley. Mainstream Water Solutions BioFiltration Water Treatment System, based on SSF technology, was used. The system included ozone pretreatment, a ZVI/sand filter, a roughing pre-filter, the SSF and biological activated carbon (BAC) filter (table 3). Figure 1 provides a process diagram of the system.

Table 3: Pilot system processes.

	ZVI/sand	Roughing	SSF	BAC
Filter Size	46 cmØ x 165 cm	46 cmØ x 165 cm	66 cmØ x 165 cm	66 cmØ x 165 cm
Media (effective size)	50/50, % (v/v) mixture of ZVI and quartz silica sand (20x40 mesh ASTM (0.45-0.55 mm))	Pea gravel (3.0-3.5 mm)	Quartz silica sand (20x40 mesh ASTM (0.45-0.55 mm))	Activated carbon (10x30 mesh U.S. (0.6-0.8 mm))
Total Media depth	88 cm	88 cm	88 cm	88 cm
Media Volume	0.15 m^3	0.15 m^3	0.30 m^3	0.30 m^3
Filter Surface Area	0.17 m^2	0.17 m^2	0.34 m^2	0.34 m^2

The ozone contacting system was configured to deliver a maximum ozone dose of 8 g/h to facilitate the oxidation of As(III) to As(V). Ozonation also oxidized other metals in the raw ground water including manganese and iron into forms that precipitate in the roughing and SSFs. Furthermore, ozone addition increases the oxidation of ZVI filings leading to rust formation, which supports As removal.

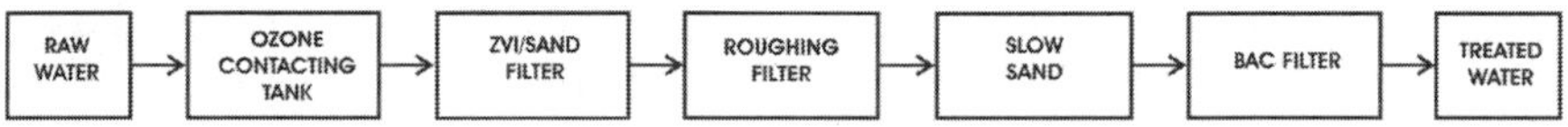

Figure 1: Pilot system processes.

The pilot project included 2 experimental trials – one with a routine flow rate of 1.26 L/min and an ozone dose of 105 mg/L. The second experiment was set up with the same flow rate (1.26 L/min; however without the addition of ozone (pre-oxidation).

3 Results

3.1 Batch study

Batch study data was fitted to both Freundlich and Langmuir adsorption isotherms and both were statistically significant at the 95% confidence interval. However the Langmuir isotherm more adequately describes As adsorption onto ZVI in both authentic ground water samples. The Langmuir isotherm curve for Kannata Valley is shown in figure 2. These results are similar to published literature where adsorption of As onto various adsorbents followed the Langmuir isotherm [9].

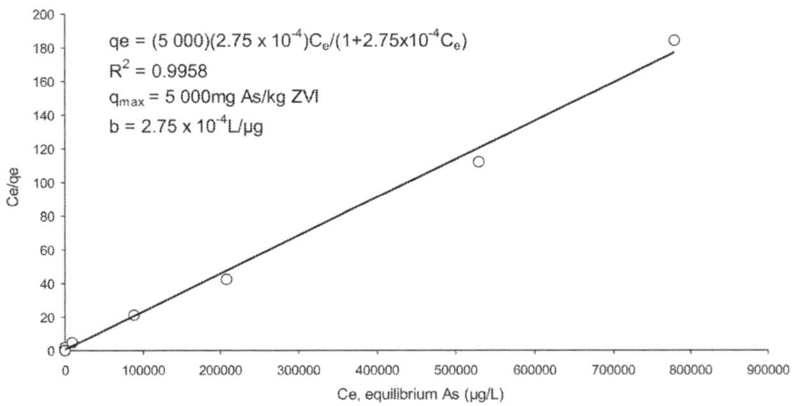

Figure 2: Langmuir adsorption isotherm for Kannata Valley ground water.

Average As removal efficiencies for Kannata Valley and Buena Vista raw water were 97 and 95%, respectively, for the selected range of As concentrations. As removal was greater than 90% in all samples up to a spiked concentration of 200,000 µg/L. A comparable binding energy constant (b) was observed for both waters suggesting similar adsorbate-adsorbent interactions. The maximum adsorption capacity, determined by the Langmuir q_{max} constant, was 5000 and 2000 mg As/kg ZVI for Kannata Valley and Buena Vista respectively and are consistent with earlier reported As removal capacities of ZVI filings [8, 10-12].

3.2 Continuous-flow column studies

3.2.1 Phase I: RO water

Nutrient-modified RO water spiked to 50 µg/L As(V) was pumped through triplicate columns containing five different ratios of ZVI/sand (v/v) ranging from 0/100 to 50/50 (Table 4). Each set of columns was examined for performance at similar loading rates. The columns were operated at 4 mL/min (0.023 m/h) for sufficient duration to achieve 149 bed volumes (total throughput volume of 173 L).

The 0/100 columns containing sand only provided minimal As removal as anticipated. All ZVI/sand columns, except those at 20/80 achieved >98% removal of influent As where the As was reduced from 50 to <1 µg/L (table 4).

A Tukey multiple comparison test was completed on the performance data. Analysis at the 95% confidence interval indicates that the As removal performance of the 20/80 columns was statistically poorer than the other ZVI/sand columns. The remaining columns performed statistically similarly to each other. Although the 20/80 columns were capable of removing As to below 10 µg/L, they did not achieve the same level of performance as the remaining columns. The reduced ZVI volume in the 20/80 columns may have resulted in dissolved As passing through the column via "null" paths without reacting with the ZVI or oxidized Fe [11].

Table 4: Summary of results from nutrient-enhanced RO water experiment.

ZVI/ Sand (v/v)	Initial As (Ave. ± St. Dev.) (µg/L)	Final As (Ave. ± St. Dev.) (µg/L)	% As Removed (Ave. ± St. Dev.)
0/100	41.70 ± 0.53	39.42 ± 2.50	5.42 ± 16.51
50/50	41.70 ± 0.53	0.64 ± 0.03	98.46 ± 0.17
40/60	41.70 ± 0.53	0.80 ± 0.08	98.09 ± 0.46
30/70	41.70 ± 0.53	0.56 ± 0.01	98.67 ± 0.07
20/80	41.70 ± 0.53	1.29 ± 0.10	92.80 ± 0.02

Operationally, variations in hydraulic resistance and, therefore, flow rate were noted, which were due primarily to plugging of pore spaces by the formation of Fe oxides where greater oxygen concentrations existed at the filter inlet [13].

3.2.2 Phase II: Kannata Valley and Buena Vista ground water

Triplicate 0/100, 40/60 and 50/50 (v/v) columns were assembled and operated at 0.023 m/h for 154 bed volumes (184 L) without achieving breakthrough. The results for this filtration run are summarized in table 5.

Table 5: Continuous-flow column performance on authentic ground water.

	Influent As (Ave. ± St. Dev.) (µg/L)	Effluent As (Ave. ± St. Dev.) (µg/L)	% As Removed (Ave. ± St. Dev.)	Influent Fe (Ave. ± St. Dev.) (mg/L)	Effluent Fe (Ave. ± St. Dev.) (mg/L
			Kannata Valley		
0/100	23.80 ± 5.11	25.50 ± 12.33	7.52 ± 443.14	0.10 ± 0.06	0.06 ± <0.01
50/50	23.80 ± 5.11	2.48 ± 0.05	89.51 ± 1.07	0.10 ± 0.06	0.03 ± <0.01
40/60	23.80 ± 5.11	2.49 ± 0.34	89.43 ± 1.64	0.10 ± 0.06	0.02 ± <0.01
			Buena Vista		
0/100	16.44 ± 0.31	16.43 ± 1.66	0.72± 93.35	0.37 ± 0.48	0.04 ± <0.01
50/50	16.44 ± 0.31	0.50 ± <0.01	96.96 ± 0.01	0.37 ± 0.48	0.03 ± <0.01
40/60	16.44 ± 0.31	0.52 ± <0.01	96.86 ± 0.13	0.37 ± 0.48	0.03 ± <0.01

The 0/100 control columns were not capable of removing As to below the pending 10 µg/L regulatory standard. In fact, these columns appear to have caused As concentration variation during the treatment.

Both the 50/50 and the 40/60 ZVI/sand column effluent water consistently contained As concentrations below 3 µg/L, which equates to >89% removal for Kannata Valley ground water. Both sets of ZVI/sand columns achieved As removal efficiency >96% thereby resulting in As at less than 0.5 µg/L in the Buena Vista water. Statistically, there was no significant difference in performance between the 50/50 and 40/60 columns within the 95% confidence interval.

The higher As removal rate observed in the columns treating Buena Vista ground water was attributed primarily to the greater concentration of As(V) species rather than the more soluble As(III) species, as well as to reduced competitive ions in the water matrix relative to Kannata Valley raw water. Although oxidation promoted the presence of As(V) in the Kannata Valley water, approximately half of the influent As was As(III). Comparatively, the majority of As in the Buena Vista raw water was As(V). Arsenate (As(V)) reacts more quickly with Fe oxides than As(III) due to the easy formation of ferric arsenate [14, 15]. Additionally, the slow filtration rate may have allowed scale formation of bicarbonate compounds on the ZVI filings limiting the corrosion required for As removal [16, 17]. Some sorption sites may also have been blocked due to particle deposition, which was more prevalent in the relatively complex Kannata Valley water matrix.

As speciation of Kannata Valley water indicated that excessive ozone application (>33 mg O^3/L) was capable of oxidizing only 40 to 60% of the As(III) to As(V). Speciation results on effluent samples also indicated that approximately 50% of the effluent As was in the As(III) form.

Despite some limitations to As removal from the Kannata Valley ground water, it was clearly noted that both the 40/60 and 50/50 ZVI/sand columns were proficient at removing both As(III) and As(V). The ability of ZVI to remove both species has been noted in previously published research [18] in which it was observed that As(III) was partially oxidized to As(V) by Fe corrosion byproducts.

Column effluent was also analysed for effluent Fe concentration to evaluate potential sloughing into the produced water. The results indicate that overall Fe concentrations were lower in the effluent than in the raw water (table 5).

3.3 Pilot study

The pilot treatment system was designed with a 50/50 ZVI/sand filter and incorporated pre-ozonation to achieve oxidation of As(III) from an initial proportion of 57-78% of total As to 16-21% of total As in the influent raw water. The experimental results with Kannata Valley ground water are displayed in table 6.

The ZVI/sand filter and SSF were successful in removing the majority of the As in the influent raw water. Total As in the effluent from the ZVI/sand filter was consistently <0.10 μg/L. The effluent from the SSF contained slightly higher

Table 6: As and Fe concentrations in raw water and after pilot system processes.

Process	As conc. (Ave. ± St, Dev) (μg/L)	Fe conc. (Ave. ± St, Dev) (mg/L)
Raw	30.76 ± 22.43	0.79 ± 0.54
ZVI/sand	<0.10 ± <0.01	<0.09 ± <0.01
SSF	2.04 ± 1.15	0.03 ± <0.01
BAC	1.68 ± 0.85	<0.09 ± <0.01

As concentration attributed to limited precipitation of complexed Fe and As in that filter. Regardless, As in the treated water from the Mainstream system never exceeded 3.9 μg/L, far below the new MAC. In addition, several samples were analysed for total dissolved Fe content throughout the pilot process to examine potential Fe leaching from the ZVI/sand filter. Results indicated total Fe concentrations below the detectable limit. From this analysis it appears that the ZVI/sand filter also removed Fe from the raw water and did not allow leaching (table 6). Complete analysis of system effluent proved overall good quality potable water production. iron and manganese were removed to well below the target objectives of 0.3 mg/L, 0.05 mg/L respectively. Ammonia was reduced to <0.25 mg/L and turbidity was consistently well below 1 NTU.

3.3.1 No oxidation of As(III)

The pilot system was operated for 7 days without ozone oxidation at a constant flow rate of 1.26 L/min. Although there was no oxidation of As species removal was still successful. The ZVI/sand filter effectively removed As(III) and As(V); and total As in the treated water never exceeded 1μg/L. The average removal of As(III) was 97% and As(V) was 99%. It was deemed likely that the slower SSF filtration rates allowed for adequate contact time and removal of As(III).

3.3.2 Filter operation

The top 3 cm of the filter bed, where DO and oxidation of ZVI were high, became cemented over time due to formation of Fe oxides. In order to restore flow a daily water scour was required. Others recommend using the lowest effective Fe:sand ratio to avoid excessive Fe hydroxide formation clogging [16]. Because the 30/70 ZVI/sand media was shown to be effective for As removal and limits the Fe:sand ratio, this was deemed the optimal media configuration for both operational and regulatory reasons.

A study of a pilot-scale filter using 30/70 ZVI/sand was completed over a short duration (14 days) to prove this conclusion. Not only does the 30/70 media mixture require less maintenance, it performed similarly to the 50/50 media mixture regarding As removal efficiency.

4 Economic feasibility

Based on the loading capacity observed in the batch isotherm study and assuming that only 25% of the Fe is available for adsorption [16], the life of the media in (6) ZVI/sand filters (based on a conceptual full-scale plant design for Kannata Valley at 720 m^3) is estimated to be approximately 40 months. The treatment cost is estimated to be <$0.01/L and includes filter installation, media, operation and maintenance, and disposal costs.

A comparison of adsorptive media reveals that ZVI/sand is economical with respect to media and treatment cost and has a satisfactory estimated life (table 7). ZVI filings can be obtained for less than $1.50/kg and the cost of manufacturing this filter is comparable to a BAC filter making it a feasible technology for Mainstream Water Solutions to utilize.

Table 7: Treatment costs and media replacement frequency for comparable adsorptive media.

Media	BV Treated	Media density (kg/m^3)	Media cost (m^3)	Treatment cost (\$/L)	Change-out frequency (months)
E33[a]	54 700	454	5 377	0.10	28.6
GFH[a]	18 300	1 281	7 769	0.42	9.6
Z33[a]	3 300	881	486	0.15	1.7
ZVI/sand[b]	25 900	1 929	1 305	<0.01	40

[a]Calculations based on 1,284m^3/day and 45.4m^3 of media [19].
[b]Estimates for Kannata Valley conceptual full-scale plant design based on bench-scale and pilot studies.

5 Conclusions

The present study concluded that ZVI/sand filtration is an effective adsorbent for As removal in potable water treatment. Column and pilot studies proved that a balance of sufficient ZVI content for both As removal and for low maintenance requirements can be achieved in the range of 30/70 ZVI/sand for the raw water sources investigated. The results also showed that ZVI can remove both As species to well below 4 μg/L without significant Fe leaching. While the use of a ZVI/sand filter for arsenic removal is an economically feasible treatment in comparison to alternative technologies, disposal of spent media as well as treatment of backwash waste prior to release must be considered.

References

[1] Bhattacharya, P., Ahmed, K., Hasan, M., Broms, S., Fogelstrom, J., Jacks, G., Sracek, O., von Bromssen, M. & Routh, J., Mobility of arsenic in groundwater in part of Brahmanbaria district, NE Bangladesh. *Managing Arsenic in the Environment from Soil to Human Health,* eds. R. Naidu, E. Smith, G. Owens, P. Bhattacharya and P. Nadebaum, CSIRO Publishing, pp. 95-116, 2006.

[2] Bhumbla, D. & Keefer, R., Arsenic mobilization and bioavailability in soils. *Arsenic in the Environment, Part I: Cycling and Characterization*, ed. J. Nriagu, John Wiley and Sons, Inc. pp. 51-82, 1994.

[3] Korte, N.C. & Fernando, Q., A review of As(III) in groundwater. *Critical Reviews in Environmental Controls,* **21**, pp. 1-39, 1991.

[4] Guidelines for Canadian Drinking Water Quality. Health Canada, 2008. www.hc-sc.gc.ca/ewh-semt/alt_formats/hecs-sesc/pdf/pubs/water-eau/sum_guide-res_recom/summary-sommaire-eng.pdf

[5] Thomas, V., *Market Assessment – Arsenic Removal Technology for Drinking Water*. Unpublished report, 2007.

[6] Pokhrel, D., Viraraghavan, T. & Braul. L., Evaluation of treatment systems for the removal of arsenic from groundwater. *Practice Periodical of*

Hazardous, Toxic and Radioactive Waste Management, **9(3)**, pp.152-157, 2005.

[7] Treatment Technologies for Arsenic Removal. U.S. Environmental Protection Agency, National Risk Management Research Laboratory, EPA/600/S-05/006 Online: www.epa.gov/nrmrl/pubs/600s05006/600s 05006.pdf

[8] Su, C. & Puls, R., Arsenate and arsenite removal by zerovalent iron: kinetics, redox transformation, and implications for in situ groundwater remediation. *Environmental Science and Technology*, **35(7)**. pp. 1487-1492, 2001.

[9] Guo, H., Stuben, D. & Berner, Z., Removal of arsenic from aqueous solution by natural siderite and hematite. *Applied Geochemistry*, **22(5)**, pp. 1039-1051, 2007.

[10] Nikolaidis, N., Dobbs, G. & Lackovic, J., Arsenic removal by zero-valent iron: field, laboratory and modeling studies. *Water Research*, **37(6)**, pp.1417-1425, 2003.

[11] Lien, H. & Wilkin, R., High-level arsenic removal from groundwater using zero-valent iron. *Chemosphere*, **59(3)**, pp.377-386, 2004.

[12] Lackovic, J., Nikolaidis, N. & Dobbs, G., Inorganic arsenic removal by zero-valent iron. *Environmental Engineering Science*, **17(1)**, pp. 29-39, 2000.

[13] Johnson, R., Tratnyek, P. Miehr, R, Thoms, R. & Bandstra, J., Reaction of hydraulic conductivity and reactivity in zero-valent iron columns by oxygen and TNT. *Ground Water Monitoring & Remediation*, **25(1)**, pp.129-136, 2005.

[14] Joshi, A. & Chaudhuri, M., Removal of Arsenic from Ground Water by Iron Oxide-Coated Sand. *Journal of Environmental Engineering,* **122(8)**, pp. 769-771, 1996.

[15] Raven, K., Jain, A. & Loeppert, R., Arsenite and arsenate adsorption on ferrihydrite: kinetics, equilibrium, and adsorption envelopes. *Journal of Environmental Science and Technology*, **32(3)**, pp. 344-349, 1998.

[16] Leupin, O. & Hug, S., Oxidation and removal of arsenic(III) from aerated groundwater by filtration through sand and zero-valent iron. *Water Research*, **39(9)**, pp.1729-1740, 2005.

[17] Leupin, O., Hug, S. & Badruzzaman, A., Arsenic removal from Bangladesh tube well water with filter columns containing zerovalent iron filings and sand. *Environmental Science and Technology*, **29(20)**, pp. 8032-8037, 2005.

[18] Su, C. & Puls, R., Significance of Iron II, III) hydroxycarbonate green rust in arsenic remediation using zerovalent iron in laboratory column tests. *Environmental Science and Technology*, **38(19)**, pp. 5224-5231, 2004.

[19] Westerhoff, P., DeHaan, M., Martindale, A. & Badruzzaman, M., Arsenic Adsorptive Media Technology Selection Strategies. *Water Quality Research Journal of Canada,* **41(2)**, pp.171-184, 2006.

Characterization of an absorbent based on *Raphanus sativus* (L. var.), a solid residue from biodiesel production, for the removal of methylene blue

D. A. Lázaro[1], C. L. Caldeira[1], M. S. S. Dantas[1], A. S. Franca[2] & M. B. Mansur[1]
[1]*Department of Metallurgical and Materials Engineering, Federal University of Minas Gerais, Brazil*
[2]*Department of Mechanical Engineering, Federal University of Minas Gerais, Brazil*

Abstract

In this study, an adsorbent produced from *Raphanus sativus* (L. var.), a residue from the biodiesel industry, was investigated for the removal of methylene blue (MB) from textile effluents aiming to: (i) obtain physical and textural characteristics of the given carbon, and (ii) understand the MB adsorption mechanism onto such carbonaceous structures. The following methods were employed: SEM-EDS microscopy, gas adsorption and Raman spectroscopy. Physical and textural characteristics of the carbon obtained by classical methodologies such as BET, BJH and SEM analyses were corroborated by the Raman analysis. For the experimental conditions used on the preparation of the activated carbon of *Raphanus sativus*, the main physical parameters measured were specific surface area (236.3 $m^2.g^{-1}$), total pore volume (0.06 $cm^3.g^{-1}$) and average pore size (1.4 nm). SEM and Raman analysis revealed that a microporous structure of carbon was obtained as the ratio between the disorder band and the graphite band was higher than unity, thus indicating that MB adsorption may occur predominantly at the surface of the adsorbent, given that the diameter of MB molecule was estimated to be nearly 1.5 nm. Raman analysis of solid MB revealed that MB molecules are predominantly found as dimmers in the solid however they are adsorbed as monomer in the *Raphanus sativus* carbon. Results evidence that MB adsorption probably occurs by bonding of monomer

WIT Transactions on Ecology and the Environment, Vol 135, © 2010 WIT Press
www.witpress.com, ISSN 1743-3541 (on-line)
doi:10.2495/WP100301

MB through the central nitrogen atom of its molecule; MB bonding through sulfur atom may also occur but at lesser intensity while MB bonding through amino nitrogen atom does not seem probable.
Keywords: carbon characterization, Raman spectroscopy, adsorption, methylene blue.

1 Introduction

The removal of synthetic dyes from textile wastewaters is extremely important to allow the reuse of water and/or the final discharge of effluent because most of such dyes are toxic, mutagenic and carcinogenic. Several methods have been developed to remove dyes from aqueous solutions in order to decrease their impact on the environment. A comparative survey of technologies including adsorption, coagulation, chemical degradation, ultrafiltration, photodegradation with ozone, electrochemical flotation and biodegradation is found elsewhere [1-6].

As long as there is no universal method available for color removal from aquatic systems, some technologies have been extensively investigated for the remediation of dyes using non-conventional low-cost sorbents based on waste materials generated by industrial and/or agricultural activities such as wheat, corn straw, olive stones, bagasse, birch wood, miscanthus, sunflower shell, pinecone, rapeseed, cotton residues, olive residues, pine rayed, eucalyptus maculata, sugar cane bagasse and many others [7]. Recently the removal of methylene blue (MB) using activated carbons based on the pressed cake of *Raphanus sativus* (L. var.), which is a residue from the biodiesel industry, was found to be technically feasible [6].

In this context, the aim of this paper is to characterize the activated carbon produced from *Raphanus sativus* (L. var.) using SEM-EDS (Scanning Electron Microscopy coupled with X-ray Energy Dispersive Spectrometry), gas adsorption and Raman spectroscopy methods in order to better understand the adsorption behavior of MB from aqueous solutions as well as of other cationic dyes produced by textile industries.

2 Experimental

2.1 Preparation of the carbon from *Raphanus sativus* (L. var.)

Activated carbon from *Raphanus sativus* (L. var.) was prepared according to the following procedure. Firstly, seeds of *Raphanus sativus* were screw-pressed (Ecirtec, Brazil) for oil extraction followed by Sohxlet extraction procedure at 60°C for 6h to remove the excess of oil still present in the cake. Samples of the cake were weighted (20 g), placed in ceramic containers and then submitted to one-step thermal activation procedure (carbonization) using an electric muffle furnace.

During the carbonization, samples were initially burned at 100°C for 10 min. Temperature was then raised at the rate of 10°C min^{-1} to 800°C; N_2 gas was injected into the furnace (N_2 flow rate = 0.5 L min^{-1}) when T ≥ 300°C in order to create an inert atmosphere. Finally, temperature was maintained constant at 800°C for 40 min. After carbonization samples were cooled down and submitted to the characterization tests described as follows. The carbonization procedure was based on preliminary tests and also on literature data for the production of adsorbents using coffee beans pressed cake as precursor material [8].

2.2 Morphological analysis and determination of textural properties

Morphological characteristics of the adsorbent surface (before and after carbonization) were carried out by Scanning Electron Microscopy (SEM) using a JEOL microscopy (model JSM6360LV). Due to the non-conductive nature of the material, samples were covered with a thin layer of gold. The qualitative composition at surface of some particles was obtained by SEM coupled with X-ray Energy Dispersive Spectrometry (EDS) using a Thermo Noram spectrometer (model Quest).

Surface area, total pore volume and average pore diameter were assessed by measuring multipoint N_2 isotherms at 77 K in an Autosorb-1 MP (Quantachrome). For such analysis, samples were outgassed at 473 K under vacuum for 3 h. The specific surface area value was calculated according to the Brunauer–Emmett–Teller (BET) equation and pore parameters (pore volume and pore diameter) were evaluated based on Barrett–Joyner–Halenda (BJH) model.

2.3 Raman spectroscopy analysis

Raman analysis was carried out in order to evaluate the structural change of the activated carbon due to the adsorption of MB using a Jobin-Yvon/Horiba Raman laser system (model Labran HR800) coupled with a transference optic microscopy. Samples of carbon (unloaded and MB loaded samples) were analyzed under the following conditions: lens 100x, laser power < 1 mW, laser acquisition time = 20s (acquisitions 5x), red laser wavelength = 632.8 nm, analysis in two points (1 μm). MB loaded carbon samples were obtaining by contacting MB solutions in 250 mL agitated Erlenmeyer flasks (100 rpm) at following conditions [6]: initial MB concentration = 500 mg L^{-1}; initial pH = 5; adsorbent dosage = 10 g L^{-1}; T = 25°C; t = 72 h. The flasks were covered with aluminium foil to prevent photodegradation. As it was not possible to get the Raman spectra for MB solutions in concentrations as those commonly used in the adsorption tests (an intense fluorescence was obtained probably due to the color of solution), the solid MB reagent was evaluated for comparative analysis under the following conditions: laser power = 0.08 mW (power was raised to reduce signal noise), wavelength = 632.8 nm, laser acquisition time = 10 s. Analytical MB, 15% hydrated, (95% purity, Labsynth, Brazil) was used to load carbon for Raman analysis.

3 Results and discussion

3.1 Morphological analysis and determination of textural properties

A typical view of produced adsorbent is shown in Fig. 1(a) (SEM micrographs at resolution of 25x).

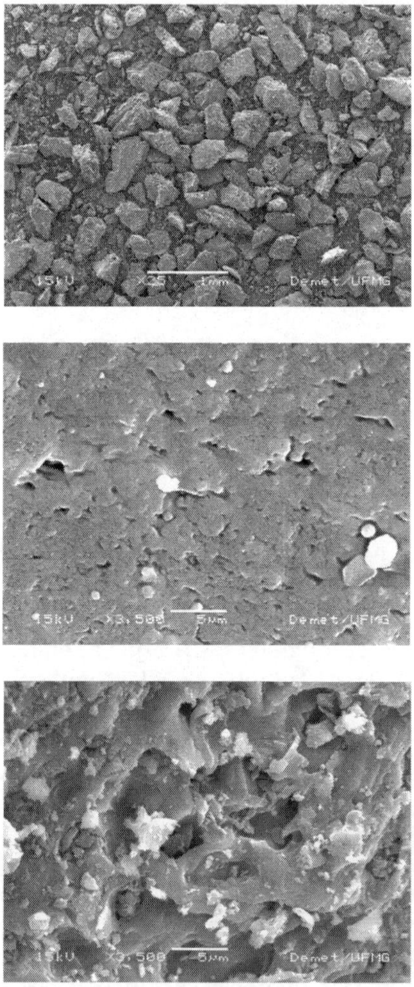

(a)

(b)

(c)

Figure 1: Typical surface morphology of *Raphanus sativus* and the produced adsorbent at different SEM micrograph resolutions: (a) Adsorbent at SEM resolution of 25x, (b) *Raphanus sativus* (raw material) at SEM resolution of 3,500x, and (c) Adsorbent at SEM resolution of 3,500x.

It can be seen that particles of homogeneous color, highly irregular shapes and different sizes were obtained; in fact, particles are smaller than 1 mm size and they could be classified into fine and coarse fractions due to low mechanical resistance of the material. In addition, particles seem to be not too porous, so adsorption is expected to occur predominantly at the surface of the carbon. Such evidence is corroborated by analyzing SEM micrographs at resolution of 3,500x as shown in Figs. 1(b) and 1(c). As expected, an increase on the porosity of *Raphanus sativus* due to carbonization was verified but it was not significant for the conditions studied. It might explain why adsorption of MB was found to occur predominantly at the surface of the adsorbent [6].

The qualitative composition given by EDS shown in Fig. 2 reveals that particles contain predominantly carbon, as well as phosphorous, calcium, magnesium and potassium. Metals such as zinc and copper were found at very low quantities while presence of gold in the carbon is due to the thin layer of metal which covered the sample to carry out SEM analysis. The presence of oxygen was found to be quite low as well.

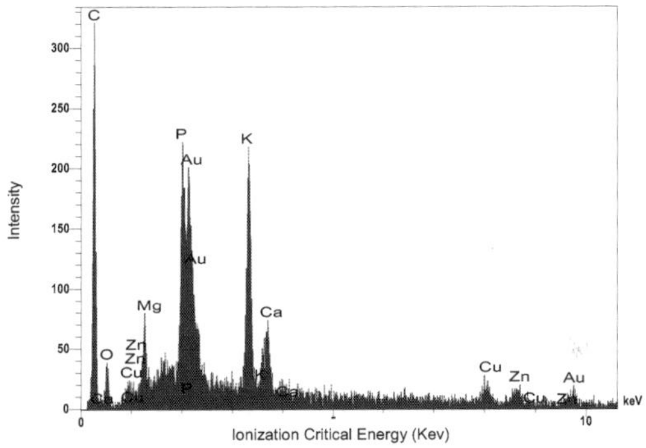

Figure 2: Qualitative composition of the activated carbon surface of *Raphanus sativus* by EDS analysis.

The following properties of the activated carbon of *Raphanus sativus* (L. var.) were estimated by N_2 gas adsorption: specific surface area = 236.3 m^2 g^{-1} (Multipoint BET equation, R^2 = 0.998), total pore volume = 0.06 cm^3 g^{-1} (BJH method) and average pore size = 1.4 nm (BJH method). Such results reveal that specific surface area and total pore volume of the produced adsorbent are low if compared to other carbons from agricultural residues. Ioannidou and Zabaniotou [7], for instance, pointed out the surface area of apricot stones activated carbon is 1190 m^2 g^{-1} (similar to some commercial activated carbons) with total pore volume of 0.5 cm^3 g^{-1} while the carbon from oat hulls has surface area ranging from 349 to 625 m^2 g^{-1} with total pore volume from 84.2 to 91.4 cm^3 g^{-1}. Such results, however, do not guarantee satisfactory MB adsorption. Ferrero [9] reported that although hazelnut shells activated carbon achieved high surface area due to thermal treatment (793 m^2 g^{-1}), the maximum adsorption capacity at equilibrium was q_{max} = 8.8 $mg.g^{-1}$, as opposed to 18.3 $mg.g^{-1}$ obtained for the adsorbent based produced in the present study [6].

The Langmuir isotherm was also obtained in the BET test (R^2 = 0.999) thus giving Langmuir surface area = 366.4 $m^2.g^{-1}$ and Langmuir constant K = 98.795. The adsorption (A) and desorption (D) isotherms for the activated carbon of *Raphanus sativus* shown in Fig. 3 reveal that an IUPAC type I isotherm was obtained. Such curves are normally encountered when adsorption is limited to at most only a few molecular layers and to microporous carbons whose pore size does not exceed a few adsorbate molecular diameters so, once the micropores are filled there is only a small fraction of the original surface exposed for continued adsorption [10].

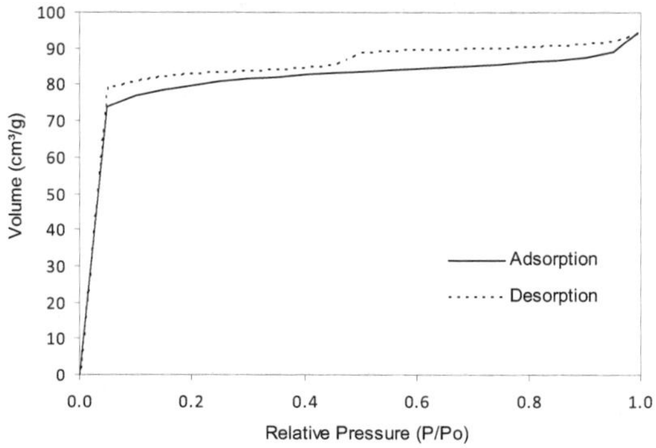

Figure 3: Adsorption (A) and desorption (D) isotherms of *Raphanus sativus* activated carbon of BET test.

3.2 Raman spectroscopy analysis

Raman spectra of the activated carbon of *Raphanus sativus* before (unloaded carbon) and after MB adsorption (loaded carbon) are shown in Fig. 4. Two characteristic bands which are typical of carbonaceous materials used as precursors for activated carbons were found at 1340 and 1590 cm^{-1} in the unloaded carbon, thus corresponding to the disorder line (D band) and the graphite line (G band), respectively; another characteristic band was identified at 2793 cm^{-1} but much softer than D and G bands. The intensity of D and G bands is an indicative of the structure of the carbon. In fact, the amount of disorganized carbons which is characterized by the intensity ratio of D to G bands (hereafter named ID/IG ratio) will affect both the pore structure and the adsorption properties of the adsorbent. Haghseresht *et al.* [11] evaluated the relationship between ID/IG ratio with total BET surface area, total pore volumes, micropore volumes, mesopore volumes and other parameters for a number of microporous carbonaceous adsorbents prepared by oxidation of coal reject in various conditions prior to heat treatment at 600^0C. They verified an increase on the adsorption capacity of the resulting carbons for organic compounds with the increase of ID/IG ratio. Also, higher amount of disorganized carbons (which is related to the increase on the amount of aromatic compounds in comparison with aliphatic carbons) indicates lesser space between the clusters of aromatic compounds, hence smaller micropores are obtained. For *Raphanus sativus* activated carbon shown in Fig. 4, ID/IG = 1.20 for the unloaded carbon, thus indicating that pores are probably too small (microporous) and corroborating previous BET observations. Fig. 4 also shows an intense peak in the MB loaded curve at 480 cm^{-1} as well as other less intense peaks that are related to the adsorption of MB; the interaction between MB and the carbon of *Raphanus sativus* is discussed as follows.

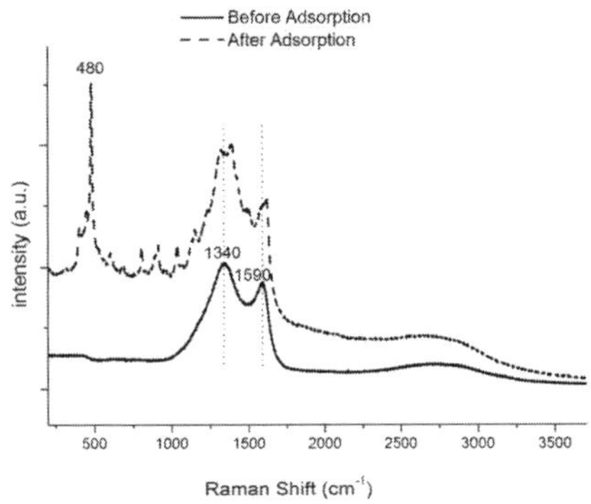

Figure 4: Raman spectra of activated carbon of *Raphanus sativus* before and after MB adsorption.

In order to evaluate how MB molecules are adsorbed in the *Raphanus sativus* carbon and also to identify the predominant adsorption mechanism involved in this separation, the MB loaded carbon spectra was subtracted from the unloaded carbon spectra; a comparison between this curve and that of solid MB spectra is shown in Fig. 5(a) as the Raman spectra of MB aqueous solutions was not possible to be carried out. In fact, the bonding between MB and carbon can be identified according to changes verified in the peak positions of both curves. In order to obtain a better visualization of the Raman spectra, the curves shown in Fig. 5(a) were split into regions according to Raman shift interval of 200 to 700 cm^{-1} and 700 to 1700 cm^{-1}, shown respectively in Fig. 5(b) and 5(c).

According to Rubim *et al.* [12], the bands identified at 448, 480 and 500 cm^{-1} as shown in Fig. 5(b) are associated to the aggregation state of the MB molecules that might change due to the contact with the carbon surface or even with pH of the aqueous solution. The presence of dimmer-MB species is evidenced by peaks around 448 and 500 cm^{-1} while the monomer-MB species can be identified at 485 cm^{-1} (δ (CNC) skeletal deformation mode). Therefore, according to Fig. 5(b), MB molecules are predominantly found as dimmers in the solid but they are adsorbed as monomer in the *Raphanus sativus* carbon. Recently, UV-vis analysis of aqueous solutions containing MB carried out by Tafulo *et al.* [13] pointed to the non existence of dimmers in the aqueous solution as referred by

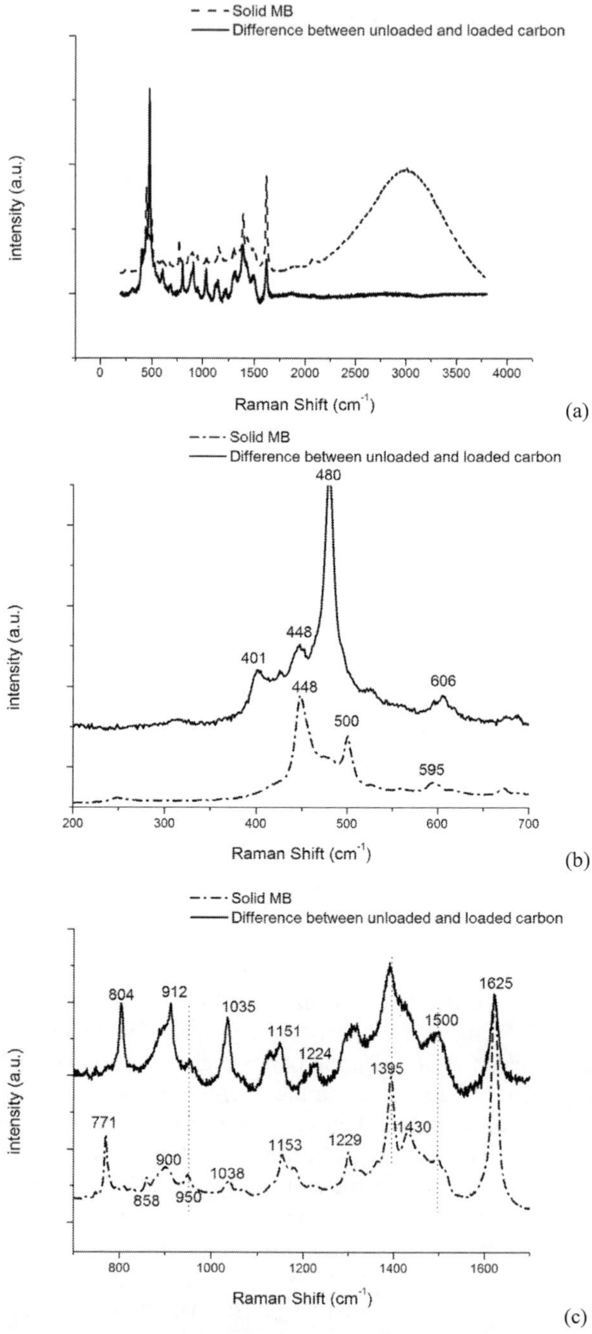

Figure 5: Raman spectra of solid MB and the spectra difference between unloaded and MB loaded carbon: (a) 0 to 4000 cm^{-1}, (b) 200 to 700 cm^{-1} and (c) 700 to 1700 cm^{-1}.

several authors and that changes on the position and shape of bands obtained by them are in fact influenced by the ionic strength of the solution. The chemical structure of MB and its possible dimerization reaction in solution is shown in Fig. 6. The peak at 401 cm^{-1} can be attributed to some deformation in the C-S-C bond of the MB species [14] and no significant change in the intensity of peaks nearby 600 cm^{-1} was considered.

Some peaks related to atoms deformation in the molecules of solid MB and adsorbed MB are shown in Fig. 5(c). Firstly, peaks related to the CH bond deformation β (C-H) are found at 1030 and 1151 cm^{-1} [15]. One peak was identified at 1038 cm^{-1}, showing low intensity in the solid MB spectra but being quite significant when MB molecules are adsorbed in the *Raphanus sativus* carbon; the other peak occurred at 1153 cm^{-1} for both curves with quite similar intensities. Other deformation on the atoms of MB molecules is related to the symmetric and asymmetric stretch of C-N bond that are found at 1390 and 1430 cm^{-1}, respectively [15]. According to Fig. 5(c), the symmetric stretch of C-N bond occurs in both solid and MB loaded carbon, but the asymmetric stretch of C-N bond has been identified in the solid only. And finally a typical large MB peak was found at 1625 cm^{-1} for both spectra which is related to the C-C bond stretch of the benzene ring of MB molecule [15-16]. It was considerably more intense in the solid MB spectra. Other non identified peaks occurred at 771, 858, 900, 950 and 1229 cm^{-1} for the solid MB spectra and at 804, 912, 1224 and 1500 cm^{-1} for the MB adsorbed in the carbon.

The identification of bands that have appeared, disappeared or just shifted after adsorption of MB molecules in comparison with those bands of solid MB may help to elucidate how MB can possibly be linked to the adsorbent based on *Raphanus sativus*. Hutchinson and Hester [16], studying the structure of the surface coat of a thionine-modified gold electrode, suggested that the coat occurred basically by thionine units linked via an amino nitrogen in a secondary amine structure. According to the molecule of MB shown in Fig. 6, there are three probable sites for this bonding: the central nitrogen atom, the central sulfur atom and/or the amino nitrogen atoms. MB bonding as monomer species through

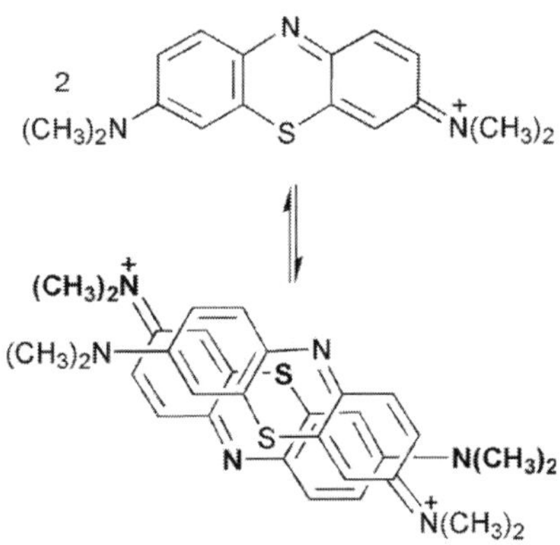

Figure 6: MB chemical structure and its possible dimmer [13].

the central nitrogen atom seems to be the most probable mechanism to occur during adsorption of MB molecules onto *Raphanus sativus* carbon because a very high intensity peak at 480 cm^{-1} was identified, which is assigned to a δ (CNC) skeletal deformation mode. MB bonding through the sulfur atom seems to occur as well but at lesser amount because intensity of peak near 405 cm^{-1}, attributed to some deformation in the C-S-C bond of the MB species, has also appeared. However, bonding through an amino nitrogen atom to a carbon atom seems less probable because symmetric stretch of C-N bond 1395 cm^{-1} was found to be not affected and the asymmetric stretch at 1430 cm^{-1} just vanished at the MB loaded curve.

4 Conclusions

The characterization of carbons is an important step for the design of adsorption processes. Main results on the characterization of the adsorbent based on *Raphanus sativus* in correlation to MB adsorption are:

- A microporous structure was verified qualitatively (SEM and Raman analysis) indicating that MB adsorption occurs predominantly at the surface of the adsorbent (surface area = 236.3 $m^2.g^{-1}$, pore volume = 0.06 $cm^3.g^{-1}$; pore size = 1.4 nm; MB diameter ≈ 1.5 nm [17]).
- Adsorption probably occurs by MB bonding of monomer species through central nitrogen atom; MB bonding through sulfur atom may occur while MB bonding through amino nitrogen atom seems improbable.

Acknowledgements

Authors are grateful to the National Institute of Science and Technology (INCT, Brazil): Mineral Resources, Water and Biodiversity for the financial support (MCT/CNPq/FNDCT/CAPES/FAPEMIG/FAPERJ/FAPESP).

References

[1] I.M. Banat, P. Nigam, D. Singh, R. Marchant, Microbial decolorization of textile-dye-containing effluents: A review, Bioresource Technol. 58 (1996) 217-227.

[2] Y.M. Slokar, M.A. Le Marechal, Methods of decoloration of textile wastewaters, Dyes Pigments 37 (1998) 335-356.

[3] T. Robinson, G. Mcmullan, R. Marchant, P. Nigam, Remediation of dyes in textile effluent: a critical review on current treatment technologies with a proposed alternative, Bioresource Technol. 77 (2001) 247-255.

[4] G. Crini, Non-conventional low-cost adsorbents for dye removal: a review, Bioresource Technol. 97 (2006) 1061-1085.

[5] A.B. Santos, F.J. Cervantes, J.B. van Lier, Review paper on current technologies for decolourisation of textile wastewaters: Perspectives for anaerobic biotechnology. Bioresource Technol. 98 (2007) 2369-2385.

[6] D.A. Lázaro, M.B. Mansur, A.S. Franca, L.S. Oliveira, S.D.F. Rocha, Performance of cold-pressed cake from *Raphanus sativus* (L. Var.) oilseeds, a solid residue from biodiesel production, as adsorbent for basic dyes, Int. J. Chem. Eng. 1 (2008) 147-160.

[7] O. Ioannidou, A. Zabaniotou, Agricultural residues as precursors for activated carbon production – A review, Renew. Sust. Energ. Rev. 11 (2007) 1966-2005.

[8] A. Nunes, A.S. Franca, L.S., Oliveira, Activated carbons from waste biomass: An alternative use for biodiesel production solid residues, Bioresource Technol. 100 (2009) 1786-1792.

[9] F. Ferrero, Dye removal by low cost adsorbents: Hazelnut shells in comparison with wood sawdust, J. Hazard. Mater. 142 (2007) 144-152.

[10] S. Lowell, J.E. Shields, Powder surface analysis area and porosity, third ed, Chapman & Hall, Great Britain, 1991.

[11] F. Haghseresht, G.Q. Lu, A.K. Whittaker, Carbon structure and porosity of carbonaceous adsorbents in relation to their adsorption properties, Carbon 37 (1999) 1491-1497.

[12] J.C. Rubim, M.H. Sousa, J.C.O. Silva, F.A. Tourinho, Raman spectroscopy as a powerful technique in the characterization of ferrofluids, Braz. J. Physics 31 (2001) 402-408.

[13] P.A.R. Tafulo, R.B. Queirós, G. González-Aguilar, On the "concentration-driven" methylene blue dimerization. Spectrochim, Acta Part A 73 (2009) 295-300.

[14] G.N. Tognalli, A. Fainstai, C. Vericat, M.R. Vela, R.C. Salvarezza, An exploring three-dimensional nanosystems with Raman spectroscopy: methylene blue adsorbed on thiol and sulfur monolayers on gold, J. Phys. Chem. B 110 (2006) 354-360.

[15] R.R. Naujok, V.D. Robert, R.M. Corn, Fluorescence and Fourier transform surface-enhanced raman scattering measurements of methylene blue adsorbed onto a sulfur-modified gold electrode, Langmuir 9 (1993) 1771-1774.

[16] K. Hutchinson, R.E. Hester, Raman spectroscopic studies of a thionine-modified electrode, J. Chem. Soc., Faraday Trans. I 80 (1984) 2053-2071.

[17] Y. Xun, Z. Shu-Ping, X. Wei, C. Hong-You, D. Xiao-Dong, L. Xin-Mei, Y. Zi-Feng, Aqueous dye adsorption on ordered mesoporous carbons, J. Colloid and Int. Sci. 310 (2007) 83-89.

Author Index